Operator Theory
Advances and Applications
Vol. 86

Editor
I. Gohberg

Functional Analysis
Vol. II

Y.M. Berezansky
Z.G. Sheftel
G.F. Us

Translated from the Russian by Peter V. Malyshev

Birkhäuser Verlag
Basel · Boston · Berlin

Yurij M. Berezansky
Institute of Mathematics
Ukrainian Academy of Sciences
Repin str. 3
252601 Kiev
Ukraine

Georgij F. Us
Mechanics and Mathematics Faculty
Kiev University
Vladimirskaya str. 64
252617 Kiev
Ukraine

Zinovij G. Sheftel
Department of Mathematics
Pedagogical Institute
Sverdlov str. 53
250038 Chernigov
Ukraine

Originally published in 1990 by Vysha Shkola, Kiev.

1991 Mathematics Subject Classification 46-XX

A CIP catalogue record for this book is available from the Library of Congress, Washington D.C., USA

Deutsche Bibliothek Cataloging-in-Publication Data
Berezanskij, Jurij M.:
Functional analysis / Y. M. Berezansky ; Z. G. Sheftel ; G. F.
Us. Transl. from the Russian by Peter V. Malyshev. - Basel ;
Boston ; Berlin : Birkhäuser
Einheitssacht.: Funkcional'nyj analiz <engl.>
NE: Šeftel, Zinovij G.:; Us, Georgij F.:
Vol. 2 (1996)
(Operator theory ; Vol. 86)
ISBN-13: 978-3-0348-9872-0 e-ISBN-13: 978-3-0348-9024-3
DOI: 10.1007/978-3-0348-9024-3
NE: GT

Softcover reprint of the hardcover 1st edition 1996
Printed on acid-free paper produced from chlorine-free pulp. TCF ∞
Cover design: Heinz Hiltbrunner, Basel

9 8 7 6 5 4 3 2 1

Contents

Volume I

Chapter 4
Measures in the Products of Spaces.
Fubini Theorem

Volume II

Chapter 12
General Theory of Unbounded Operators in Hilbert Spaces

Most of the operators encountered in mathematical physics are unbounded. As a rule, they are constructed by using the operation of differentiation. In this chapter, we present general principles of the theory of unbounded operators in complex Hilbert spaces.

1 Definition of an Unbounded Operator. The Graph of an Operator

1.1 Definitions

First, we consider a simple example of an unbounded operator in the Hilbert space $H = L_2((a,b)) = L_2$ of functions $f(x)$ square summable with respect to the Lebesgue measure on a bounded interval (a,b). An operator A acting on L_2 is defined on the linear set $C^1([a,b]) \subseteq L_2$ by setting

$$C^1([a,b]) \ni f \mapsto Af = \frac{df}{dx} = f'. \tag{1.1}$$

Clearly, this operator is linear. At the same time, it is not bounded: Indeed, let $f_n(x) = e^{inx}$ $(n \in \mathbb{N})$, then, for any $n \in \mathbb{N}$, we have $\|Af_n\|_{L_2} = \|\inf f_n\|_{L_2} = n\|f_n\|_{L_2}$. Therefore, the inequality $\|Af_n\|_{L_2} \leq c\,\|f\|_{L_2}$ $\quad (f \in C^1([a,b]))$ with some $c > 0$ is impossible.

In this example, the operator A was defined not on the whole Hilbert space $H = L_2$ but on its dense linear subset $C^1([a,b])$. Below, we show that this situation is in a certain sense general, namely an operator defined on the whole H and satisfying a certain fairly general requirement is automatically bounded (see Theorem 3.4). Let H be a given Hilbert space. Consider a linear set $\mathcal{D}(A) \subseteq H$. In this set, we define a linear mapping $\mathcal{D}(A) \ni f \mapsto Af \in H$ (i.e., a mapping for which $A(\lambda f + \mu g) = \lambda Af + \mu Ag$ for all $f, g \in \mathcal{D}(A)$ and $\lambda, \mu \in \mathbb{C}$).

This mapping is called a linear operator with the domain of definition (or simply domain) $\mathcal{D}(A)$. We say that two linear operators A and B are equal if their domains and actions coincide, i.e., $A = B$ if both $\mathcal{D}(A) = \mathcal{D}(B)$ and $Af = Bf$ $(f \in \mathcal{D}(A))$.

The domain of an operator A is always denoted by $\mathcal{D}(A)$ (one can also encounter the notation dom (A)). Generally speaking, this domain may be not dense in H. Its denseness or nondenseness will always be clear from the situation or will be specially indicated.

Let A and B be two operators such that $\mathcal{D}(A) \subseteq \mathcal{D}(B)$ and $Af = Bf$ $(f \in \mathcal{D}(A))$. *In this case, the operator B is called an extension of the operator A, A is called a restriction of the operator B, and we write $A \subseteq B$ and $B \restriction \mathcal{D}(A) = A$.*

Example

1.1 As in (1.1), we assume that $H = L_2((a,b)) = L_2$, $\mathcal{D}(A_k) = C^k([a,b])$, and $L_2 \supset \mathcal{D}(A_k) \ni f \mapsto f' \in L_2$ $(k \in \mathbb{N})$. It is clear that $A_1 \supseteq A_2 \supseteq \dots$; $\mathcal{D}(A_k)$ is dense in L_2 $(\forall k \in \mathbb{N})$.

The operations with linear operators are defined in a natural way. One should only take care of the domains of the operators that are encountered. Thus, let A and B be two operators acting on H and let $\lambda \in \mathbb{C}$. We set

(i) $(\lambda A)f = \lambda(Af) \quad (f \in \mathcal{D}(\lambda A) = \mathcal{D}(A))$;

(ii) $(A+B)f = Af + Bf \quad (f \in \mathcal{D}(A+B) = \mathcal{D}(A) \cap \mathcal{D}(B))$;

(iii) $(AB)f = A(Bf) \quad (f \in \mathcal{D}(AB) = \{f \in \mathcal{D}(B) \mid Bf \in \mathcal{D}(A)\})$. (1.2)

In other words, the domains in (1.2) are constructed as the largest possible for a given operation. In this case, of course, $\mathcal{D}(A+B)$ and $\mathcal{D}(AB)$ may be not dense in H, though $\mathcal{D}(A)$ and $\mathcal{D}(B)$ are dense. One must always keep in mind the possibility of this situation and, if necessary, require in addition that $\mathcal{D}(A+B)$ or $\mathcal{D}(AB)$ be dense.

Recall that the range of an operator A, i.e., the linear set of vectors of the form Af, where $f \in \mathcal{D}(A)$, is denoted by $\mathcal{R}(A)$ (or ran (A)). Assume that A establishes a one-to-one correspondence between $\mathcal{D}(A)$ and $\mathcal{R}(A)$. Then there exists an inverse operator A^{-1} defined by the equality $A^{-1}(Af) = f \quad (f \in \mathcal{D}(A))$. Thus, $\mathcal{D}(A^{-1}) = \mathcal{R}(A)$ and $\mathcal{R}(A^{-1}) = \mathcal{D}(A)$. A criterion for the existence of a one-to-one correspondence can be formulated as follows: The kernel of the operator A must be equal to zero, i.e., $\mathrm{Ker} A = \{f \in \mathcal{D}(A) \mid Af = 0\} = 0$.

The inverse operator just defined is usually called the algebraically inverse operator because it is customary to say that the operator A^{-1} is inverse to A if $\mathcal{D}(A^{-1}) = \mathcal{R}(A) = H$ and A^{-1} is bounded.

1.2 Graphs of Operators

The study of unbounded operators is connected with serious difficulties caused by the necessity of simultaneous investigation of the action of an operator A and its domain $\mathcal{D}(A)$. This is why the technique of graphs of operators proves to be a very convenient tool in the theory of unbounded operators.

The graph of an operator A acting on H is introduced quite naturally. Consider the orthogonal sum $H \oplus H$ of pairs $\langle f, g \rangle$, where $f, g \in H$. Recall that the linear operations with these pairs are defined "coordinatewise" and that their scalar product is introduced as follows:

$$(\langle f_1, g_1 \rangle, \langle f_2, g_2 \rangle)_{H \oplus H} = (f_1, f_2)_H + (g_1, g_2)_H \quad (f_1, f_2, g_1, g_2 \in H). \tag{1.3}$$

The set

$$\Gamma_A = \{\langle f, Af\rangle \in H \oplus H \mid f \in \mathcal{D}(A)\}. \tag{1.4}$$

is called the graph Γ_A *of the operator* A.

This definition and the fact that A is linear imply that Γ_A is a linear set in $H \oplus H$. It is natural to ask whether any linear set L in $H \oplus H$ is the graph of a certain operator (possibly with a nondense domain). In the case of a one-dimensional real space H, any straight line, which passes through the origin of coordinates in the plane $\mathbb{R}^2 = H \oplus H$ but does not coincide with the ordinate, is the graph of an operator.

It is clear that a similar condition can be formulated in the general case. Indeed, a linear set $L \subset H \oplus H$ is the graph of an operator A acting on H if and only if, for any f such that $\langle f, g\rangle \in L$, the coordinate g is determined uniquely (i.e., "the second coordinate of the vectors from L is uniquely determined if the first coordinate is given"). In this case, we set $\mathcal{D}(A) = \{f \in H \mid \langle f, g\rangle \in L\}$ and $Af = g$. It follows from the linearity of L that the set $\mathcal{D}(A)$ and the operator A are linear; $L = \Gamma_A$. One can also say that a linear set $L \subset H \oplus H$ is the graph of an operator if the inclusion $\langle 0, h\rangle \in L$ implies that $h = 0$. If the first coordinates of the vectors $\langle f, g\rangle \in L$ form a dense set in H, then A is densely defined.

We stress that the definition of an operator A in H (i.e., the definition of its domain and action) is equivalent to the definition of its graph Γ_A. The relation $A \subseteq B$ is equivalent to the inclusion $\Gamma_A \subseteq \Gamma_B$.

To simplify the study of operators by using their graphs, it is convenient to introduce the following two operators acting on $H \oplus H$:

$$\begin{aligned} H \oplus H \ni \langle f, g\rangle &\mapsto U\langle f, g\rangle = \langle g, f\rangle \in H \oplus H, \\ H \oplus H \ni \langle f, g\rangle &\mapsto O\langle f, g\rangle = \langle -g, f\rangle \in H \oplus H. \end{aligned} \tag{1.5}$$

These operators are isometric. Indeed, according to (1.3), we have

$$\begin{aligned} (U\langle f_1, g_1\rangle, U\langle f_2, g_2\rangle)_{H\oplus H} &= (\langle g_1, f_1\rangle, \langle g_2, f_2\rangle)_{H\oplus H} \\ &= (g_1, g_2)_H + (f_1, f_2)_H = (\langle f_1, g_1\rangle, \langle f_2, g_2\rangle)_{H\oplus H}, \\ (O\langle f_1, g_1\rangle, O\langle f_2, g_2\rangle)_{H\oplus H} &= (\langle -g_1, f_1\rangle, \langle -g_2, f_2\rangle)_{H\oplus H} \\ &= (-g_1, -g_2)_H + (f_1, f_2)_H = (\langle f_1, g_1\rangle, \langle f_2, g_2\rangle)_{H\oplus H} \end{aligned}$$

$$(f_1, f_2, g_1, g_2 \in H).$$

By virtue of (1.5), $\mathcal{R}(U) = H \oplus H$ and $\mathcal{R}(O) = H \oplus H$; therefore, U and O are unitary operators. Definition (1.5) immediately implies that

$$U^2 = \mathbb{1}, \quad O^2 = -\mathbb{1}, \quad OU = -UO. \tag{1.6}$$

The graphs of operators obey the well-known "school" rule for the construction of the graphs of inverse functions, namely, the original graph must be symmetrically reflected in the bisectrix of the first coordinate angle. In the case under consideration, this rule takes the following form:

Theorem 1.1. *Let A be an operator with, in general, nondense domain. In order that the algebraically inverse operator A^{-1} exist, it is necessary and sufficient that the set $U\Gamma_A$ be the graph of a certain operator. Furthermore,*

$$\Gamma_{A^{-1}} = U\Gamma_A. \tag{1.7}$$

Proof. Assume that A^{-1} exists and $\langle f, g\rangle \in \Gamma_A$, i.e., $f \in \mathcal{D}(A)$ and $g = Af$. Then $g \in \mathcal{D}(A^{-1})$ and $f = A^{-1}g$, i.e., $U\langle f, g\rangle = \langle g, f\rangle \in \Gamma_{A^{-1}}$. In other words, $U\Gamma_A \subset \Gamma_{A^{-1}}$. By applying the operator U to the last equality in view of the first equality in (1.6), we arrive at the opposite inclusion. This proves equality (1.7).

It remains to show that if $U\Gamma_A$ is the graph of a certain operator, then the operator A^{-1} exists. But $U\Gamma_A$ consists of the vectors $\langle g, f\rangle$ with $f \in \mathcal{D}(A)$ and $g = Af$. The assertion that the first coordinate g of this vector determines its second coordinates f uniquely is equivalent to the existence of A^{-1}. □

Exercises

1.1. Let A, B, and C be linear operators in H. Prove that

(a) $A + B = B + A$; (b) $(A + B) + C = A + (B + C)$; (c) $0A \subseteq 0$;

(d) $(AB)C = A(BC)$; (e) $(A + B)C = AC + BC$;

(f) $A(B + C) \supseteq AB + AC$.

Give an example of operators A, B, and C for which the inclusion in (f) is strict.

1.2. Let A^{-1} and B^{-1} be the operators algebraically inverse to operators A and B. Then $(AB)^{-1} = B^{-1}A^{-1}$. Prove this.

1.3. Let $A \in \mathcal{L}(H)$. We say that an operator A commutes with an operator B if $AB \subseteq BA$ (notation: $A \smile B$). Prove the following statements:

(a) $A \smile B_1, A \smile B_2 \Rightarrow A \smile (B_1 + B_2), A \smile B_1B_2$;

(b) $A_1 \smile B, A_2 \smile B \Rightarrow (A_1 + A_2) \smile B, A_1A_2 \smile B$;

(c) if the algebraically inverse B^{-1} exists and $A \smile B$, then $A \smile B^{-1}$.

1.4. Give examples of operators acting on H and such that

(a) $(\mathcal{D}(A))^{\sim} = H, \mathcal{D}(A^2) = \{0\}$;

(b) $(\mathcal{D}(A))^{\sim} = H$, $(\mathcal{D}(B))^{\sim} = H$, $\mathcal{D}(A + B) = \{0\}$.

1.5. Let $(e_n)_{n=1}^\infty$ be an orthonormal basis in H and let $\varphi \in H \setminus \text{l.s.}\big((e_n)_{n=1}^\infty\big)$. Consider the operator T defined on $\mathcal{D}(T) = \text{l.s.}\big(\varphi, (e_n)_{n=1}^\infty\big)$ by setting

$$T\Big(\lambda\varphi + \sum_{i=1}^{n} \lambda_i e_i\Big) = \lambda\varphi.$$

Show that the closure of the graph of the operator T is not a graph for any linear operator in H.

2 Closed and Closable Operators. Differential Operators

We introduce two important classes of unbounded operators for which it is possible to construct a fairly detailed theory (without additional restrictions, the notion of unbounded operator is too general).

2.1 Closed Operators

First, we give three equivalent definitions of a closed operator A acting on H.

(1) *An operator A is closed if its graph Γ_A is closed in $H \oplus H$.*

(2) *An operator A is closed if, for any sequence $(f_n)_{n=1}^\infty \subseteq \mathcal{D}(A)$, the facts that $f_n \to f \in H$ and $Af_n \to g \in H$ as $n \to \infty$ imply that $f \in \mathcal{D}(A)$ and $Af = g$.*

(This definition can be regarded as a somewhat weakened version of the standard definition of a continuous operator; indeed, if the operator is continuous, then $Af_n \to g$ automatically.)

(3) In the domain $\mathcal{D}(A)$ of an operator A, we introduce the so-called *graph scalar product*

$$(f, g)_{\Gamma_A} = (f, g)_H + (Af, Ag)_H \quad (f, g \in \mathcal{D}(A)). \tag{2.1}$$

The operator A is closed if $\mathcal{D}(A)$ is a complete space with respect to the graph scalar product.

Also note that the norm corresponding to (2.1) is called the *norm of a graph*. For this norm, we have

$$\|f\|_{\Gamma_A}^2 = \|f\|_H^2 + \|Af\|_H^2 \qquad (f \in \mathcal{D}(A)). \tag{2.2}$$

Theorem 2.1. *The definitions of closed operators given above are equivalent.*

Proof.

(1) $\Rightarrow$ (2). The fact that $f_n \to f$ and $Af_n \to g$ in H means that $(\langle f_n, Af_n\rangle)_{n=1}^\infty \subseteq \Gamma_A$ converges in $H \oplus H$ to the point $\langle f, g\rangle$, which must belong to Γ_A in view of the assumed closeness of Γ_A in $H \oplus H$. This means that $f \in \mathcal{D}(A)$ and $Af = g$.

(2) $\Rightarrow$ (3). Let $(f_n)_{n=1}^{\infty} \subseteq \mathcal{D}(A)$ be a fundamental sequence with respect to $\|\cdot\|_{\Gamma_A}$. It follows from (2.2) that $(f_n)_{n=1}^{\infty}$ and $(Af_n)_{n=1}^{\infty}$ are fundamental sequences in H. Let f and g be the limits of these sequences. By virtue of (2), $f \in \mathcal{D}(A)$ and $Af = g$. It is easy to see that $\|f_n - f\|_{\Gamma_A} \to 0$ as $n \to \infty$. Indeed, for any $\varepsilon > 0$, there exists $N = n(\varepsilon)$ such that $\|f_n - f_m\|_{\Gamma_A} < \varepsilon$ for $n, m > N$. In view of relation (2.2) for the norm $\|\cdot\|_{\Gamma_A}$, we can pass to the limit as $m \to \infty$ in the last inequality arriving, as a result, at the required relation $\|f_n - f\|_{\Gamma_A} \leq \varepsilon$ $(n > N)$.

(3) $\Rightarrow$ (1). Let $\Gamma_A \ni \langle f_n, Af_n\rangle \to \langle f, g\rangle$ as $n \to \infty$. It follows from (2.2) that the sequence $(f_n)_{n=1}^{\infty}$ is fundamental in the norm of the graph and, according to (3), it converges in this norm to a vector $h \in \mathcal{D}(A)$. But then, by virtue of (2.2), $\|f_n - h\|_H \to 0$ and $\|Af_n - Ah\|_H \to 0$. Since the limits are unique, we have $h = f$ and $Ah = g$, whence $\langle f, g\rangle \in \Gamma_A$. □

Example

2.1 Each operator A_k $(k \in \mathbb{N})$ appearing in Example 1.1 is not closed. For example, consider the operator A_1. It is not closed because the fact that the sequences $f_n \in \mathcal{D}(A_1) = C^1([a,b])$ and f_n' converge in L_2 to $f \in L_2$ and $g \in L_2$, respectively, does not imply that $f \in C^1([a,b])$. The last statement can be proved, e.g., as follows: We set $[a,b] = [-1,1]$, $f(x) = |x|$, and $f_n(x) = \left(S_{\frac{1}{n}} f\right)(x)$ $(x \in [-1,1],\ n \in \mathbb{N})$, where the averaging operator S_ε is given by the formula

$$(S_\varepsilon f)(x) = \frac{1}{2\varepsilon}\int_{x-\varepsilon}^{x+\varepsilon} f(\xi)\,d\xi \quad (\varepsilon > 0; x \in \mathbb{R}) \tag{2.3}$$

(the integrand is extended to the outside of $[-1,1]$ as zero); the required properties of this sequence follow from the properties of averaging operators (see Lemma 11.1.1 and Exercise 11.1.11).

2.2 Closable Operators

If an operator A acting on H is not closed, then, at first sight, it is always possible to construct its "closure", i.e., to extend A to a closed operator $\tilde{A}$ by adding to its domain $\mathcal{D}(A)$ all vectors $f \in H$, for which one can find a sequence $(f_n)_{n=1}^{\infty} \subseteq \mathcal{D}(A)$ such that $f_n \to f$ as $n \to \infty$ and there exists $\lim_{n\to\infty} Af_n = g$. On the added vectors f, we naturally set $\tilde{A}f = g$. However, this procedure is, generally speaking, incorrect: $\tilde{A}f$ may depend on a sequence $(f_n)_{n=1}^{\infty}$ that approximates f.

(1) *We say that an operator A admits a closure $\tilde{A}$ (or is closable) if the procedure outlined above is correct (for all $f \in H$ that admit a required approximating sequence $(f_n)_{n=1}^{\infty}$).*

Instead of this definition, which has a somewhat descriptive nature, it is more convenient to use the following two definitions:

(2) *An operator A is closable if the fact that* $\lim_{n\to\infty} f_n = 0$ *and* $\lim_{n\to\infty} Af_n = h \in H$ *for a sequence $(f_n)_{n=1}^{\infty} \subset \mathcal{D}(A)$ implies that $h = 0$.*

(3) *An operator A is closable if the closure $\tilde{\Gamma}_A$ of its graph is the graph of some operator.*

Theorem 2.2. *The definitions of closable operators given above are equivalent.*
Proof.

(1) $\Rightarrow$ (2). This is evident since, according to definition (1), we have $h = \tilde{A}0 = 0$.

(2) $\Rightarrow$ (1). Let a vector $f \in H$ be such that there exist $f_n', f_n'' \in \mathcal{D}(A)$ $(n \in \mathbb{N})$ for which $f_n' \to f, f_n'' \to f, Af_n' \to g'$, and $Af_n'' \to g''$ as $n \to \infty$. It is necessary to show that $g' = g''$. But this follows from definition (2) if we set $f_n = f_n' - f_n''$.

(2) $\Rightarrow$ (3): $\tilde{\Gamma}_A \subseteq H \oplus H$ is a linear set. Let $\langle 0, h\rangle \in \tilde{\Gamma}_A$. The last inclusion means that there exists a sequence $(f_n)_{n=1}^{\infty}, f_n \in \mathcal{D}(A)$, such that $f_n \to 0$ and $Af_n \to h$ as $n \to \infty$. But then, by virtue of (2), $h = 0$ and this means that $\tilde{\Gamma}_A$ is a graph.

(3) $\Rightarrow$ (2). This is an obvious inversion of the reasoning used in the previous item. □

Assume that an operator A admits a closure $\tilde{A}$. We stress once again that the values of $\tilde{A}f$ are computed as follows: We consider all $f \in H$, for which one can find sequences $(f_n)_{n=1}^{\infty}, f_n \in \mathcal{D}(A)$, such that $\lim_{n\to\infty} f_n = f$ and the limit $\lim_{n\to\infty} Af_n = g$ exists. These vectors f constitute the domain of $\tilde{A}$ and $\tilde{A}f = g$.

It follows from this construction that $\tilde{A}$ is a linear closed operator. Note that

$$\Gamma_{\tilde{A}} = \tilde{\Gamma}_A. \tag{2.4}$$

If the original operator A is continuous, then $\tilde{A}$ is a continuous operator defined in $\mathcal{D}(\tilde{A}) = (\mathcal{D}(A))^{\sim}$. In this case, the procedure of closing coincides with the standard procedure of extension of operators by continuity.

REMARK 2.1. Assume that the operator A admits a closure. Then $\mathcal{D}(\tilde{A}) = (\mathcal{D}(A))^{\sim\Gamma_A}$, where $\sim \Gamma_A$ denotes the closure in norm of the graph. This immediately follows from the comparison of the definition of $\tilde{A}$ with relation (2.1).

REMARK 2.2. Let $A \subseteq B$ and let the operator B be closable. Then the operator A is also closable and $\tilde{A} \subseteq \tilde{B}$. Indeed, $A \subseteq B \Leftrightarrow \Gamma_A \subseteq \Gamma_B$. Hence, $\tilde{\Gamma}_A \subseteq \tilde{\Gamma}_B = \Gamma_{\tilde{B}}$, i.e., $\tilde{\Gamma}_A$ is a part of a graph $\Gamma_{\tilde{B}}$ and, therefore, is a graph. It follows from the last inclusion and (2.4) that $\tilde{A} \subseteq \tilde{B}$. □

Note that, in this section, we did not assume that the domain of the operator A is dense in H.

Example

2.2. (A nonclosable operator.) Let $H = L_2((a,b)) = L_2$, $\mathcal{D}(A) = C([a,b])$, and $(Af)(x) = f(a)$. Clearly, one can always construct a sequence of functions $(f_n)_{n=1}^{\infty}$, $f_n \in C([a,b])$ such that $f_n(a) = 1$ and $\|f_n\|_{L_2} \to 0$ as $n \to \infty$. Thus, we have $Af_n = h \equiv 1 \neq 0$ $(n \in \mathbb{N})$ and $f_n \to 0$. Therefore, according to definition (2), the operator A is not closable.

2.3 Differential Operators

We now introduce a class of operators admitting a closure, namely, differential operators, which are quite important for modern mathematical physics.

Let G be a bounded domain in the N-dimensional space $\mathbb{R}^N$ of points $x = (x_1, \dots, x_N)$. For simplicity, its boundary ∂G is assumed to be smooth, i.e., every sufficiently small part of the boundary is described by an equation of the form $x_j = \varphi_j(x_1, \dots, x_{j-1}, x_{j+1}, \dots, x_N)$, where j is the number of a selected variable, which is specified separately for each part of the boundary ($j = 1, \dots, N$), and φ_j is an l times continuously differentiable function of the other variables ($l \in \mathbb{Z}_+ \cup \infty$). We say that boundaries of this sort form a class C^l. In what follows, we assume that l is sufficiently large (to meet the requirements of the corresponding calculations).

Consider a differential expression $\mathcal{L}$ of the form

$$(\mathcal{L}u)(x) = \sum_{|\alpha| \le r} a_\alpha(x)(D^\alpha u)(x) \quad (x \in G). \tag{2.5}$$

Here, we use the following notation (see Section 6.7):

$$D^\alpha = D_1^{\alpha_1} \dots D_N^{\alpha_N}, \quad \alpha = (\alpha_1, \dots \alpha_N), \quad \alpha_1, \dots \alpha_N \in \mathbb{Z}_+,$$
$$|\alpha| = \alpha_1 + \dots + \alpha_n, \quad \text{and} \quad D_k = \frac{\partial}{\partial x_k} \quad (k = 1, \dots N). \tag{2.6}$$

Thus, (2.5) is a general linear rth order differential expression with complex-valued coefficients $a_\alpha(x)$. Assume that they are sufficiently smooth or, more precisely, that $a_\alpha \in C^{|\alpha|}(G)$ $(|\alpha| \le r)$.

Note that notation (2.6) is frequently used in what follows.

The differential expression $\mathcal{L}$ of the form (2.5) is not an operator because we did not indicate the classes (spaces) of functions to which this expression is applied. One should only assume that the function u in (2.5) is sufficiently smooth such that the derivatives appearing in $\mathcal{L}$ are meaningful.

It is useful to introduce a *differential expression $\mathcal{L}^+$ formally adjoint to $\mathcal{L}$ (or adjoint in Lagrange's sense).* To define this expression, we consider two functions u and v sufficiently smooth in G and assume that one of these functions vanishes in a neighbourhood of the boundary of the domain G. Integrating by parts, we obtain (the terms without integrals vanish)

$$\begin{aligned}\int_G (\mathcal{L}u)(x)\overline{v(x)}dx &= \sum_{|\alpha| \le r} \int_G a_\alpha(x)(D^\alpha u)(x)\overline{v(x)}dx \\ &= \sum_{|\alpha| \le r} (-1)^\alpha \int_G u(x)\overline{(D^\alpha \overline{a_\alpha(\cdot)}v(\cdot))(x)}dx \\ &= \int_G u(x)\, \overline{(\mathcal{L}^+ v)(x)}dx, \end{aligned} \tag{2.7}$$

$$(\mathcal{L}^+ v)(x) = \sum_{|\alpha| \le r} (-1)^{|\alpha|} D^\alpha (\overline{a_\alpha(\cdot)}v(\cdot))(x). \tag{2.8}$$

By differentiating $\overline{a_\alpha v}$ according to the Leibniz formula, we can write $\mathcal{L}^+$ in the form (2.5) but this notation is not used in what follows. Clearly, by virtue of the arbitrariness of u and v in (2.7), the expression $\mathcal{L}^+$ is uniquely defined by (2.7).

A differential expression $\mathcal{L}$ is called formally selfadjoint (or Hermitian) if $\mathcal{L} = \mathcal{L}^+$.

The Schrödinger expression

$$(\mathcal{L}u)(x) = -\sum_{j=1}^{N}(D_j^2 u)(x) + q(x)u(x) = -(\Delta u)(x) + q(x)u(x) \qquad (2.9)$$

with a real-valued potential $q = a_0$ is an important example of expressions of this sort. Another example:

$$(\mathcal{L}u)(x) = -i(D_k u)(x) \quad (k = 1, \dots, N \text{ is fixed}).$$

Certainly, the definition presented above remains true for unbounded regions (in particular, for $G = \mathbb{R}^N$). One should only impose an additional requirement that one of the functions in (2.7) (u or v) must vanish in a neighbourhood of ∞.

First, we associate expressions $\mathcal{L}$ of the form (2.5) with the so-called minimal operators acting on the Hilbert space $H = L_2(G) = L_2$ (G is either bounded or not). For this purpose, we denote by $C_0^l(G)$ $(l \in \mathbb{Z}_+ \cup \infty)$ the linear set of finite l times continuously differentiable functions, i.e., the collection of functions u vanishing in a certain (depending on u) neighbourhood of the boundary of G (or of ∞ if the region is unbounded). Clearly, $C_0^l(G)$ is dense in L_2. In L_2, we introduce an operator L' by the formula

$$L_2 \supset C_0^r(G) = \mathcal{D}(L') \ni f \mapsto \mathcal{L}f \in L_2. \qquad (2.10)$$

Theorem 2.3. *The operator L' given by (2.10) admits a closure. This closure $L = \tilde{L}'$ is called the minimal operator corresponding to the expression $\mathcal{L}$ in the region G.*

Proof. Let $f_n \in \mathcal{D}(L') = C_0^r(G)$ $(n \in \mathbb{N})$ be such that $f_n \to 0$ and $L'f_n = \mathcal{L}f_n \to h$ in L_2 as $n \to \infty$. It is necessary to prove that $h = 0$. Let $v \in C_0^r(G)$. By virtue of (2.7),

$$(h, v)_{L_2} = \lim_{n\to\infty} (\mathcal{L}f_n, v)_{L_2} = \lim_{n\to\infty} (f_n, \mathcal{L}^+ v)_{L_2} = 0 \qquad (2.11)$$

(clearly, $\mathcal{L}^+ v \in L_2$). In view of the arbitrariness of v and denseness of $C_0^r(G)$ in L_2, we conclude from (2.11) that $h = 0$. □

REMARK 2.3. It follows from Corollary 2.1 that the domain of the minimal operator L coincides with the complement of $C_0^r(G)$ with respect to the scalar product

$$(f, g)_{\Gamma_L} = (f, g)_{L_2} + (\mathcal{L}f, \mathcal{L}g)_{L_2} \qquad (f, g \in C_0^r(G)). \qquad (2.12)$$

In some cases, this complement can be described more explicitly (see Section 16.1).

REMARK 2.4. Denote by L'_k $(k = r, r+1, \dots, \infty)$ the operator defined by relation (2.10) but on $\mathcal{D}(L'_k) = C_0^r(G)$ (thus, $L' = L'_r$). Clearly, $L'_r \supseteq L'_{r+1} \supseteq \dots$. At the same time, for any $k = r+1, r+2, \dots, \infty$, we have $\tilde{L}'_k = \tilde{L}'_r = L$ (therefore, in the definition of L (2.10), one can replace $C_0^r(G)$ by $C_0^k(G)$ with the indicated k). In fact, since $\tilde{L}'_r \supseteq \tilde{L}'_{r+1} \supseteq \dots$, it suffices to show that $\mathcal{D}(\tilde{L}'_r) \subseteq \mathcal{D}(\tilde{L}'_\infty)$. According to Remark 2.3, this is true if

$$(\forall f \in C_0^r(G))\ (\forall \varepsilon > 0)\ (\exists \varphi \in C_0^\infty(G)):\ \|f - \varphi\|_{L_2} < \varepsilon$$

and $\|\mathcal{L}f - \mathcal{L}\varphi\|_{L_2} < \varepsilon$. But these relations follow from the denseness of $C_0^\infty(G)$ in $C_0^r(G)$ in the norm of the space $C^r(G)$ for bounded G. □

Exercises

2.1. Let A be a closed operator in H. Prove that

(a) $\mathrm{Ker}A$ is a subspace;

(b) if $B \in \mathcal{L}(H)$, then $A + B$ is closed.

Are these statements true for a nonclosed operator A?

2.2. Let A be a closed operator in H. Is it true that

(a) $\mathcal{D}(A)$ is a closed set;

(b) $\mathcal{R}(A)$ is a closed set?

2.3. Let an operator A be bounded on $\mathcal{D}(A)$. Prove that A is closed if and only if $(\mathcal{D}(A))^\sim = \mathcal{D}(A)$.

2.4. Let A be a linear operator in H satisfying the conditions:

(i) $(\mathcal{R}(A))^\sim = \mathcal{R}(A)$;

(ii) $(\exists m > 0)(\forall x \in \mathcal{D}(A)):\ \|Ax\| \geq m\|x\|$.

Prove that A is closed.

2.5. Let A be a linear operator in H satisfying the condition $\mathrm{Ker}A = \{0\}$. Consider the following additional assertions:

(i) A is a closed operator;

(ii) $(\mathcal{R}(A))^\sim = H$;

(iii) $\mathcal{R}(A)$ is a closed set;

(iv) $(\exists m > 0)(\forall x \in \mathcal{D}(A)):\ \|Ax\| \geq m\|x\|$.

Prove that

(a) Assertion (iv) follows from assertions (i)–(iii);

(b) (i) follows from (ii)–(iv);

(c) (iii) follows from (i) and (iv).

2.6. Is the operator of multiplication by the independent variable $(Ax)(t) = tx(t)$ defined on the following domains in $L_2(\mathbb{R})$ closed: (a) $C_0^\infty(\mathbb{R})$; (b) $L_2(\mathbb{R}, (1+t^2)dt)$?

2.7. Denote by $AC([0,1])$ the set of all absolutely continuous functions on $[0,1]$ whose derivatives belong to $L_2([0,1])$. Is the operator of differentiation $(Ax)(t) = ix'(t)$ defined on the following domains in $L_2([0,1])$ closed:

(a) $AC([0,1])$;

(b) $\{x \in AC([0,1]) \mid x(0) = 0\}$;

(c) $\{x \in AC([0,1]) \mid x(0) = x(1) = 0\}$;

(d) $\{x \in AC([0,1]) \mid x(1) = e^{i\alpha}x(0)\}$, where $\alpha \in \mathbb{R}$ is fixed?

2.8. Prove that the operator of multiplication defined on $C_0^\infty(\mathbb{R})$ by a function $a \in C(\mathbb{R})$ defined on $L_2(\mathbb{R})$ is closable. Find its closure.

3 The Adjoint Operator

3.1 Definition and Properties of the Adjoint Operator

Recall that the operator A^* adjoint to a bounded operator A acting on the Hilbert space H is defined by the equality (see Section 8.4)

$$(Af, g)_H = (f, A^*g)_H \quad (f, g \in H). \tag{3.1}$$

In the case where the operator A is unbounded, the definition of A^* is more complicated, since it is necessary to take care of the domains of definition of operators. Thus, let A be an operator in H with a dense domain $\mathcal{D}(A)$. Consider a vector $g \in H$, for which one can find $g^* \in H$ such that

$$(Af, g)_H = (f, g^*)_H \quad (\forall f \in \mathcal{D}(A)). \tag{3.2}$$

It follows from the linearity of the scalar product that the collection of all vectors g of this sort (denoted somewhat arbitrarily by $\mathcal{D}(A^*)$) forms a linear set and

$$(\lambda g + \mu h)^* = \lambda g^* + \mu h^* \quad (g, h \in \mathcal{D}(A^*);\ \lambda, \mu \in \mathbb{C}).$$

The vector g^* is uniquely defined for a given g. Indeed, assume that, parallel with (3.2), there exists another representation $(Af, g)_H = (f, g^{*'})_H$ $(f \in \mathcal{D}(A))$. Then $(f, g^* - g^{*'})_H = 0$ $(f \in \mathcal{D}(A))$ and $g^{*'} = g^*$ because $\mathcal{D}(A)$ is dense in H.

For $g \in \mathcal{D}(A^*)$, we set $A^*g = g^*$. The operator thus defined is linear in view of the argument presented above. It is called the operator *adjoint* to A. In this case, we can also write equality (3.1) and, moreover, $\mathcal{D}(A^*)$ consists of all g such that relation (3.2) is satisfied (this relation can also be expressed in another form, namely $g \in \mathcal{D}(A^*) \iff$ a functional $\mathcal{D}(A) \ni f \mapsto l_g(f) = (Af, g)_H \in \mathbb{C}$ is continuous).

REMARK 3.1. The operator adjoint to an operator A is well-defined if and only if $\mathcal{D}(A)$ is dense in H. Indeed, it remains to show that if there exists an operator A^* connected with A by equality (3.1) (where $f \in \mathcal{D}(A)$ and $g \in \mathcal{D}(A^*)$) and uniquely defined for given A, then $\mathcal{D}(A)$ is dense in H. But this is obvious. Indeed, if $\mathcal{D}(A)$ is not dense, then, for given g, the vector A^*g is defined by (3.1) ambiguously (up to an arbitrary vector $h \perp \mathcal{D}(A)$). □

We have not studied yet the domain of the operator A^*. It is clear that 0 belongs to it (however, it may happen that $\mathcal{D}(A^*) = \{0\}$, see Example 3.2). Below, we clarify the relationship between the denseness of $\mathcal{D}(A^*)$ in H and the closability of A. Here, we dwell upon some important decompositions of $H \oplus H$ connected with A^*.

Lemma 3.1. *Assume that $\mathcal{D}(A)$ is dense in H and, therefore, A^* exists. Then*

$$\Gamma_{A^*} = (O\Gamma_A)^{\perp} = (H \oplus H) \ominus (O\Gamma_A) \tag{3.3}$$

where O is the operator defined in (1.5)

Proof. Let $\langle g, A^*g\rangle \in \Gamma_{A^*}$. This means that $g \in \mathcal{D}(A^*)$ and the equality $(Af, g)_H = (f, A^*g)_H$ $(f \in \mathcal{D}(A))$ holds. In view of (1.5) and (1.3), this gives

$$\begin{aligned}(\langle g, A^*g\rangle, O\langle f, Af\rangle)_{H\oplus H} &= (\langle g, A^*g\rangle, \langle -Af, f\rangle)_{H\oplus H} \\ &= -(g, Af)_H + (A^*g, f)_H = 0 \quad (f \in \mathcal{D}(A)).\end{aligned} \tag{3.4}$$

Since $\langle g, A^*g\rangle \perp O\langle f, Af\rangle$, we have $\Gamma_{A^*} \subseteq (O\Gamma_A)^{\perp}$.

Conversely, if $\langle g, h\rangle \in (O\Gamma_A)^{\perp}$, then we can read (3.4) in the reverse order and conclude that

$$0 = (\langle g, h\rangle, O\langle f, Af\rangle)_{H\oplus H} = -(g, Af)_H + (h, f)_H$$

for any $f \in \mathcal{D}(A)$. In other words, $g \in \mathcal{D}(A^*)$ and $h = g^* = A^*g$, i.e., $\langle g, h\rangle \in \Gamma_{A^*}$. Thus, $(O\Gamma_A)^{\perp} \subseteq \Gamma_{A^*}$. This proves (3.3). □

Equality (3.3) is equivalent to the following orthogonal decomposition:

$$H \oplus H = \Gamma_{A^*} \oplus (O\Gamma_A)^{\sim}. \tag{3.5}$$

We stress that (3.5) contains the closure $(O\Gamma_A)^{\sim}$ of the linear set $O\Gamma_A$. However, if the operator A is closed, then Γ_A and, hence, $O\Gamma_A$ are closed in $H \oplus H$ (O is a unitary operator). Therefore, (3.5) turns into the decomposition

$$H \oplus H = \Gamma_{A^*} \oplus O\Gamma_A. \tag{3.6}$$

The following assertion (converse to Lemma 3.1) is true:

Lemma 3.2. *Assume that an operator A acting on H is such that the subspace $(H \oplus H) \ominus (O\Gamma_A) = (O\Gamma_A)^{\perp}$ is the graph of an operator. Then $\mathcal{D}(A)$ is dense in H and, therefore, the operator A^* exists and its graph satisfies equality (3.3).*

Proof. If $f \in \mathcal{D}(A)$, then the vectors from $O\Gamma_A \subseteq H \oplus H$ have the form $\langle -Af, f\rangle$ and the vectors $\langle g, h\rangle \in (O\Gamma_A)^{\perp}$ satisfy the relation

$$0 = (\langle -Af, f\rangle, \langle g, h\rangle)_{H\oplus H} = (f, h)_H - (Af, g)_H \quad (f \in \mathcal{D}(A)). \tag{3.7}$$

Assume the contrary, i.e., let $\mathcal{D}(A)$ be not dense in H and $0 \neq h \perp \mathcal{D}(A)$. The vector $\langle 0, h\rangle \in H \oplus H$ satisfies (3.7) and, therefore, belongs to $(O\Gamma_A)^{\perp}$. Since $h \neq 0$, this vector cannot belong to any graph. Hence, $(O\Gamma_A)^{\perp}$ is not a graph but this contradicts the assumption. □

Theorem 3.1. *Let A be an operator with a dense domain acting on H and let A^* be its adjoint operator. Then*

(i) *the operator A^* is closed;*

(ii) *if A admits a closure, then $(\tilde{A})^* = A^*$ (the operation of closure does not affect the adjoint operator);*

(iii) *assume that $(\mathcal{R}(A))^\sim = H$ and the operator A^{-1} algebraically inverse to A exists; then the operator $(A^*)^{-1}$ exists and*

$$(A^{-1})^* = (A^*)^{-1}; \tag{3.8}$$

(iv) *if B is an operator of the same type as A, then*

$$B \supseteq A \Rightarrow A^* \supseteq B^*; \tag{3.9}$$

(v) *let an operator B be of the same type as A and let, in addition, $\mathcal{D}(A+B)$ be dense in H; then*

$$(A+B)^* \supseteq A^* + B^*; \tag{3.10}$$

(vi) *let an operator B be of the same type as A and let, in addition, $\mathcal{D}(BA)$ be dense in H; then*

$$(BA)^* \supseteq A^* B^*. \tag{3.11}$$

Proof.

(i) The fact that A^* is closed follows from (3.3) because orthogonal complements are always closed.

(ii) By using (3.3) and (2.4), we obtain

$$\Gamma_{(\tilde{A})^*} = (O\Gamma_{\tilde{A}})^\perp = ((O\Gamma_A)^\sim)^\perp = (O\Gamma_A)^\perp = \Gamma_{A^*}$$

and, therefore, $(\tilde{A})^* = A^*$.

(iii) The operator $(A^{-1})^*$ exists because $\mathcal{D}(A^{-1}) = \mathcal{R}(A)$ is dense in H. Let us construct its graph by using relations (3.3), (1.7), and (1.5). We have

$$\begin{aligned}\Gamma_{(A^{-1})^*} &= (O\Gamma_{A^{-1}})^\perp = (OU\Gamma_A)^\perp = (-UO\Gamma_A)^\perp \\ &= (H \oplus H) \ominus (UO\Gamma_A) \quad = U((H \oplus H) \ominus (O\Gamma_A)) = U\Gamma_{A^*}.\end{aligned}$$

Thus, $U\Gamma_{A^*}$ is the graph of an operator. Therefore, by virtue of Theorem 1.1, the operator $(A^*)^{-1}$ exists and $\Gamma_{(A^*)^{-1}} = U\Gamma_{A^*} = \Gamma_{(A^{-1})^*}$. This means that $(A^*)^{-1} = (A^{-1})^*$.

(iv) Since $B \supseteq A$, we have $\Gamma_B \supseteq \Gamma_A$, $O\Gamma_B \supseteq O\Gamma_A$, and, according to (3.3), $\Gamma_{B^*} = (O\Gamma_B)^\perp \subseteq (O\Gamma_A)^\perp = \Gamma_{A^*}$. Consequently, $B^* \subseteq A^*$.

(v) Let $g \in \mathcal{D}(A^* + B^*) = \mathcal{D}(A^*) \cap \mathcal{D}(B^*)$. Then we can write the following equalities:

$$(Af, g)_H = (f, A^* g)_H \quad (f \in \mathcal{D}(A)),$$
$$(Bf, g)_H = (f, B^* g)_H \quad (f \in \mathcal{D}(B)).$$

Assume that $f \in \mathcal{D}(A) \cap \mathcal{D}(B) = \mathcal{D}(A+B)$. By adding these equalities, we obtain

$$((A+B)f, g)_H = (f, A^*g + B^*g)_H = (f, (A^* + B^*)g)_H \quad (f \in \mathcal{D}(A+B)).$$

But this equality means that $g \in \mathcal{D}((A+B)^*)$ and $(A+B)^*g = (A^*+B^*)g$. Hence, $A^* + B^* \subseteq (A+B)^*$.

(vi) Let $g \in \mathcal{D}(A^*B^*) = \{g \in \mathcal{D}(B^*) \,|\, B^*g \in \mathcal{D}(A^*)\}$. Then, for any $f \in \mathcal{D}(BA) = \{f \in \mathcal{D}(A) \,|\, Af \in \mathcal{D}(B)\}$, we can write

$$(BAf, g)_H = (Af, B^*g)_H = (f, A^*B^*g)_H \tag{3.12}$$

(the first equality in (3.12) is a consequence of the fact that $Af \in \mathcal{D}(B)$ and $g \in \mathcal{D}(B^*)$; the second equality follows from the inclusions $f \in \mathcal{D}(A)$ and $B^*g \in \mathcal{D}(A^*)$). Relation (3.12) means that $g \in \mathcal{D}((BA)^*)$ and $(BA)^*g = A^*B^*g$. Therefore, $A^*B^* \subseteq (BA)^*$. □

In the case where at least one operator in (3.10) and (3.11) is bounded, they turn into equalities. More precisely, we have the following theorem:

Theorem 3.2. *Let A be an operator acting on H with a dense domain and let $B \in \mathcal{L}(H)$. Then*

$$(A+B)^* = A^* + B^* \quad \textit{and} \quad (BA)^* = A^*B^*. \tag{3.13}$$

Proof. Let us prove the first relation in (3.13). By virtue of (3.10), it suffices to show that $(A+B)^* \subseteq A^*+B^*$. Let $g \in \mathcal{D}((A+B)^*)$, i.e., $((A+B)f, g)_H = (f, g^*)_H$ with some $g^* \in H$ (equal to $(A+B)^*g$) for all $f \in \mathcal{D}(A+B) = \mathcal{D}(A)$. But

$$((A+B)f, g)_H = (Af, g)_H + (f, B^*g)_H.$$

Therefore, it follows from the previous equality that

$$(Af, g)_H = ((A+B)f, g)_H - (f, B^*g)_H = (f, g^* - B^*\, g)_H \quad (f \in \mathcal{D}(A)).$$

Hence, $g \in \mathcal{D}(A^*)$, $A^*g = g^* - B^*g = (A+B)^*g - B^*g$, and, consequently, $(A+B)^* \subseteq A^* + B^*$.

The second relation in (3.13) follows from (3.11) and the inclusion $(BA)^* \subseteq A^*B^*$, which we are now going to prove. Let $g \in \mathcal{D}((BA)^*)$, i.e., $(BAf, g)_H = (f, g^*)_H$ with some $g^* \in H$ (equal to $(BA)^*g$) for all $f \in \mathcal{D}(BA) = \mathcal{D}(A)$. But $(BAf, g)_H = (Af, B^*g)_H$; therefore, the last equality can be rewritten in the form

$$(Af, B^*g)_H = (BAf, g)_H = (f, g^*)_H \quad (f \in \mathcal{D}(A)),$$

whence $B^*g \in \mathcal{D}(A^*)$ and $A^*B^*g = g^* = (BA)^*g$. This implies that $g \in \mathcal{D}(A^*B^*) = \{g \in H \,|\, B^*g \in \mathcal{D}(A^*)\}$ and $(A^*B^*)g = (BA)^*g$. Thus, $(BA)^* \subseteq A^*B^*$. □

3.2 The Second Adjoint Operator

The following theorem generalizes the equality $(A^*)^* = A$, which is trivial for operators $A \in \mathcal{L}(H)$.

Theorem 3.3 (on the second adjoint operator). *Let A be an operator in H with a dense domain of definition. Assume that A admits a closure. Then the second adjoint operator $(A^*)^*$ exists and satisfies the equality*

$$(A^*)^* = \tilde{A}. \tag{3.14}$$

Conversely, assume that A has a dense domain of definition and the operator $(A^*)^*$ exists. Then A admits a closure and (3.14) holds.

Proof. First, we assume that A is closed, i.e., $\tilde{A} = A$. Then, according to (3.6), we have $H \oplus H = \Gamma_{A^*} \oplus O\Gamma_A$. By applying the unitary operator O to this equality, in view of the relation $O^2 = -\mathbb{I}$ (see (1.6)), we obtain $H \oplus H = O\Gamma_{A^*} \oplus \Gamma_A$. Hence,

$$(O\Gamma_{A^*})^{\perp} = \Gamma_A, \tag{3.15}$$

i.e., this orthogonal complement is the graph of an operator. According to Lemma 3.2, $\mathcal{D}(A^*)$ is dense in H and, therefore, $(A^*)^*$ exists. Since A^* is closed, according to (3.6), we conclude that $H \oplus H = \Gamma_{(A^*)^*} \oplus O\Gamma_{A^*}$. By comparing this formula with (3.15), we get $\Gamma_{(A^*)^*} = \Gamma_A$ and this implies that (3.14) holds for closed A.

Now assume that A admits a closure $\tilde{A}$. Let us apply the already proved part of the theorem with A replaced by $\tilde{A}$. According to (3.14), we obtain $((\tilde{A})^*)^* = \tilde{A}$. At the same time, $(\tilde{A})^* = A^*$, and this means that (3.14) follows from the last equality.

Let us prove the converse statement of the theorem: $\exists (A^*)^* \Rightarrow A$ is closable. Since O is a unitary operator in $H \oplus H$, we have $(O\Gamma_A)^{\sim} = O\tilde{\Gamma}_A$ and, therefore, equality (3.5) can be rewritten in the form $H \oplus H = \Gamma_{A^*} \oplus O\tilde{\Gamma}_A$. By applying the operator O to this equality and using the fact that $O^2 = -\mathbb{I}$, we arrive at the equality

$$H \oplus H = O\Gamma_{A^*} \oplus \tilde{\Gamma}_A. \tag{3.16}$$

In addition, decomposition (3.6) rewritten for the closed operator A^* implies that $H \oplus H = \Gamma_{(A^*)^*} \oplus O\Gamma_{A^*}$. Both this decomposition and (3.16) contain the component $O\Gamma_{A^*}$; hence, $\tilde{\Gamma}_A = \Gamma_{(A^*)^*}$. This means that the closure of the graph of the operator A is a graph, i.e., A is closable. □

Examples

3.1 It follows from Theorem 3.3 that the operator A appearing in Example 2.2 has no second adjoint operator $(A^*)^*$. Let us find $\mathcal{D}(A^*)$. Equality (3.2) now means that

$$f(a) \int_a^b \overline{g(x)}\, dx = \int_a^b f(x)\overline{g^*(x)}\, dx \tag{3.17}$$

for all $f \in C([a, b])$ with some $g^* \in L_2$. Consider a sequence of functions $f_n \in C([a, b])$ $(n \in \mathbb{N})$ such that, $f_n(a) = 1$ for any n and $\|f_n\|_{L_2} \to 0$ as $n \to \infty$.

By substituting f_n for f in (3.17) and passing there to the limit, we arrive at the equality $(1, g)_{L_2} = 0$. But then it follows from (3.17) that the equality $(f, g^*)_{L_2} = 0$ holds for all $f \in C([a, b])$ and, consequently, $g^* = 0$. Thus, in this case, $\mathcal{D}(A^*)$ is not dense in L_2 and consists of all functions from L_2 that are orthogonal to 1; A^* is the null operator defined in this $\mathcal{D}(A^*)$.

3.2 It is clear that the operator A in Example 3.1 can be modified so that $\mathcal{D}(A^*) = \{0\}$. As in Examples 2.2 and 3.1, we set $H = L_2((a, b)) = L_2$ and fix a sequence of functions, $(\varphi_j)_{j=1}^{\infty}$, $\varphi_j \in C([a, b])$, total in L_2 and such that the series $\sum_{j=1}^{\infty} |\varphi_j(x)|$ is uniformly convergent in $[a, b]$. We also fix a sequence $(a_j)_{j=1}^{\infty}$ of distinct points $a_j \in [a, b]$ and set $\mathcal{D}(A) = C([a, b])$ and $(Af)(x) = \sum_{j=1}^{\infty} f(a_j)\varphi_j(x)$ for all $f \in \mathcal{D}(A)$. This series is uniformly convergent. Therefore, the operator A is well-defined in L_2 and its domain is dense. Equality (3.2) now means that (cf. (3.17))

$$\sum_{j=1}^{\infty} f(a_j) \int_a^b \varphi_j(x)\overline{g(x)}\, dx = \int_a^b f(x)\overline{g^*(x)}\, dx \tag{3.18}$$

for any $f \in C([a, b])$. We fix $j = j_0$ and consider a sequence of uniformly bounded functions $f_n \in C([a, b])$ $(n \in \mathbb{N})$ such that $f_n(a_{j_0}) = 1$ for all n and $f_n(x) = 0$ for $x \notin \left(a_{j_0} - \frac{1}{n}, a_{j_0} + \frac{1}{n}\right)$. By substituting f_n for f in (3.18) and passing to the limit as $n \to \infty$, we easily arrive at the equality $(\varphi_{j_0}, g)_{L_2} = 0$. In view of the arbitrariness of $j_0 \in \mathbb{N}$, we have $g = 0$, i.e., $\mathcal{D}(A^*) = \{0\}$.

It is clear that the operators adjoint to the differential operators introduced at the end of Section 2 exist. At the same time, it is not easy to describe their action (because g appearing in (3.2) is nothing but an element of L_2 and one cannot *a priori* expect that it is smooth). We consider this problem in Section 16.2.

3.3 The Closed Graph Theorem

The following theorem is quite important for the investigation of the problems under consideration:

Theorem 3.4 (Banach closed graph theorem). *Let A be a closed operator acting on a Hilbert space H and defined on the whole H, i.e., $\mathcal{D}(A) = H$. Then A is necessarily bounded.*

Proof. First, we prove that A^* is bounded in $\mathcal{D}(A^*)$ (in this case, it certainly exists). Assume the contrary. Then there exists a sequence $(g_n)_{n=1}^{\infty} \subset \mathcal{D}(A^*)$, $\|g_n\|_H = 1$, such that $\|A^* g_n\|_H \to \infty$ as $n \to \infty$.

For any $n \in \mathbb{N}$, consider the functionals

$$l_n(f) = (f, A^* g_n)_H \quad (f \in H); \quad \|l_n\| = \|A^* g_n\|_H \xrightarrow[n\to\infty]{} \infty. \tag{3.19}$$

On the other hand, by virtue of the equality

$$(Af, g)_H = (f, A^* g)_H \qquad (f \in \mathcal{D}(A) = H,\ g \in \mathcal{D}(A^*))$$

we have $l_n(f) = (Af, g_n)_H$ and, therefore ,

$$|l_n(f)| \leq \|Af\|_H = c \quad (f \in H, n \in \mathbb{N}).$$

According to the Banach-Steinhaus theorem (7.7.1), the norms $\|l_k\|$ are bounded in n but this contradicts (3.19).

Since A is closed, $\mathcal{D}(A^*)$ is dense in H by virtue of Theorem 3.3 and, hence, $\tilde{A}^* \in \mathcal{L}(H)$. The same theorem implies that

$$A = (A^*)^* = (\tilde{A}^*)^* \in \mathcal{L}(H). \qquad \square$$

Thus, unbounded operators which are encountered quite often (closed or admitting a closure) cannot be defined in the whole H because they must have nontrivial domains. (This was first mentioned in Section 1.)

REMARK 3.2. The Banach closed graph theorem can be derived from the Banach inverse operator theorem (Theorem 8.3.4). For this purpose, in a subspace $\Gamma_A \subset H \oplus H$, we define an operator $\Gamma_A \ni \langle f, Af\rangle \mapsto P_1\langle f, Af\rangle = f \in H$. This is a linear continuous operator that maps Γ_A onto the whole H in a one-to-one manner. According to the Banach inverse operator theorem, the inverse operator $P_1^{-1}\colon H \mapsto \Gamma_A$ exists; moreover, it is linear and continuous. Further, the operator $P_2\langle f, Af\rangle = Af$ acting from Γ_A into H is also linear and continuous. But this means that $A = P_2P_1^{-1}$ is a linear continuous operator in H.

Exercises

3.1. Let $(e_n)_{n=1}^\infty$ be an orthonormal basis in H and let $(\alpha_n)_{n=1}^\infty \subset \mathbb{C}$ be a fixed sequence. We set

$$Ax = \sum_{n=1}^{\infty} \alpha_n x_n e_n \quad \text{for} \quad x \in \mathcal{D}(A) = \text{l.s.}\left((e_n)_{n=1}^\infty\right).$$

Find A^*.

3.2. Find A^* for the following operators acting on $L_2([0,1])$:

(a) $(Ax)(t) = x(t^2)$, $\mathcal{D}(A) = \{x \in L_2([0,1]) \mid \int_0^1 |x(t^2)|^2 dt < \infty\}$;

(b) $(Ax)(t) = tx(0)$, $\mathcal{D}(A) = C([0,1])$.

3.3. Find adjoint operators for the operators introduced in Exercises 2.6–2.8.

3.4. Let A be the operator of differentiation $(Ax)(t) = x'(t)$ defined on the following domains in $L_2([0,\infty))$:

(a) $\mathcal{D}_1 = C_0^\infty([0,\infty))$;

(b) $\mathcal{D}_2 = \{x \mid x(0) = 0,\ \exists x' \in L_2([0,\infty))\}$.

Find A^*.

3.5. Find the operator adjoint to the Laplace operator $\Delta = D_1^2 + \cdots + D_N^2$ in $L_2(\mathbb{R}^N)$ with the domain $C_0^\infty(\mathbb{R}^N)$.

3.6. Let A be a linear operator densely defined in H. Prove that $\mathcal{R}(A)^\perp = \operatorname{Ker} A^*$. Check whether $(\operatorname{Ker} A)^\perp = (\mathcal{R}(A^*))^\sim$.

3.7. Give an example of an operator A acting on l_2 such that $\mathcal{D}(A^*) = \{0\}$.

4 Defect Numbers of General Operators

Here, we introduce some notions related to the problem of invertibility of operators.

4.1 Deficient Subspaces

Consider an operator A acting on a Hilbert space H with domain $\mathcal{D}(A)$ (which may be not dense in H; at the same time, the situations where $\mathcal{D}(A) = H$ and A is bounded are possible).

A point $z \in \mathbb{C}$ is called a point of regular type for the operator A if there exists $c_z > 0$ such that

$$\|(A - z\mathbb{I})f\|_H \geq c_z\|f\|_H \qquad (f \in \mathcal{D}(A)). \tag{4.1}$$

Denote $(A - z\mathbb{I})f = g$. Then (4.1) means that the inverse operator $\mathcal{R}(A - z\mathbb{I}) \ni g \mapsto (A - z\mathbb{I})^{-1}g \in \mathcal{D}(A) \subseteq H$ exists and is continuous. According to the definition in Section 8.8, it is natural to say that a *point z of regular type is regular if $\mathcal{R}(A - z\mathbb{I}) = H$*. Thus, the notion of points of regular type generalizes the notion of regular points.

Let us establish several simple properties of regular-type points.

(i) For a given operator A, the set of points of regular type is open.

Indeed, let z_0 be a point of regular type for the operator A. It is necessary to indicate a neighbourhood that consists only of points of regular type. For $z \in \mathbb{C}$ and $f \in \mathcal{D}(A)$, we have

$$\begin{aligned}\|(A - z\mathbb{I})f\|_H &= \|(A - z_0\mathbb{I})f - (z - z_0)f\|_H \\ &\geq \|(A - z_0\mathbb{I})f\|_H - |z - z_0|\,\|f\|_H \\ &\geq c_{z_0}\|f\|_H - |z - z_0|\,\|f\|_H = (c_{z_0} - |z - z_0|)\|f\|_H.\end{aligned}$$

Hence, for z from the circle $|z - z_0| < \frac{c_{z_0}}{2}$, we have

$$\|(A - z\mathbb{I})f\|_H \geq \frac{c_{z_0}}{2}\|f\|_H.$$

This means that all z of this sort are points of regular type. □

(ii) Let A be closed and let $z \in \mathbb{C}$ be a point of regular type. Then $\mathcal{R}(A - z\mathbb{I})$ is a subspace (i.e., $\mathcal{R}(A - z\mathbb{I})$ is closed). Conversely, let z be a point of regular type and let $\mathcal{R}(A - z\mathbb{I})$ be a subspace. Then A is closed.

Indeed, assume that $g_n \in \mathcal{R}(A - z\mathbb{I})$ and $g_n \to g \in H$ as $n \to \infty$. It is necessary to prove that $g \in \mathcal{R}(A - z\mathbb{I})$. Consider $f_n \in \mathcal{D}(A)$ such that $(A - z\mathbb{I})f_n = g_n$. For any $n, m \in \mathbb{N}$, we get

$$\|g_n - g_m\|_H = \|(A - z\mathbb{I})(f_n - f_m)\|_H \geq c_z\|f_n - f_m\|_H,$$

whence we conclude that the sequence $(f_n)_{n=1}^{\infty}$ is fundamental. Let $f = \lim_{n\to\infty} f_n$. Then $Af_n = g_n + zf_n \xrightarrow[n\to\infty]{} g + zf$. Since A is closed, $f \in \mathcal{D}(A)$ and $Af = g + zf$, i.e., $g \in \mathcal{R}(A - z\mathbb{I})$. Let us prove the converse statement. Let $\mathcal{D}(A) \ni f_n \to f$ and $Af_n \to g$ as $n \to \infty$. Since $\mathcal{R}(A - z\mathbb{I})$ is closed, there exists $u \in \mathcal{D}(A)$ such that $Af_n - zf_n \to (A - z\mathbb{I})u = h \in \mathcal{R}(A - z\mathbb{I})$. At the same time,

$$\|f_n - u\|_H \le c_z^{-1}\|(A - z\mathbb{I})(f_n - u)\| = c_z^{-1}\|Af_n - zf_n - h\|_H \to 0 \quad \text{as } n \to \infty.$$

Then $f = u \in \mathcal{D}(A)$ and

$$Af = (A - z\mathbb{I})f + zf = h + zf = \lim_{n\to\infty} \big((Af_n - zf_n) + zf_n\big) = g. \qquad \square$$

(iii) Assume that the operator A admits a closure $\tilde{A}$. Every point z of regular type for the operator A is also a point of regular type for $\tilde{A}$. Furthermore,

$$\mathcal{R}(\tilde{A} - z\mathbb{I}) = (\mathcal{R}(A - z\mathbb{I}))^{\sim}. \tag{4.2}$$

Certainly, z is also a point of regular type for $\tilde{A}$; this immediately follows from the definition of $\tilde{A}$ if we pass to the limit in inequality (4.1). Taking into account that $A \subseteq \tilde{A}$, where the operator $\tilde{A}$ is closed, and assertion (ii), we conclude that $\mathcal{R}(A - z\mathbb{I}) \subseteq \mathcal{R}(\tilde{A} - z\mathbb{I}) \Rightarrow (\mathcal{R}(A - z\mathbb{I}))^{\sim} \subseteq \mathcal{R}(\tilde{A} - z\mathbb{I})$.

Let us prove the inverse inclusion. Let $g \in \mathcal{R}(\tilde{A} - z\mathbb{I})$ and $g = (\tilde{A} - z\mathbb{I})f$ ($f \in \mathcal{D}(\tilde{A})$). According to the definition of $\tilde{A}$, there exists $(f_n)_{n=1}^{\infty} \subset \mathcal{D}(A)$ such that $f_n \to f$ and $Af_n \to \tilde{A}f$ as $n \to \infty$. But then $\mathcal{R}(A - z\mathbb{I}) \ni (A - z\mathbb{I})f_n \to g$ and, therefore, $g \in (\mathcal{R}(A - z\mathbb{I}))^{\sim}$. Thus, $\mathcal{R}(\tilde{A} - z\mathbb{I}) \subseteq (\mathcal{R}(A - z\mathbb{I})^{\sim}$. $\square$

Let $z \in \mathbb{C}$ be a point of regular type for the considered operator A.

The subspace $N_z = H \ominus (\mathcal{R}(A - z\mathbb{I})) = (\mathcal{R}(A - z\mathbb{I}))^{\perp}$ is called the deficient subspace of the operator A corresponding to z.

Hence, we can write the decomposition

$$H = (\mathcal{R}(A - z\mathbb{I}))^{\sim} \oplus N_z. \tag{4.3}$$

If the operator A admits a closure or is closed, then, according to (iii) and (ii), decomposition (4.3) can be rewritten in the form

$$H = (\mathcal{R}(\tilde{A} - z\mathbb{I})) \oplus N_z \quad \text{or} \quad H = \mathcal{R}(A - z\mathbb{I}) \oplus N_z, \tag{4.4}$$

respectively.

Let us describe deficient subspaces in a somewhat different way.

By analogy with Section 8.8, we say that *φ is an eigenvector of the operator B with a domain $\mathcal{D}(B)$ if $0 \ne \varphi \in \mathcal{D}(B)$ and $B\varphi = \lambda\varphi$ with some $\lambda \in \mathbb{C}$, which is called the eigenvalue corresponding to the eigenvector φ.*

The set $\Phi(\lambda)$ which consists of 0 and all eigenvectors corresponding to the same eigenvalue λ is linear. It is clear that if B is closed, then the set $\Phi(\lambda)$ is also closed. We say *that $\Phi(\lambda)$ is the eigensubspace corresponding to λ.*

(iv) Assume that the domain of A is dense and, therefore, A^* exists. Then the deficient subspace N_z coincides with the eigensubspace of the operator A^* corresponding to the eigenvalue $\overline{z}$.

Indeed, let $\varphi \in N_z$. Then

$$(\forall f \in \mathcal{D}(A)): ((A - z\mathbb{I})f, \varphi)_H = 0 \Rightarrow (Af, \varphi)_H = (f, \overline{z}\varphi)_H \Rightarrow \varphi \in \mathcal{D}(A^*)$$

and $A^*\varphi = \overline{z}\varphi$. Therefore, $N_z \subseteq \Phi(\overline{z})$ where $\Phi(\overline{z})$ denotes the indicated eigensubspace. Conversely, if $A^*\varphi = \overline{z}\varphi$, then

$$(\forall f \in \mathcal{D}(A)): (zf, \varphi)_H = (f, A^*\varphi)_H = (Af, \varphi)_H,$$

whence $\varphi \perp \mathcal{R}(A - z\mathbb{I})$, i.e., $\Phi(\overline{z}) \subseteq N_z$. □

4.2 Defect Numbers

Let us formulate the main result of this section. Consider an operator A whose domain $\mathcal{D}(A)$ may be not dense. Let z be a point of regular type. It is obvious that deficient subspaces N_z corresponding to different z's are different. At the same time, a remarkable property of these spaces is that the dimension $\dim N_z$ is invariant under changes of z. More precisely, the following theorem is true:

Theorem 4.1 (Krasnoselsky-Krein). *Let A be a closed operator in H. Then $n_z = \dim N_z$ is invariant under the changes of z within a connected component of the set of points z of regular type for the operator A. Thus, every component G of this sort can be associated with a fixed number n_z, where $z \in G$. This number is called the defect number of the operator A (in the component G).*

Proof. First, we localize the problem. Below, it will be shown that, for every point z_0 of regular type, one can indicate a neighbourhood $U(z_0)$ that consists of regular-type points and $\dim N_z = \dim N_{z_0}$ for any $z \in U(z_0)$. This would be enough to prove the theorem. In fact, let $z_1, z_2 \in G$, where G is a connected component of the open set of regular-type points. We connect z_1 and z_2 by a closed rectifiable curve $\gamma \subset G$ and, for every point $z \in \gamma$, consider a neighbourhood $U(z) \subseteq G$, which exists according our assumption. Then we select a finite subcovering from the covering of γ by these neighbourhoods. For z from each of this neighbourhoods, $\dim N_z$ remains unchanged. Thus, by passing from the point z_1 to z_2 step by step, we finally conclude that $\dim N_{z_1} = \dim N_{z_2}$.

Let us return to the main part of the proof and find the required neighbourhood $U(z_0)$. Assume the contrary. Then one can find a sequence $(z_n)_{n=1}^{\infty}$ of points z_n of regular type such that $\lim_{n\to\infty} z_n = z_0$ and, at the same time, $\dim N_{z_n} \neq \dim N_{z_0} (n \in \mathbb{N})$. The following two situations may occur:

(a) the sequence $(z_n)_{n=1}^{\infty}$ contains a subsequence (it is also denoted by $(z_n)_{n=1}^{\infty}$) such that $\dim N_{z_n} < \dim N_{z_0}$ $(n \in \mathbb{N})$;

(b) $(\forall n \in \mathbb{N})$: $\dim N_{z_n} > \dim N_{z_0}$.

Consider case (a). Denote by $P_{N_{z_0}}$ the orthogonal projector onto the subspace N_{z_0}. The corresponding image $P_{N_{z_0}}N_{z_n}$ is such that

$$\dim(P_{N_{z_0}}N_{z_n}) \le \dim N_{z_n} < \dim N_{z_0}$$

(the operation of projection cannot increase the dimension of a subspace). Therefore, one can choose $0 \neq g_n \in N_{z_0} \ominus P_{N_{z_0}}N_{z_n} \quad (n \in \mathbb{N})$.

It is easy to show that $g_n \perp N_{z_n}$. Indeed, let $h \in N_{z_n}$ and $h = h_1 + h_2$ ($h_1 = P_{N_{z_0}}h$) be the decomposition of this vector by the projector $P_{N_{z_0}}$. Then

$$(g_n, h)_H = (g_n, h_1)_H + (g_n, h_2)_H = 0$$

because the first term vanishes in view of the fact that $g_n \in N_{z_0} \ominus P_{N_{z_0}}N_{z_n}$ and $h \in N_{z_0}$ and the second term is zero since $h_2 \perp N_{z_0}$.

For a closed operator A, we can write a decomposition of the form (4.4) (the second formula)

$$H = (\mathcal{R}(A - z_n\mathbb{1})) \oplus N_{z_n} \quad (n \in \mathbb{N}). \tag{4.5}$$

Since $g_n \perp N_{z_n}$, it follows from (4.5) that $g_n \in \mathcal{R}(A - z_n\mathbb{1})$, i.e., there exists $f_n \in \mathcal{D}(A)$, $f_n \neq 0$, such that $g_n = (A - z_n\mathbb{1})f_n$. By multiplying f_n by a scalar, we can guarantee that $\|f_n\|_H = 1$. In addition, $g_n \in N_{z_0}$; therefore, $g_n \perp \mathcal{R}(A - z_0\mathbb{1})$. In particular, $g_n \perp (A - z_0\mathbb{1})f_n$. Thus, we can write

$$((A - z_n\mathbb{1})f_n, (A - z_0\mathbb{1})f_n)_H = 0 \quad (n \in \mathbb{N}). \tag{4.6}$$

The left-hand side of (4.6) can be transformed as follows:

$$0 = ((A - z_n\mathbb{1})f_n, (A - z_0\mathbb{1})f_n)_H = ((A - z_0\mathbb{1})f_n - (z_n - z_0)f_n,$$
$$(A - z_0\mathbb{1})f_n)_H = \|(A - z_0\mathbb{1})f_n\|_H^2 - (z_n - z_0)(f_n, (A - z_0\mathbb{1})f_n)_H,$$

whence

$$\begin{aligned}\|(A - z_0\mathbb{1})f_n\|_H^2 &\le |z_n - z_0||f_n, (A - z_0\mathbb{1})f_n)_H| \\ &\le |z_n - z_0|\|f_n\|_H\|(A - z_0\mathbb{1})f_n\|_H, \\ \|(A - z_0\mathbb{1})f_n\|_H &\le |z_n - z_0|\|f_n\|_H \\ &(\|f_n\|_H = 1, n \in \mathbb{N}).\end{aligned}$$

Since $z_n \to z_0$ as $h \to \infty$, the last inequality contradicts the fact that z_0 is a point of regular type for A.

Case (b) can be investigated similarly. Consider the projector $P_{N_{z_n}}$ onto the subspace N_{z_n}. Then

$$\dim(P_{N_{z_n}}N_{z_0}) \le \dim N_{z_0} < \dim N_{z_n}.$$

We choose $0 \neq g_n \in N_{z_n} \ominus P_{N_{z_n}} N_{z_0}$. As above, $g_n \perp N_{z_0} (n \in \mathbb{N})$. We now use the decomposition $H = (\mathcal{R}(A - z_0 \mathbb{I})) \oplus N_{z_0}$. Since $g_n \in \mathcal{R}(A - z_0 \mathbb{I})$, one can find $f_n \in \mathcal{D}(A)$ such that $g_n = (A - z_0 \mathbb{I}) f_n$ and we can assume that $\|f_n\|_H = 1$ $(n \in \mathbb{N})$. In addition, $g_n \in N_{z_n}$; therefore, $g_n \perp \mathcal{R}(A - z_n \mathbb{I})$ and, in particular, $g_n \perp (A - z_n \mathbb{I}) f_n$. Relation (4.6) takes the form

$$((A - z_0 \mathbb{I}) f_n, (A - z_n \mathbb{I}) f_n)_H = 0, \quad \|f_n\|_H = 1 \quad (n \in \mathbb{N}).$$

By using this equality, we arrive at a contradiction just as in case (a), where we have used (4.6). □

REMARK 4.1. Theorem 4.1 has the same form in the case of closable operators A. More precisely, in (iii), we have shown that if z is a point of regular type for A, then it is also a point of regular type for $\tilde{A}$ and equality (4.2) holds. This equality implies that the operators A and $\tilde{A}$ have the same deficient subspace N_z, the same connected components of the set of points of regular type, and the same defect numbers that correspond to these components.

REMARK 4.2. The quantity $\dim N_z$ can also be regarded as the dimension of the subspace N_z in the sense of Section 8.10, i.e., as the cardinality of an orthonormal basis in N_z. In this case, Theorem 4.1 remains valid but its proof becomes slightly more complicated. Clearly, this remark is meaningful only for nonseparable spaces.

Note that Theorem 4.1 also holds for nonclosable operators. (We do not present the proof of this assertion.)

5 Hermitian and Selfadjoint Operators. General Theory

As we have already seen in Chapters 8 and 10, the class of bounded operators acting on a Hilbert space H contains the important subclass of selfadjoint operators, i.e., the operators A for which

$$(Af, g)_H = (f, Ag)_H \qquad (f, g \in H)$$

or, in other words, $A^* = A$. In the case of unbounded operators, these operators play a similar or even more important role. Note that this definition can be generalized to the case of unbounded operators either on the basis of the first equality presented above or on the basis of the second equality. As a result, we obtain two different classes of operators.

5.1 Hermitian Operators

Let A be an operator in H with a dense domain. A is called Hermitian if

$$(Af, g)_H = (f, Ag)_H \quad (f, g \in \mathcal{D}(A)). \tag{5.1}$$

An operator A with a dense domain is called selfadjoint if

$$A^* = A. \tag{5.2}$$

Let A be an Hermitian operator, i.e., (5.1) is true. This equality means that g belongs to $\mathcal{D}(A^*)$ and $A^*g = Ag$, i.e., $A \subseteq A^*$. It is also clear that the last inclusion yields (5.1). Thus, for an operator A with a dense domain, the fact that it is Hermitian is equivalent to the inclusion

$$A \subseteq A^*, \tag{5.3}$$

while its selfadjointness is equivalent to equality (5.2). Note that the theory of spectral decompositions presented in Chapter 10 admits a generalization just to the case of selfadjoint operators. At the same time, the theory of more general Hermitian operators is much more primitive. (Note that the Hermitian property can also be defined by (5.1) for operators with nondense domains; however, these operators are mostly outside the scope of our interests.)

An Hermitian operator A always admits a closure; this follows from inclusion (5.3) and the fact that A^* is closable (see Remark 2.2). Its closure $\tilde{A}$ is also an Hermitian operator. Indeed, $\tilde{A} \subseteq \tilde{A}^* = A^* = (\tilde{A})^*$, i.e., (5.3) is satisfied.

Let us give a useful definition:

An operator A is called essentially selfadjoint if its closure $\tilde{A}$ is selfadjoint.

Lemma 5.1. *Any $z \in \mathbb{C} \setminus \mathbb{R}$ is a point of regular type for an arbitrary Hermitian operator.*

Proof. We have

$$\begin{aligned} \|(A - z\mathbb{I})f\|_H^2 &= ((A - x\mathbb{I})f - iyf, (A - x\mathbb{I})f - iyf)_H \\ &= \|(A - x\mathbb{I})f\|_H^2 + iy((A - x\mathbb{I})f, f)_H - iy(f, (A - x\mathbb{I})f)_H + y^2\|f\|_H^2 \\ &\geq y^2\|f\|_H^2 \end{aligned} \tag{5.4}$$

for any $f \in \mathcal{D}(A)$ and $z = x + iy \quad (x, y \in \mathbb{R})$ (here, we have used the fact that the operator $A - x\mathbb{I}$ is Hermitian. It follows from (5.4) that

$$\|(A - z\mathbb{I})f\|_H \geq |\mathrm{Im} z| \, \|f\|_H \quad (f \in \mathcal{D}(A)), \tag{5.5}$$

i.e., inequality (4.1) is true with $c_z = |\operatorname{Im} z|$. □

Hence, the set of regular-type points of Hermitian operator has two connected components — the upper and lower half planes. According to Theorem 4.1, these half planes have defect numbers $m = \dim N_z$ ($\mathrm{Im} z > 0$) and $n = \dim N_z$ ($\mathrm{Im} z < 0$); the pair (m, n) is called the deficiency index of the operator A. Clearly, $\tilde{A}$ and A have the same deficiency indices.

As in the case of bounded operators, one can easily prove that *eigenvalues of Hermitian operators are always real.*

Indeed, if A is an Hermitian operator and $A\varphi = \lambda\varphi$ for some $0 \neq \varphi \in \mathcal{D}(A)$, then

$$\lambda(\varphi, \varphi)_H = (A\varphi, \varphi)_H = (\varphi, A\varphi)_H = \overline{\lambda}(\varphi, \varphi)_H,$$

whence it follows that $\lambda \in \mathbb{R}$. □

5.2 Criterion of Selfadjointness

To establish the selfadjointness of an operator one can use e.g., the following criterion:

Theorem 5.1. *A closed Hermitian operator is selfadjoint if and only if its defect numbers are equal to zero, i.e.,* $m = n = 0$. *In other words, an Hermitian operator* A *is selfadjoint if the equalities*

$$\mathcal{R}(A - z_1 \mathbb{I}) = H, \quad \text{and} \quad \mathcal{R}(A - z_2 \mathbb{I}) = H \tag{5.6}$$

hold for some $z_1, z_2 \in \mathbb{C}$, *where* $\mathrm{Im}\, z_2 < 0$ *and* $\mathrm{Im}\, z_1 > 0$. *Conversely, if* A *is selfadjoint, equalities (5.6) hold for any* z_1 *and* z_2 *of the indicated type.*

Proof. *Necessity.* Let $A = A^*$. It is necessary to prove that $(\forall z \in \mathbb{C} \setminus \mathbb{R})\colon N_z = \{0\}$. However, by virtue of statement (iv) in Section 4, N_z coincides with the eigensubspace of the operator A^* that corresponds to a nonreal eigenvalue $\overline{z}$. The assumption $N_z \neq \{0\}$ contradicts the already established fact that the eigenvalues of Hermitian operators are real.

Sufficiency. Prove that $A^* \subseteq A$. The required equality $A^* = A$ is a consequence of this inclusion and (5.3). Fix $z \in \mathbb{C} \setminus \mathbb{R}$. Since $m = n = 0$, we have $N_z = N_{\overline{z}} = \{0\}$. Let $g \in \mathcal{D}(A^*)$. Then $(A^* - z\mathbb{I})g \in H$ and, by virtue of the equality $N_z = \{0\}$, there exists $f \in \mathcal{D}(A)$ such that $(A - z\mathbb{I})f = (A^* - z\mathbb{I})g$. But $A \subseteq A^*$; therefore, $Af = A^*f$ and the last equality can be rewritten in the form $(A^* - z\mathbb{I})f = (A^* - z\mathbb{I})g$ or

$$A^*(f - g) = z(f - g). \tag{5.7}$$

Equality (5.7) means that $f - g$ is an eigenvector of the operator A^* corresponding to z. According to statement (iv) in Section 4, $f - g \in N_{\overline{z}} = \{0\}$, whence $g = f \in \mathcal{D}(A)$. Thus, $\mathcal{D}(A^*) \subseteq \mathcal{D}(A)$ and this implies that $A^* \subseteq A$.

It remains to prove the last part of the theorem connected with equalities (5.6). This would be just a reformulation of the first part if we manage to show that the validity of (at least) one relation in (5.6) guarantees that A is closed. This can be easily proved by direct calculations (see Section 4, (ii)) but one can also apply Theorem 1.1. Indeed, the first equality in (5.6) means that the operator $(A-z_1\mathbb{I})^{-1}$ exists, is bounded and defined in the whole H (and, therefore, closed). Then, according to (1.7), the graph $\Gamma_{A-z_1\mathbb{I}} = U\Gamma_{(A-z_1\mathbb{I})^{-1}}$ is closed, i.e., the operator $A - z_1\mathbb{I}$ and, consequently, the operator A are closed. □

Corollary 5.1. *Let A be an Hermitian operator (generally speaking, nonclosed). It is essentially selfadjoint provided that its defect numbers are equal to zero.*

5.3 Semibounded Operators

Assume that an Hermitian operator A is such that at least one of its points of regular type lies on the real axis. Then the set of all points of regular type of the operator A is connected and, therefore, its defect numbers are equal, i.e., $m = n$ (note that this nothing but a sufficient (not necessary) condition for the operator A to have equal defect numbers).

Semibounded operators constitute an important class of Hermitian operators with this property.

Let A be an operator with dense domain $\mathcal{D}(A)$. Assume that there exists $\alpha \in \mathbb{R}$ such that

$$(Af, f)_H \geq \alpha\|f\|_H^2 \quad (f \in \mathcal{D}(A)). \tag{5.8}$$

The operator A is called semibounded (below), and the number α is called a vertex of A (it is clear that the definition of α (for given A) is ambiguous).

The operators semibounded above are defined similarly; the only difference is that (5.8) must be replaced by the inverse inequality. If an operator A is semibounded, then $-A$ is semibounded above.

A semibounded operator with a vertex $\alpha = 0$ is called nonnegative (cf. Section 8.5.3).

The fact that $(Af, f)_H$ is real and the polarization identity imply that $(Af, g)_H = (f, Ag)_H \quad (f, g \in \mathcal{D}(A))$, i.e., A is an Hermitian operator (this formula was first considered in Section 8.5; in Section 14.8, we return to these problems once again in connection with the study of bilinear forms). If A is semibounded and admits a closure $\tilde{A}$, then it is evident that $\tilde{A}$ is also semibounded and has the same vertex.

Lemma 5.2. *Let A be a semibounded operator with a vertex $\alpha \in \mathbb{R}$. Any $z \in \mathbb{R} \setminus [\alpha, +\infty)$ is a point of regular type for this operator.*

Proof. We set $\varepsilon = \alpha - z > 0$. Then, for any $f \in \mathcal{D}(A)$, we can write

$$\begin{aligned}\|(A-z\mathbb{I})f\|_H^2 &= ((A-\alpha\mathbb{I})f+\varepsilon f, (A-\alpha\mathbb{I})f+\varepsilon f)_H \\ &= \|(A-\alpha\mathbb{I})f\|_H^2 + \varepsilon((A-\alpha\mathbb{I})f, f)_H + \varepsilon(f, (A-\alpha\mathbb{I})f)_H + \varepsilon^2\|f\|_H^2 \\ &\geq \varepsilon^2\|f\|_H^2.\end{aligned}$$

Here, we have used the fact that (5.8) is equivalent to the inequality

$$((A-\alpha\mathbb{I})f, f) = (f, (A-\alpha\mathbb{I})f)_H \geq 0.$$

□

Theorem 5.2. *Let A be a closed semibounded operator with a vertex $\alpha \in \mathbb{R}$. It has equal defect numbers. In order for this operator to be selfadjoint, it is sufficient that*

$$\mathcal{R}(A - z\mathbb{I}) = H \tag{5.9}$$

for some $z \in \mathbb{C} \setminus [\alpha, +\infty)$.

Proof. The fact that the defect numbers m and n are equal follows from Lemma 5.2. The fact that each $z \in \mathbb{C} \setminus [\alpha, +\infty)$ is a point of regular type is a consequence of Lemmas 5.1 and 5.2. Therefore, (5.9) is equivalent to the equality $m = n = 0$. It remains to apply Theorem 5.1. □

Let $A = A^*$. Recall that *a point $z \in \mathbb{C}$ is called a regular point of the operator A if the inverse operator $R_z = (A - z\mathbb{I})^{-1}$ exists and is bounded and defined on the whole H (this operator is called the resolvent of A).*

Thus, a point z is regular if it is of regular type and $\mathcal{R}(A - z\mathbb{I}) = H$. As in the case of bounded operators, one can easily prove that the set of regular points is open in $\mathbb{C}$ and the Hilbert identity

$$R_z - R_\zeta = (z - \zeta)R_z R_\zeta. \tag{5.10}$$

holds for any two regular points z and ζ. To prove this assertion, one must simply repeat the calculations carried out in Subsection 8.8.

The spectrum of a selfadjoint operator A is defined as the complement of the set of its regular points in $\mathbb{C}$.

Thus, the spectrum $S(A)$ is a closed, generally speaking, unbounded set concentrated on the real axis. If, in addition, the operator A is semibounded with a vertex α, then $S(A) \subseteq [\alpha, +\infty)$ (see Lemmas 5.1 and 5.2).

Examples

5.1 Let $G \subseteq \mathbb{R}^N$ be a bounded (or unbounded) domain considered in Section 2, let $H = L_2(G) = L_2$, and let $\mathcal{L}$ be a differential expression (2.5) with coefficients $a_\alpha \in C^{|\alpha|}(G)$. If $\mathcal{L}$ is formally selfadjoint, i.e., $\mathcal{L}^+ = \mathcal{L}$, then the minimal operator L corresponding to $\mathcal{L}$ is Hermitian. Indeed, for the operator L' (see (2.10)), this immediately follows from (2.7) and we recall that $L = \tilde{L}'$. In particular, the Schrödinger operator, i.e., the minimal operator generated by expression (2.9), is Hermitian.

However, the operator L is, as a rule, not selfadjoint. The problem of establishing the conditions under which this operator or its extensions (corresponding to various boundary conditions on ∂G) are selfadjoint is quite complicated and, at the same time, very important. We dwell upon this problem in Section 16.4. Here, we restrict ourselves to the investigation of the following example:

5.2. Let G be a bounded region in $\mathbb{R}^N$ and let $\mathcal{L} = \mathcal{L}^+$ be a formally selfadjoint expression (2.5) with constant coefficients a_α which is not identically equal to zero. Then the minimal operator L associated with this expression is Hermitian but not selfadjoint. (Note that the condition of formal selfadjointness now has the following form: $(\forall\alpha\colon |\alpha| \le r) : a_\alpha = (-1)^{|\alpha|}\overline{a_\alpha}$, i.e., the coefficient a_α is real for even $|\alpha|$ and imaginary for odd $|\alpha|$; see (2.8).)

Indeed, consider an exponential function $\mathbb{R}^N \ni x = (x_1, \dots, x_N) \mapsto \varphi(x) = \exp(x,\zeta)_{\mathbb{R}^N} \in \mathbb{C}$, where $\zeta = (\zeta_1, \dots, \zeta_N) \in \mathbb{C}^N$ is a given vector. Since G is bounded, $(\forall\zeta)\colon 0 \ne \varphi \in L_2$. At the same time, integrating by parts, we obtain

$$((L' - z\mathbb{I})f, \varphi)_{L_2} = ((\mathcal{L} - z\mathbb{I})f, \varphi)_{L_2} = (f, (\mathcal{L} - \overline{z}\mathbb{I})\varphi)_{L_2} \tag{5.11}$$

for $f \in C_0^r(G)$ and $z \in \mathbb{C}\setminus\mathbb{R}$. But

$$(\mathcal{L}\exp(\,\cdot\,,\zeta)_{\mathbb{R}^N})(x) = \mathcal{L}[\zeta]\exp(x,\zeta)_{\mathbb{R}^N}, \tag{5.12}$$

$$\mathcal{L}[\zeta] = \sum_{|\alpha|\le r} a_\alpha\zeta^\alpha \quad (\zeta^\alpha = \zeta_1^{\alpha_1}\dots\zeta_N^{\alpha_N};\ x \in \mathbb{R}^N,\ \zeta \in \mathbb{C}^N).$$

Thus, if ζ satisfies the equation $\mathcal{L}[\zeta] = \overline{z}$, then $(\mathcal{L} - \overline{z}\mathbb{I})\varphi = 0$ and (5.11) implies that $\mathcal{R}(L - z\mathbb{I}) = (\mathcal{R}(L' - z\mathbb{I}))^\sim$ is not dense in L_2. Since $\mathcal{L} \ne 0$, there exists $\zeta \in \mathbb{C}^N$ such that $\operatorname{Im}(\mathcal{L}[\zeta]) \ne 0$. We set $z = \overline{\mathcal{L}[\zeta]} \in \mathbb{C}\setminus\mathbb{R}$. Then $N_z \ne \{0\}$ and the operator L is not selfadjoint.

This reasoning fails in the case of unbounded G (generally speaking, in this case, $\varphi \notin L_2$). Furthermore, in what follows, we prove that L is always selfadjoint if $G = \mathbb{R}^N$ (see Section 14.4).

In conclusion, we note that the reader may encounter in the literature other names of the classes of operators considered in this section. Thus, Hermitian operators are called symmetric, while selfadjoint operators are called hypermaximal or Hermitian.

Exercises

5.1. Let A and B be Hermitian operators with the same domain $\mathcal{D}$ and let $\alpha, \beta \in \mathbb{R}$. Prove that $\alpha A + \beta B$ is an Hermitian operator.

5.2. Let A be an Hermitian operator whose range $\mathcal{R}(A)$ is dense in H. Prove that

(a) the algebraic inverse A^{-1} exists;

(b) the operator A^{-1} is Hermitian.

5.3. Consider the operators appearing in Exercises 2.6 and 2.7. Are they Hermitian?

Find necessary and sufficient conditions for the operators in Exercises 2.8 and 3.1 to be Hermitian.

5.4. Let A be an Hermitian operator and let (m, n) be its deficiency index. Prove the following assertions:

(a) if $m = 0$ and $n > 0$, then $S(A) = \{z \in \mathbb{C} \mid \operatorname{Im} z \le 0\}$;

(b) if $m > 0$ and $n = 0$, then $S(A) = \{z \in \mathbb{C} \mid \operatorname{Im} z \ge 0\}$;

(c) if $m > 0$ and $n > 0$, then $S(A) = \mathbb{C}$.

5.5. Find the deficiency indices for

(a) the operators introduced in Exercise 2.7;

(b) the operator $(Ax)(t)) = ix'(t)$ defined in $L_2([0,\infty))$ on the domains $\mathcal{D}_1$ and $\mathcal{D}_2$ from Exercise 3.4;

(c) the same operator as in (b) but defined on the set $C_0^\infty((-\infty,0])$ in $L_2((-\infty,0])$.

5.6. Let A_n be an Hermitian operator in H_n with domain $\mathcal{D}(A_n)$. In the space $\mathcal{H} = \oplus_{n=1}^\infty H_n$, we consider a domain $\mathcal{D}$ of vectors $x = (x_1, x_2, \ldots)$ such that $x_n \in \mathcal{D}(A_n)$ and all x_n (except finitely many vectors) are equal to zero. Prove that

(a) the operator $A = \sum_{n=1}^\infty A_n$ defined on $\mathcal{D}(A) = \mathcal{D}$ in $\mathcal{H}$ is Hermitian;

(b) the defect numbers of the operator A are given by the equalities $m = \sum_{k=1}^\infty m_k$ and $n = \sum_{k=1}^\infty n_k$, where (m_k, n_k) is the deficiency index of A_k $(k \in \mathbb{N})$.

5.7. On the basis of the results obtained in Exercises 5.5 and 5.6, construct an Hermitian operator with arbitrary preassigned defect numbers $m, n \in \mathbb{Z}_+$.

5.8. Let A be an Hermitian operator. Show that the following conditions are equivalent:

(i) A is essentially selfadjoint;

(ii) A^* has no nonreal eigenvalues;

(iii) $(\forall z \in \mathbb{C} \setminus \mathbb{R})\colon (\mathcal{R}(A - z\mathbb{1}))^\sim = H$;

(iv) $(\exists z_1\colon \operatorname{Im} z_1 > 0)\ (\exists z_2\colon \operatorname{Im} z_2 < 0)\colon (\mathcal{R}(A - z_1\mathbb{1}))^\sim = (\mathcal{R}(A - z_2\mathbb{1}))^\sim = H$.

5.9. Are the operators considered in Exercises 2.6, 2.7, 5.5(b) and (c), and 3.5 essentially selfadjoint? Are they selfadjoint?

5.10. Assume that the conditions for the operators considered in Exercises 2.8 and 3.1 to be Hermitian are satisfied (see Exercise 5.3). Are these operators essentially selfadjoint? Are they selfadjoint?

5.11. Establish a criterion of essential selfadjointness of a semibounded operator.

5.12. Let A_1 and A_2 be the operators of multiplication by t in $L_2(\mathbb{R})$ with domains $\mathcal{D}(A_1)$ and $\mathcal{D}(A_2)$. It is known that A_1 and A_2 are essentially selfadjoint. Is it possible that $\mathcal{D}(A_1) \cap \mathcal{D}(A_2) = \{0\}$?

5.13. Find two linear subsets $\mathcal{D}_1$ and $\mathcal{D}_2$, dense in $L_2(\mathbb{R})$ such that $\mathcal{D}_1 \cap \mathcal{D}_2 = \{0\}$, the operator of multiplication by t is essentially selfadjoint in $\mathcal{D}_1$, and the operator of multiplication by t^2 is essentially selfadjoint in $\mathcal{D}_2$.

5.14. Let A be an Hermitian operator with domain $\mathcal{D}(A)$. Let $\mathcal{D}_1 \subseteq \mathcal{D}(A)$ be a dense linear subset of H. Assume that the operator $A_1 = A \restriction \mathcal{D}_1$ is essentially selfadjoint. Prove that A is also essentially selfadjoint and $\tilde{A} = \tilde{A}_1$.

6 Isometric and Unitary Operators. Cayley Transformation

6.1 Defect Numbers of Isometric Operators

Isometric and unitary operators have already been studied in Section 8.5. Here, we present some additional facts about these operators and clarify their relationship to Hermitian and selfadjoint operators. As before, all operators considered below act on a Hilbert space H.

An operator U acting from the subspace $\mathcal{D}(U) \subseteq H$ to the subspace $\mathcal{R}(U) \subseteq H$ is called isometric if

$$(Uf, Ug)_H = (f, g)_H \qquad (f, g \in \mathcal{D}(U)). \tag{6.1}$$

This operator is called unitary if, in addition, $\mathcal{D}(U) = \mathcal{R}(U) = H$.

We stress that the isometric operator U is necessarily continuous; therefore, it seems useless to study this operator in a nonclosed linear set $\mathcal{D}(U)$ because it is always possible to pass to the investigation of $\tilde{U}$ in $(\mathcal{D}(U))^{\sim} = \mathcal{D}(\tilde{U})$ closing it by continuity. Thus, we always assume that $\mathcal{D}(U)$ and $\mathcal{R}(U)$ are subspaces and the operator U is closed. Further, recall (see Section 8.5) that equality (6.1) holds for all $f, g \in \mathcal{D}(U)$ if it holds for $g = f \in \mathcal{D}(U)$, i.e., if the operator under consideration preserves the norm.

Let us describe the regions that contain only points of regular type for an isometric operator.

Lemma 6.1. *Every $z \in \mathbb{C}, |z| \neq 1$, is a point of regular type of an isometric operator.*

Proof. Let U be an isometric operator and let $|z| < 1$. Then

$$\|(U - z\mathbb{1})f\|_H \geq \|Uf\|_H - |z|\,\|f\|_H = (1 - |z|)\|f\|_H.$$

Similarly, for $|z| > 1$,

$$\|(U - z\mathbb{1})f\|_H \geq |z|\,\|f\|_H - \|Uf\|_H = (|z| - 1)\|f\|_H. \qquad \square$$

Thus, the structure of the set of regular-type points for isometric operators characterized by the presence of two connected components $\{z \in \mathbb{C} \mid |z| > 1\}$ and $\{z \in \mathbb{C} \mid |z| < 1\}$ is similar to that observed for Hermitian operators. According to Theorem 4.1, their defect numbers are m and n, respectively; (m, n) is the deficiency index of the operator U.

Let us prove the following theorem similar to Theorem 5.1:

Theorem 6.1. *An isometric operator U is unitary if and only if its defect numbers are equal to zero, i.e., $m = n = 0$.*

Proof. The proof of Theorem 6.1 immediately follows from the following useful relations:

$$m = \dim(H \ominus \mathcal{D}(U)) \quad \text{and} \quad n = \dim(H \ominus \mathcal{R}(U)). \tag{6.2}$$

The second relation is evident: $n = \dim N_z$ for $|z| < 1$ and, in particular, $n = \dim N_0$, where $N_0 = H \ominus \mathcal{R}(U)$.

To prove the first relation, we consider the operator U^{-1} with $\mathcal{D}(U^{-1}) = \mathcal{R}(U)$ and $\mathcal{R}(U^{-1}) = \mathcal{D}(U)$. Clearly, this operator exists and is isometric. Let n_1 be its second defect number; then, according to the second formula in (6.2) applied to U^{-1}, we can write

$$\dim(H \ominus \mathcal{R}(U^{-1} - z\mathbb{I})) = n_1 = \dim(H \ominus \mathcal{D}(U))$$

for any $z \in \mathbb{C}$ such that $|z| < 1$.

It remains to show that $\mathcal{R}(U^{-1} - z\mathbb{I}) = \mathcal{R}(U - z^{-1}\mathbb{I})(0 < |z| < 1)$. We have

$$\begin{aligned}\mathcal{R}(U^{-1} - z\mathbb{I}) = (U^{-1} - z\mathbb{I})\mathcal{D}(U^{-1}) = (U^{-1} - z\mathbb{I})\mathcal{R}(U) = (\mathbb{I} - zU)\mathcal{D}(U) \\ = (U - z^{-1}\mathbb{I})\mathcal{D}(U) = \mathcal{R}(U - z^{-1}\mathbb{I}).\end{aligned}$$ □

6.2 Direct Cayley Transformation

The analysis of the facts presented in Section 5 and Subsection 6.1 reveal an analogy between Hermitian and selfadjoint operators on the one hand, and isometric and unitary operators on the other hand. This analogy appears not by chance. Indeed, one can indicate a transformation that transforms these classes into each other (it is called the *Cayley transformation*). Let us study its properties.

In the trivial case of a one-dimensional space H, all linear operators are operators of multiplication by complex numbers and the Cayley transformation is defined as the classical linear-fractional transformation given by the formula

$$\mathbb{C} \ni a \mapsto \frac{a - \bar{z}}{a - z} = u \in \mathbb{C}, \tag{6.3}$$

where $z \in \mathbb{C} \setminus \mathbb{R}$ is a fixed number. Mapping (6.3) transforms $\mathbb{R}$ into the unit circle. Real numbers serve as an analogue of Hermitian operators, while the numbers on the unit circle are an analogue of isometric operators. Therefore, one may expect that the classes of operators introduced in Section 5 turn into the classes of operators from Section 6 under the action of the corresponding generalization of (6.3). (Note that generalizations of this sort are well known in the theory of matrices.)

Thus, let H be a Hilbert space and let A be a closed Hermitian operator in H whose domain $\mathcal{D}(A)$ may be not dense in H. We fix $z \in \mathbb{C}$ with $\operatorname{Im} z > 0$.

Consider a vector $g \in \mathcal{R}(A - z\mathbb{I})$ of the form $g = (A - z\mathbb{I})f$, where $f \in \mathcal{D}(A)$. We construct a mapping $g \mapsto (A - \bar{z}\mathbb{I})f = Ug$. This definition is correct because

f is uniquely determined for given g in view of estimate (5.5). It is also clear that U is a linear operator with domain $\mathcal{R}(A - z\mathbb{I})$ and range $\mathcal{R}(A - \bar{z}\mathbb{I})$. We have

$$g = (A - z\mathbb{I})f, \quad Ug = (A - \bar{z}\mathbb{I})f \quad (f \in \mathcal{D}(A)); \tag{6.4}$$

$$\mathcal{D}(U) = \mathcal{R}(A - z\mathbb{I}), \quad \text{and} \quad \mathcal{R}(U) = \mathcal{R}(A - \bar{z}\mathbb{I}). \tag{6.5}$$

It is clear that relations (6.4) can be rewritten in a more concise form similar to (6.3), namely,

$$Ug = (A - \bar{z}\mathbb{I})\,(A - z\mathbb{I})^{-1}g. \tag{6.6}$$

However, relation (6.6) is less convenient than (6.4) because it is necessary to indicate in what sense the inverse operator exists and take care of the domains of definition.

The operator U *is called the Cayley transform of an operator* A. Let us establish its simple properties.

(i) The Cayley transform of a closed Hermitian operator is an isometric operator.

In fact, it follows from (6.4) that $(\forall f_1, f_2 \in \mathcal{D}(A))$:

$$\begin{aligned}(Ug_1, Ug_2)_H &= ((A - z\mathbb{I})f_1, (A - z\mathbb{I})f_2)_H \\ &= (Af_1, Af_2)_H - \bar{z}(Af_1, f_2)_H - z(f_1, Af_2)_H + |z|^2(f_1, f_2)_H,\end{aligned}$$

$$\begin{aligned}(g_1, g_2)_H &= ((A - \bar{z}\mathbb{I})f_1, (A - \bar{z}\mathbb{I})f_2)_H \\ &= (Af_1, Af_2)_H - z(Af_1, f_2)_H - \bar{z}(f_1, Af_2)_H + |z|^2(f_1, f_2)_H.\end{aligned}$$

Since A is an Hermitian operator, the right-hand sides of these equalities coincide. Therefore, $(Ug_1, Ug_2)_H = (g_1, g_2)$ $(g_1, g_2 \in \mathcal{D}(U))$. □

(ii) Let $m(A)$, $n(A)$ and $m(U)$, $n(U)$ be the defect numbers of the operators A and U, respectively. Then

$$m(A) = m(U) \quad \textit{and} \quad n(A) = n(U). \tag{6.7}$$

These relations can be obtained by comparing equalities (6.2) with (6.5) (recall that $\operatorname{Im} z > 0$). □

(iii) The Cayley transform of a selfadjoint operator is a unitary operator.

This fact follows from relations (6.7). □

(iv) Let $B \supseteq A$ be the closed Hermitian extension of an operator A. Then its Cayley transform V is an isometric extension of the operator U.

This fact is an immediate consequence of relations (6.4) and (i). □

6.3 Inverse Cayley Transformation

Let us study the inverse Cayley transformation. To do this, we express the operator A in terms of U by using relations (6.4). Subtracting the first equality in (6.4) from the second one, we get

$$(U - \mathbb{I})g = (z - \overline{z})f \qquad (f \in \mathcal{D}(A)). \tag{6.8}$$

Further, by multiplying the second (first) equality in (6.4) by z ($\overline{z}$) and subtracting one inequality from the other, we obtain

$$(zU - \overline{z}\mathbb{I})g = (z - \overline{z})Af \qquad (f \in \mathcal{D}(A)). \tag{6.9}$$

Equalities (6.8) and (6.9) can be rewritten in the form

$$f = \frac{1}{z - \overline{z}}(U - \mathbb{I})g, \quad Af = \frac{1}{z - \overline{z}}(zU - \overline{z}\mathbb{I})g. \tag{6.10}$$

Assume now that there exists an isometric operator U acting on H. Then, on the vectors f of the form $f = (z - \overline{z})^{-1}(U - \mathbb{I})g$, where $g \in \mathcal{D}(U)$, we define an operator A by setting $Af = (z - \overline{z})^{-1}(zU - \overline{z}\mathbb{I})g$. This definition is correct and the operator A thus defined is linear provided that

$$\operatorname{Ker}(U - \mathbb{I}) = \{0\}. \tag{6.11}$$

Suppose that condition (6.11) is satisfied. The operator A constructed as a result is called the inverse Cayley transform of the operator U and we have

$$\mathcal{D}(A) = \mathcal{R}(U - \mathbb{I}), \quad \mathcal{R}(A) = \mathcal{R}(zU - \overline{z}\mathbb{I}). \tag{6.12}$$

On a somewhat formal level, the operator A can be expressed in terms of the operator U by a relation similar to (6.6), which is, in fact, its inversion. Indeed,

$$Af = (zU - \overline{z}\mathbb{I})(U - \mathbb{I})^{-1}f. \tag{6.13}$$

Let us establish some properties of the inverse Cayley transformation.

(v) The inverse Cayley transform of an isometric operator is a closed Hermitian operator.

In fact, it follows from (6.10) that ($\forall g_1, g_2 \in \mathcal{D}(U)$):

$$\begin{aligned}(Af_1, f_2)_H &= \left(\frac{1}{z - \overline{z}}(zU - \overline{z}\mathbb{I})g_1, \frac{1}{z - \overline{z}}(U - \mathbb{I})g_2\right)_H \\ &= \frac{1}{|z - \overline{z}|^2}\big(z(Ug_1, Ug_2)_H - z(Ug_1, g_2)_H - \overline{z}(g_1, Ug_2)_H + \overline{z}(g_1, g_2)_H\big),\end{aligned}$$

$$(f_1, Af_2)_H = \left(\frac{1}{z-\bar{z}}(U-\mathbb{I})g_1, \frac{1}{z-\bar{z}}(zU-\bar{z}\mathbb{I})g_2\right)_H$$
$$= \frac{1}{|z-\bar{z}|^2}\big(z(Ug_1, Ug_2)_H - z(Ug_1, g_2)_H - \bar{z}(g_1, Ug_2) + z(g_1, g_2)_H\big).$$

Since U is isometric, these equalities imply that $(Af_1, f_2)_H = (f_1, Af_2)_H$ ($f_1, f_2 \in \mathcal{D}(A)$), i.e., A is Hermitian.

This operator is closed. Indeed, let $(f_n)_{n=1}^{\infty}$, $f_n \in \mathcal{D}(A) = \mathcal{R}(U-\mathbb{I})$, be such that $f_n \to f$ and $Af_n \to h$ as $n \to \infty$. We have

$$f_n = (z-\bar{z})^{-1}(U-\mathbb{I})g_n \quad \text{and} \quad Af_n = (z-\bar{z})^{-1}(zU-\bar{z}\mathbb{I})g_n,$$

where $g_n \in \mathcal{D}(U)$. This enables us to find g_n and, according to (6.4), we have $g_n = (A - z\mathbb{I})f_n$ and $Ug_n = (A-\bar{z}\mathbb{I})f_n$. We pass to the limit in these relations and conclude that $\exists g = \lim g_n = h - zf$ and $Ug = h - \bar{z}f$, where $g \in \mathcal{D}(U)$ (since $\mathcal{D}(U)$ is closed). By expressing f and h in the last two equations in terms of g, we arrive at relations (6.10) with Af replaced by h. But this means that $f \in \mathcal{D}(A)$ and $h = Af$. □

(vi) Defect numbers of the operators U and A satisfy equalities (6.7).

In fact, it follows from (6.10) that $(A - z\mathbb{I})f = g$ and $(A - \bar{z}\mathbb{I})f = Ug$. Therefore, $\mathcal{R}(A - z\mathbb{I}) = \mathcal{D}(U)$ and $\mathcal{R}(A - \bar{z}\mathbb{I}) = \mathcal{R}(U)$. By using these equalities and (6.2), we arrive at (6.7) (recall that $\operatorname{Im} z > 0$). □

(vii) The inverse Cayley transform of a unitary operator is a selfadjoint operator provided that its domain $\mathcal{D}(A) = \mathcal{R}(U-\mathbb{I})$ is dense in H.

This follows from relations (6.7). The denseness of $\mathcal{D}(A)$ in H is necessary for the existence of A^* and, therefore, must be postulated. □

It is evident that a statement similar to (iv) is also true, i.e., $V \supseteq U \Rightarrow B \supseteq A$. However, in this case it is necessary to require that the isometric operator V should satisfy condition (6.11). Since, in what follows, we study only densely defined Hermitian operators, it is more convenient to use the following lemma:

Lemma 6.2. *If $\mathcal{R}(U-\mathbb{I})$ is dense in H, then* $\operatorname{Ker}(U-\mathbb{I}) = \{0\}$.

Proof. Let $h \in \operatorname{Ker}(U-\mathbb{I})$, i.e., $Uh = h$ ($h \in \mathcal{D}(U)$). Then, for any $g \in \mathcal{D}(U)$, by virtue of the isometry of U, we obtain

$$((U-\mathbb{I})g, h)_H = (Ug, h)_H - (g, h)_H = (Ug, Uh)_H - (g, h)_H = 0.$$

Since $(\mathcal{R}(U-\mathbb{I}))^{\sim} = H$, this relation implies that $h = 0$. □

It is now convenient to formulate an analogue of (iv) in the following form:

(viii) Let $V \supseteq U$ be the isometric extension of an isometric operator U such that $\mathcal{R}(U - \mathbb{I})$ is dense in H. Then the operator V possesses the inverse Cayley transform B, which is the closed Hermitian extension of a closed Hermitian operator A.

Indeed, according to Lemma 6.2, the operator U satisfies condition (6.11) and, hence, it is possible to construct the operator A. But $V \supseteq U$. Therefore, $\mathcal{R}(V - \mathbb{I}) \supseteq \mathcal{R}(U - \mathbb{I})$ is also dense in H and, thus, in view of the same lemma, one can construct the operator B. The relation $B \supseteq A$ follows from (6.12) and (v). □

Finally, by comparing the structures (6.4) and (6.10) of the direct and inverse Cayley transforms, we arrive at the following conclusion:

(ix) Let us construct the inverse Cayley transform of an operator U, which is, in turn, the Cayley transform of a given operator A. As a result, we obtain A, i.e., $A \mapsto U \mapsto A$. Similarly, $U \mapsto A \mapsto U$.

Exercises

6.1. Find the Cayley transforms of the operators considered in Exercise 2.6.

6.2. Assume that the operators introduced in Exercises 2.8 and 3.1 are Hermitian (see Exercise 5.3). Find their Cayley transforms.

6.3. Let U be a unitary operator acting on the space l_2 of sequences $(x_k)_{k=-\infty}^{\infty}$ according to the rule $(Ux)_k = x_{k-1}$ $(k \in \mathbb{Z})$. Find its inverse Cayley transform.

7 Extensions of Hermitian Operators to Selfadjoint Operators

In the next chapter, we show that the important theory of expansions in eigenvectors of selfadjoint compact operators (see Section 10.1) can be generalized to the case of selfadjoint operators but not to the case of Hermitian operators. In this connection, it is quite important to study the problem of extension of an arbitrary given Hermitian operator to a selfadjoint operator. For the minimal operator generated by the formally selfadjoint differential expression $\mathcal{L}$ (see Examples 5.1 and 5.2), this problem can be reduced, roughly speaking, to making the definition of the operator by formula (2.10) meaningful not only in the set of finite functions but on a larger set of functions satisfying certain homogeneous boundary conditions on the boundary ∂G.

7.1 The Construction of Extensions

The construction of extensions of Hermitian operators can be reduced, by using the Cayley transformation investigated in Section 6, to the theory of extensions of isometric operators, which looks very simple from the geometric point of view.

Below, we assume that the defect numbers m, n of the operators acting on a Hilbert space H take the values $0, 1, \ldots$, or ∞ (see Section 4). This is indeed true if H is separable. For general H, the numbers m, n are cardinals (see Remark 4.2). All the results presented below remain true in this case but their proofs must be somewhat modified according to the algorithm described in Section 7.10.

Theorem 7.1. *Let U be an isometric operator in H with domain $\mathcal{D}(U)$, range $\mathcal{R}(U)$, and the defect numbers $m = \dim\,(H \ominus \mathcal{D}(U)) > 0$ and $n = \dim(H \ominus \mathcal{R}(U)) > 0$. Fix $k \leq \min\,(m, n)$, choose two k-dimensional subspaces $F \subseteq H \ominus \mathcal{D}(U)$ and $G \subseteq H \ominus \mathcal{R}(U)$, and construct an isometric operator W acting from the whole F to the whole G, i.e., $\mathcal{D}(W) = F$ and $\mathcal{R}(W) = G$. The orthogonal sum*

$$V = U \oplus W, \quad \mathcal{D}(V) = \mathcal{D}(U) \oplus \mathcal{D}(W), \quad \mathcal{R}(V) = \mathcal{R}(U) \oplus \mathcal{R}(W)$$

is an isometric extension of the operator U. All possible isometric extensions of this operator can be obtained by using the same procedure for all possible k, F, G, and W.

Proof. The proof immediately follows from the definition and properties of orthogonal sums of subspaces (see Section 7.9), from the definition of extensions of operators, and from the definition of isometric operators. □

Corollary 7.1. *If at least one defect number (m or n) of an operator U is equal to zero, then this operator has no nontrivial isometric extensions.*

Corollary 7.2. *In order that the operator U admit unitary extensions, it is necessary and sufficient that its defect numbers be equal, i.e., $m = n$. To construct a unitary extension, one must set $F = H \ominus \mathcal{D}(U)$ and $G = H \ominus \mathcal{R}(U)$ and take an isometric operator W with $\mathcal{D}(W) = F$ and $\mathcal{R}(W) = G$.*

In the case where one of the numbers m and n is equal to zero, and the other one is positive, it is natural to say that U is a maximal operator. From this point of view, unitary operators may be called hypermaximal.

These simple facts lay the foundation of the theory of extensions of isometric operators to isometric or unitary operators. We now consider the case of Hermitian operators.

Let A be a closed Hermitian operator with dense domain and let $B \supseteq A$ be its closed Hermitian extension. Since $B^* \subseteq A^*$ and $A \subseteq A^*$, we arrive at the following chain:

$$A \subseteq B \subseteq B^* \subseteq A^*. \tag{7.1}$$

If B is selfadjoint, relation (7.1) turns into

$$A \subseteq B \subseteq A^*. \tag{7.2}$$

Let us clarify the conditions under which, for a given operator A, one can construct an operator B satisfying (7.1) or (7.2) and describe the set of such B.

Theorem 7.2.   *Let A be a closed Hermitian operator with dense domain and defect numbers m and n. In order that A admit nontrivial closed Hermitian extensions $B \supseteq A$, it is necessary and sufficient that m and n be positive. In order that A admit a selfadjoint extension $B = B^* \supseteq A$, it is necessary and sufficient that its defect numbers be equal, i.e., $m = n$.*

These extensions are constructed as follows: Fix a point $z \in \mathbb{C}$ with $\operatorname{Im} z > 0$. In the deficient subspaces $N_z = H \ominus \mathcal{R}(A - z\mathbb{1})$ and $N_{\bar{z}} = H \ominus \mathcal{R}(A - \bar{z}\mathbb{1})$ whose dimensions are $m > 0$ and $n > 0$, respectively, we choose subspaces $F \subseteq N_z$ and $G \subseteq N_{\bar{z}}$ of the same dimension and construct an isometric operator W that maps the whole $F = \mathcal{D}(W)$ onto the whole $G = \mathcal{R}(W)$. Let U be the Cayley transform of the operator A, $\mathcal{D}(U) = \mathcal{R}(A - z\mathbb{1})$, and $\mathcal{R}(U) = \mathcal{R}(A - \bar{z}\mathbb{1})$. Consider an isometric operator

$$ V = U \oplus W, \quad \mathcal{D}(V) = \mathcal{D}(U) \oplus \mathcal{D}(W), \quad \mathcal{R}(V) = \mathcal{R}(U) \oplus \mathcal{R}(W). $$

The inverse Cayley transform B of the operator V is a closed Hermitian extension of the operator A. By taking all possible combinations of F, G, and W (for fixed z), we exhaust the set of closed Hermitian extensions B of the operator A.

If $m = n$, then, in particular, we can set $F = N_z$ and $G = N_{\bar{z}}$. In this case, the operator $B \supseteq A$ is a selfadjoint extension of A. By taking all possible W, we exhaust the set of selfadjoint extensions B of the operator A.

Proof.   The proof of this theorem is, in fact, contained in the procedure of constructing the extension B described above. One should only take into account the properties (i)–(ix) of the Cayley transformation presented in Section 6 and Theorem 7.1. In this case, it is necessary to take into account the fact that the operator A is densely defined. Hence, by virtue of the first formula in (6.10) applied to the transition $A \mapsto U$, the set $\mathcal{R}(U - \mathbb{1})$ is dense in H. Therefore, the condition of (viii) is satisfied and it is possible to construct the inverse Cayley transform B of the operator $V = U \oplus W \supseteq U$ that extends A. It is also clear that this procedure exhausts the set of all closed Hermitian extensions $B \supseteq A$. Indeed, according to (iv), any extension of this sort gives after passing to its Cayley transform V, an isometric extension of U described by Theorem 7.1.   □

It is obvious that, if we change the parametric point $z \in \mathbb{C}$, $\operatorname{Im} z > 0$, then the objects F, G, and W necessary to get the same extension $B \supseteq A$ would also change. It is not difficult to deduce formulas that describe these changes but we do not want to study this problem here.

It follows from Theorem 7.2 that, in the case where one of the numbers m or n is equal to zero and the other one is positive, the operator A does not have closed Hermitian extensions in H. In this case, it is called maximal. If $m = n = 0$, then this operator is selfadjoint or hypermaximal (this terminology now becomes clear). Also note that if B is an Hermitian extension of A, then $\tilde{B}$ is a closed Hermitian extension of this operator. Therefore, the theory presented above also describes nonclosed Hermitian extensions of the operator A.

Here, we do not consider extensions of the operator A that lead out of the space H. We only note that if $m \neq n$, then one can "equalize" the defect numbers by finding a proper embedding of H into a broader Hilbert space $\hat{H}$. Then we can treat the operator A as acting on $\hat{H}$ and construct its extensions. Thus, we can set $\hat{H} = H \oplus H$, $\hat{A} = A \oplus (-A)$, and $\mathcal{D}(\hat{A}) = \mathcal{D}(A) \oplus \mathcal{D}(A)$. It is now not difficult to show that the operator $\hat{A}$ has equal defect numbers $m+n$, $m+n$. The extensions of this operator are, in a certain sense, extensions of the original operator A.

7.2 Von Neumann Formulas

The theory of extensions presented above and the results in Section 6 are due mainly to von Neumann. We consider two "analytic" formulas related to this theory ((i) and (ii) below), which are called the first and second von Neumann formulas, respectively.

(i) We consider a formula that describes the action of the operator A^*. First, it is necessary to recall that a linear set $L \subseteq H$ is called the direct sum of linear sets $L_1, \ldots, L_n \subseteq H$ if, for any $f \in L$, one can write a representation $f = f_1 + \cdots + f_n$, where $f_j \in L_j$, and this representation is unique (in other words, $0 = f_1 + \cdots + f_n \Rightarrow f_1 = \cdots = f_n = 0$). Denote this direct sum as follows:

$$L = L_1 \dotplus L_2 \dotplus \cdots \dotplus L_n. \tag{7.3}$$

Let A be a closed Hermitian operator in H with a dense domain and let $z \in \mathbb{C} \setminus \mathbb{R}$ be fixed. Then

$$\mathcal{D}(A^*) = \mathcal{D}(A) \dotplus N_z \dotplus N_{\bar{z}}. \tag{7.4}$$

Thus, according to (7.4), any $g \in \mathcal{D}(A^*)$ admits a unique decomposition

$$g = f + h_z + h_{\bar{z}} \tag{7.5}$$
$$(f = f(g) \in \mathcal{D}(A), \quad h_z = h_z(g) \in N_z, \quad h_{\bar{z}} = h_{\bar{z}}(g) \in N_{\bar{z}})\,.$$

The *result of the action of the operator* A^* upon vector (7.5) can now be determined in a very simple way. Indeed, since $A^* \supseteq A$, it follows from (iv) in Section 4 that

$$A^*g = Af + \bar{z}h_z + zh_{\bar{z}}. \tag{7.6}$$

The proof of relation (7.4) can be reduced to the proof of the existence of decomposition (7.5) and its uniqueness. Let us prove that this decomposition exists.

Let $g \in \mathcal{D}(A^*)$. Then, according to the decomposition $H = \mathcal{R}(A - z\mathbb{I}) \oplus N_z$, the vector $(A^* - z\mathbb{I})g \in H$ can be written in the following form:

$$(A^* - z\mathbb{I})g = (A - z\mathbb{I})f + (\bar{z} - z)h_z. \tag{7.7}$$

(It is convenient to denote the components of $(A^* - z\mathbb{I})g$ in the corresponding subspaces as indicated above.) The vectors f and h_z in (7.7) are just the first two components of (7.5). To prove this, it suffices to show that the vector $g - f - h_z \in \mathcal{D}(A^*)$, where f and h_z are taken from (7.7), belongs to $N_{\bar{z}}$, i.e., that $g - f - h_z$ is an eigenvector of the operator A^* with the eigenvalue z. By virtue of (7.7), we have

$$\begin{aligned} A^*(g - f - h_z) &= A^*g - A^*f - A^*h_z \\ &= (A^* - z\mathbb{I})g + zg - Af - \bar{z}h_z \\ &= (A - z\mathbb{I})f + (\bar{z} - z)h_z + zg - Af - \bar{z}h_z \\ &= z(g - f - h_z). \end{aligned}$$

Let us prove that decomposition (7.5) is unique. Assume that the following decomposition is true:

$$0 = f + h_z + h_{\bar{z}} \quad (f \in \mathcal{D}(A), h_z \in N_z, h_{\bar{z}} \in N_{\bar{z}}). \tag{7.8}$$

Consider the action of the operator A^* upon (7.8). We have

$$\begin{aligned} 0 &= A^*f + A^*h_z + A^*h_{\bar{z}} = Af + \bar{z}h_z + zh_{\bar{z}} \\ &= (A - z\mathbb{I})f + \bar{z}h_z + z(h_{\bar{z}} + f) \\ &= (A - z\mathbb{I})f + \bar{z}h_z + z(-h_z) \\ &= (A - z\mathbb{I})f + (\bar{z} - z)h_z. \end{aligned} \tag{7.9}$$

But $(A - z\mathbb{I})f \in \mathcal{R}(A - z\mathbb{I})$, $(\bar{z} - z)h_z \in N_z$, and $\mathcal{R}(A - z\mathbb{I}) \oplus N_z = H$. Therefore, it follows from (7.9) that $(A - z\mathbb{I})f = 0$ and $(\bar{z} - z)h_z = 0$. Hence, $f = 0$ (z is a point of regular type for A) and $h_z = 0$. But this means that relation (7.8) turns into $h_{\bar{z}} = 0$. ☐

(ii) Let us describe the action of a closed Hermitian extension B of an operator A by using decomposition (7.4). We fix $z \in \mathbb{C}$ with Im $z > 0$. Let W be the operator associated with the extension B according to Theorem 7.2, $\mathcal{D}(W) = F \subseteq N_z$, and $\mathcal{R}(W) = G \subseteq N_{\bar{z}}$.

The set $\mathcal{D}(B)$ admits a decomposition

$$\mathcal{D}(B) = \mathcal{D}(A) \dotplus (W - \mathbb{I})F, \tag{7.10}$$

i.e., for all $g \in \mathcal{D}(B) \subseteq \mathcal{D}(A^*)$, decomposition (7.5) takes the form

$$g = f - h_z + Wh_z \quad (f \in \mathcal{D}(A),\ h_z \in F \subseteq N_z,\ Wh_z \in WF \subseteq N_{\bar{z}}). \tag{7.11}$$

Since $B \subseteq A^*$, the action of B upon the vector g is defined by (7.6), namely,

$$Bg = A^*G = Af - zh_z + zWh_z. \tag{7.12}$$

Proof. According to the first formula in (6.12) applied to $V = U \oplus W$, we obtain

$$\begin{aligned}\mathcal{D}(B) = \mathcal{R}(V - \mathbb{1}) &= \mathcal{R}(U - \mathbb{1}) \dotplus \mathcal{R}(W - \mathbb{1}) \\ &= \mathcal{D}(A) \dotplus \mathcal{R}(W - \mathbb{1}) = \mathcal{D}(A) \dotplus (W - \mathbb{1})F.\end{aligned} \tag{7.13}$$

(In spite of the fact that $\mathcal{R}(V) = \mathcal{R}(U) \oplus \mathcal{R}(W)$, if the identity operator is subtracted as in (7.13), then we can write there nothing but the direct sum because the sets $\mathcal{R}(U - \mathbb{1})$ and $\mathcal{R}(W - \mathbb{1})$ are, generally speaking, not orthogonal.) □

In conclusion, we present one more theorem of von Neumann related to the results discussed in Section 3.2.

Theorem 7.3. *Let A be a closed densely defined operator acting on H. Then the operator A^*A is selfadjoint and nonnegative.*

Proof. Recall that, according to the definition of the product of operators, $\mathcal{D}(A^*A) = \{f \in \mathcal{D}(A) \,|\, Af \in \mathcal{D}(A^*)\}$. Therefore, $(A^*Af, f)_H = (Af, Af)_H \geq 0$ for $f \in \mathcal{D}(A^*A)$, i.e., the operator A^*A is nonnegative.

Let us prove that it is selfadjoint. By virtue of the closeness of A and Theorem 3.3, we have $(A^*)^* = A$. Consequently, relation (3.6) written for A^* (instead of A) yields

$$H \oplus H = \Gamma_A \oplus O\Gamma_{A^*}. \tag{7.14}$$

For all $h \in H$, the vector $\langle h, 0\rangle$ can be decomposed, according to (7.14), as follows: There exist $f \in \mathcal{D}(A)$ and $g \in \mathcal{D}(A^*)$ such that

$$\langle h, 0\rangle = \langle f, AF\rangle + \langle -A^*g, g\rangle \Leftrightarrow h = f - A^*g, \;\; 0 = Af + g. \tag{7.15}$$

Let us find the vector g from the second equality on the right-hand side of (7.15) and insert it in the first equality. This gives

$$h = f + A^*Af = (\mathbb{1} + A^*A)f \quad (f \in \mathcal{D}(A), Af \in \mathcal{D}(A^*)) \tag{7.16}$$

for all $h \in H$.

Below, we prove that $\mathcal{D}(A^*A)$ is dense in H. Equality (7.16) means that $\mathcal{R}(A^*A + \mathbb{1}) = H$. Therefore, according to Theorem 5.2, this operator is selfadjoint (in this case, $\alpha = 0$ and $z = -1 \in \mathbb{C} \setminus [0, \infty)$).

Hence, it remains to show that $(\mathcal{D}(A^*A))^\sim = H$. Assume the opposite, i.e., that $(\mathcal{D}(A^*A))^\sim \neq H$. Then there exists $0 \neq h \in H$ such that $h \perp \mathcal{D}(A^*A)$. By virtue of (7.16), one can indicate $f \in \mathcal{D}(A^*A)$ for which $f + A^*Af = h$ and, therefore,

$$0 = (h, f)_H = (f + A^*Af, f)_H = \|f\|_H^2 + (A^*Af, f)_H = \|f\|_H^2 + \|Af\|_H^2.$$

Consequently, $f = 0 \Rightarrow h = f + A^*Af = 0$, but this is absurd. □

It is worth noting that a similar theorem also holds for the operator A^*A. To prove it, one must replace A by A^* in Theorem 7.3.

Exercises

7.1. Prove that a selfadjoint operator A has no Hermitian extensions other than A.

7.2. Let A be an Hermitian operator. Prove that any Hermitian extension of A is a restriction of the operator A^*.

7.3. Let A be a closed Hermitian operator that admits a selfadjoint extension. Check whether it admits a closed Hermitian extension B which has no selfadjoint extensions.

7.4. A mapping $I\colon H \to H$ that preserves the norm is called involution if it is antilinear, i.e.,

$$I(\lambda x + \mu y) = \bar{\lambda} I x + \bar{\mu} I y$$

for all $\lambda, \mu \in \mathbb{C}$ and $x, y \in H$ and $I^2 = \mathbb{1}$. Let A be an Hermitian operator such that $I\colon \mathcal{D}(A) \to \mathcal{D}(A)$ and $AI = IA$. Prove that the defect numbers of this operator A are equal.

7.5. Prove the assertion on the "equalization" of the defect numbers formulated at the end of Subsection 1.

7.6. Prove that all selfadjoint extensions of the operator of differentiation in $L_2([0,1])$ introduced in Exercise 2.7(c) are described in Exercise 2.7(d), where $\alpha \in [0, 2\pi)$, and that different $\alpha \in [0, 2\pi)$ correspond to different selfadjoint extensions.

7.7. By using the result established in Exercise 7.4, show that differential operators with real-valued coefficients have equal defect numbers.

Chapter 13
Spectral Decompositions of Selfadjoint, Unitary, and Normal Operators. Criteria of Selfadjointness

Here, we present a convenient form of the spectral theorem for a selfadjoint operator acting on a finite-dimensional Hilbert space H. This theorem has already been studied in Section 10.1.

Thus, let H be a Hilbert space with dimensionality $\dim H = n < \infty$ and let A be a selfadjoint operator that acts on it. Let $\lambda_1 < \lambda_2 < \ldots < \lambda_m$ $(m \leq n)$ be eigenvalues of this operator. Each λ_k is associated with an eigensubspace $\Phi(\lambda_k)$ that consists of all eigenvectors of the operator A with the eigenvalue $\lambda_k : \Phi(\lambda_k) = \{\varphi \in H | A\varphi = \lambda_k \varphi\}$; its dimensionality is $N(\lambda_k) = \dim \Phi(\lambda_k) \leq n$. In $\Phi(\lambda_k)$, one can choose an orthonormal basis $\varphi_1(\lambda_k), \ldots, \varphi_{N(\lambda_k)}(\lambda_k)$ (clearly, this choice is not unique).

The spectral theorem for A states that the vectors $\varphi_\alpha(\lambda_k)$ with different k and α form an orthonormal basis in H. Thus, for any $f \in H$, we can write its decomposition in this basis, namely,

$$f = \sum_{k=1}^{m} \sum_{\alpha=1}^{N(\lambda_k)} (f, \varphi_\alpha(\lambda_k))_H \, \varphi_\alpha(\lambda_k). \tag{0.1}$$

Since $A\varphi_\alpha(\lambda_k) = \lambda_k \varphi_\alpha(\lambda_k)$, decomposition (0.1) enables us to rewrite the action of the operator A in the diagonal form

$$Af = \sum_{k=1}^{m} \sum_{\alpha=1}^{N(\lambda_k)} \lambda_k \, (f, \varphi_\alpha(\lambda_k))_H \, \varphi_\alpha(\lambda_k). \tag{0.2}$$

Denote by $P(\lambda_k)$ the (orthogonal) projector onto the subspace $\Phi(\lambda_k)$. Then

$$P(\lambda_k) f = \sum_{\alpha=1}^{N(\lambda_k)} (f, \varphi_\alpha(\lambda_k))_H \, \varphi_\alpha(\lambda_k),$$

and, therefore, we can rewrite equalities (0.1) and (0.2) in the following form:

$$\mathbb{1} = \sum_{k=1}^{m} P(\lambda_k), \qquad A = \sum_{k=1}^{m} \lambda_k P(\lambda_k). \tag{0.3}$$

One can ask a natural question: Is it possible to generalize the formulas (0.1) and (0.2), or (in the other notation) (0.3) for the case of selfadjoint operators acting on an arbitrary Hilbert space H? We already know that even the spectrum of a bounded selfadjoint operator A is, as a rule, not discrete and the eigenvectors $\varphi_\alpha(\lambda)$ not necessarily exist (for more details, see Section 8.8). Therefore, it is impossible to generalize these formulas directly by "passing to the limit as $n \to \infty$".

But relations (0.3) can be rewritten in the form of the following "integrals with respect to the operator-valued measure" E:

$$\mathbb{I} = \int_{-\infty}^{\infty} dE(\lambda), \quad A = \int_{-\infty}^{\infty} \lambda dE(\lambda);$$

$$\mathbb{R} \ni \alpha \mapsto E(\alpha) = \sum_{\lambda_j \in \alpha} P(\lambda_j) \in \mathcal{L}(H). \tag{0.4}$$

It turns out that relations (0.4) can be generalized to the case of general selfadjoint, even unbounded, operators, and this is the main aim of the present chapter. We begin with the investigation of an operator-valued measure E that will appear in a generalization of this sort — the resolution of the identity.

1 The Resolution of the Identity and its Properties

1.1 The Resolution of the Identity

A resolution of the identity is a projector-valued measure. Therefore, we first prove one of the properties of projectors in addition to what has been said about them in Section 8.5.

Lemma 1.1. *Let P_{G_1} and P_{G_2} be projectors onto the subspaces $G_1, G_2 \subseteq H$, respectively. The sum $P = P_{G_1} + P_{G_2}$ is a projector if and only if $G_1 \perp G_2$, i.e., $P_{G_1}P_{G_2} = 0$ (or, equivalently, $P_{G_2}P_{G_1} = 0$). In this case, P is a projector onto $G_1 \oplus G_2 : P = P_{G_1 \oplus G_2}$.*

Proof. *Necessity.* Let $P = P_{G_1} + P_{G_2}$, then $P^2 = P$ or

$$\begin{aligned} P_{G_1} + P_{G_2} = (P_{G_1} + P_{G_2})^2 &= P_{G_1}^2 + P_{G_1}P_{G_2} + P_{G_2}P_{G_1} + P_{G_2}^2 \\ &= P_{G_1} + P_{G_1}P_{G_2} + P_{G_2}P_{G_1} + P_{G_2}; \end{aligned} \tag{1.1}$$

$$P_{G_1}P_{G_2} + P_{G_2}P_{G_1} = 0. \tag{1.2}$$

Consider $f \in H$. It is necessary to prove that $g = P_{G_2}P_{G_1}f = 0$. By virtue of (1.2), we have

$$P_{G_1}g = P_{G_1}P_{G_2}P_{G_1}f = -P_{G_2}(P_{G_1})^2 f = -P_{G_2}P_{G_1}f = -g.$$

Hence,

$$g = -P_{G_1}g. \tag{1.3}$$

By applying the operator P_{G_1} to equality (1.3), we conclude that $P_{G_1} g = 0$. Consequently, $g = 0$, i.e., $P_{G_2} P_{G_1} = 0$.

Sufficiency. Since $P_{G_2} P_{G_1} = 0 \Leftrightarrow G_1 \perp G_2 \Leftrightarrow P_{G_1} P_{G_2} = 0$, by virtue of (1.1), we have $P^2 = P$. It is also clear that P is selfadjoint. Thus, this operator is a projector onto the subspace $G = \{f \in H | Pf = f\}$ (see Section 8.5). But the equality $f = Pf = P_{G_1} f \oplus P_{G_2} f$ means that $f \in G_1 \oplus G_2$. Thus, $G = G_1 \oplus G_2$. □

We now proceed to the definition of a general resolution of the identity. Let R be an abstract space and let $\mathfrak{R}$ be a σ-algebra of its subsets. In other words, we define a measurable space $\langle R, \mathfrak{R} \rangle$. Furthermore, we introduce a Hilbert space H.

An operator-valued function $\mathfrak{R} \ni \alpha \mapsto E(\alpha) \in \mathcal{L}(H)$ is called a resolution of the identity (on R) provided that

(a) for any $\alpha \in \mathfrak{R}$, $E(\alpha)$ is a projector in H; $E(\emptyset) = 0$, and $E(R) = \mathbb{I}$.

(b) E is countably additive, i.e., for an arbitrary sequence $(\alpha_j)_{j=1}^{\infty}$ that consists of disjoint sets from $\mathfrak{R}$, the equality

$$E\left(\bigcup_{j=1}^{\infty} \alpha_j\right) = \sum_{j=1}^{\infty} E(\alpha_j) \tag{1.4}$$

holds, where the series converges in the sense of the strong convergence of operators (i.e., on every vector $f \in H$ in a norm of H).

Theorem 1.1. *A resolution of the identity possesses the property of orthogonality, namely*

$$E(\alpha)E(\beta) = E(\alpha \cap \beta) \quad (\alpha, \beta \in \mathfrak{R}). \tag{1.5}$$

It is clear that, for an ordinary scalar measure E, a property of the form (1.5) holds only in trivial cases.

Proof. First, let $\alpha \cap \beta = \emptyset$. Then, according to (b), $E(\alpha \cup \beta) = E(\alpha) + E(\beta)$ and, by virtue of Lemma 1.1, $E(\alpha)\ E(\beta) = 0$. Thus, property (1.5) takes place, since $E(\alpha \cap \beta) = E(\emptyset) = 0$.
Consider the general case. We set $\gamma = \alpha \cap \beta$. Then $\alpha = (\alpha \setminus \gamma) \cup \gamma$, $\beta = (\beta \setminus \gamma) \cup \gamma$ are the decompositions of α and β into disjoint sets from $\mathfrak{R}$ and, according to the already proved result, $E(\alpha \setminus \gamma) E(\gamma) = 0$, $E(\beta \setminus \gamma) E(\gamma) = 0$, and $E(\alpha \setminus \gamma) = 0$. Taking into account these equalities and (b), we get

$$\begin{aligned} E(\alpha)E(\beta) &= \Big(E(\alpha \setminus \gamma) + E(\gamma)\Big)\Big(E(\beta \setminus \gamma) + E(\gamma)\Big) \\ &= E(\alpha \setminus \gamma)E(\beta \setminus \gamma)E(\gamma) + E(\gamma)E(\beta \setminus \gamma) + E^2(\gamma) = E(\gamma). \end{aligned}$$

□

Corollary 1.1. *It follows from (1.5) that the operators $E(\alpha)$ $(\alpha \in \mathfrak{R})$ are commuting.*

REMARK 1.1. In condition (b), the strong convergence of series (1.4) can be replaced by weak convergence.

Indeed, independently of the type of convergence of series (1.4), the finite additivity of E follows from (b) (all α_j, except finitely many, are replaced by $\emptyset$). The proof of Theorem 1.1 was based only on the use of finite additivity of E; therefore, Theorem 1.1 remains true if we understand condition (b) in the sense of the weak convergence of series (1.4). But then, for any $f \in H$, the vectors $E(\alpha_j)f$ in (1.4) are mutually orthogonal, and we know that, for the series which are formed of mutually orthogonal vectors, the notions of strong and weak convergences are equivalent (see Exercise 7.10.5). □

Let $\mathfrak{R} \ni \alpha \mapsto E(\alpha)$ be a resolution of the identity. We fix $f \in H$. Then the set function

$$\mathfrak{R} \ni \alpha \mapsto \rho_{f,f}(\alpha) = \big(E(\alpha)f, f\big)_H = \|E(\alpha)f\|_H^2 \geq 0, \tag{1.6}$$

is, clearly, a nonnegative finite measure on $\mathfrak{R}$ (in (1.6), we have used the fact that $\big(E(\alpha)\big)^2 = E(\alpha)$, $\big(E(\alpha)\big)^* = E(\alpha)$). For fixed $f, g \in H$, the set function

$$\mathfrak{R} \ni \alpha \mapsto \rho_{f,g}(\alpha) = \big(E(\alpha)f, g\big)_H \in \mathbb{C} \tag{1.7}$$

is a complex-valued measure (charge) on $\mathfrak{R}$. It is a linear combination of measures (1.6) by virtue of the polarizational formula (see Section 8.5). Measures (1.6) and (1.7) are frequently used in what follows.

Let us present the simplest examples of a resolution of the identity.

Examples

1.1. Let $R = \mathbb{R}$, let $(\lambda_k)_{k=1}^{\infty}$ be a sequence of points from $\mathbb{R}$ (possibly, finite), and let $(P_k)_{k=1}^{\infty}$ be a sequence of projectors in H such that $P_j P_k = \delta_{jk} P_j$, $\sum_{k=1}^{\infty} P_k = \mathbb{1}$. As $\mathfrak{R}$, we take an arbitrary σ-algebra of subsets of $\mathbb{R}$ that contains one-point sets (for example, the Borel σ-algebra $\mathfrak{B}(\mathbb{R})$) and set ($\forall \alpha \in \mathfrak{R}$):

$$E(\alpha) = \sum_{\lambda_k \in \alpha} P_k. \tag{1.8}$$

It is easy to see that the resolution of the identity (0.4) constructed in the introduction is an example of a resolution of the identity of this sort.

1.2. Let $\langle R, \mathfrak{R} \rangle$ be a measurable space and let μ be a measure in this space (possibly, infinite). We set $H = L_2(R, \mathfrak{R}, d\mu) = L_2$ and consider in this space the operator of multiplication $E(\alpha)$ by the indicator $\chi_\alpha(\cdot)$ of the set $\alpha \in \mathfrak{R}$, i.e.,

$$\Big(E(\alpha)f\Big)(\lambda) = \chi_\alpha(\lambda) f(\lambda) \quad (f \in L_2). \tag{1.9}$$

The boundedness of χ_α yields the boundedness of operator (1.9); the fact that χ_α is real-valued implies that it is selfadjoint, and the equality $(\chi_\alpha(\lambda))^2 = \chi_\alpha(\lambda)(\lambda \in R)$ gives the equality $E^2(\alpha) = E(\alpha)$. Thus, $E(\alpha)$ is a projector. It projects onto the subspace of those $f \in L_2$ for which $\chi_\alpha(\lambda)f(\lambda) = f(\lambda)$ for μ-almost all $\lambda \in R$ (i.e., for $f \in L_2$ that vanish μ-almost everywhere for $\lambda \notin \alpha$).

The set function $\mathfrak{R} \ni \alpha \mapsto E(\alpha)$ is a resolution of the identity in L_2. However, the countable additivity of (1.4) is not completely evident in this case. But, in view of Remark 1.1, this problem reduces to the validity of the equality $(\forall f, g \in L_2)$:

$$\int\limits_{\cup_{j=1}^{\infty}\alpha_j} f(\lambda)\overline{g(\lambda)}d\mu(\lambda) = \sum_{j=1}^{\infty}\int\limits_{\alpha_j} f(\lambda)\overline{g(\lambda)}d\mu(\lambda),$$

which follows from the fact that $f(\lambda)\overline{g(\lambda)} \in L_1(R, \mathfrak{R}, d\mu)$. Note that equality (1.5) is now reflected in the relation $\chi_\alpha(\lambda)\chi_\beta(\lambda) = \chi_{\alpha\cap\beta}(\lambda)(\alpha, \beta \in \mathfrak{R}; \lambda \in R)$.

Example 1.2 represents, in fact, the general form of a resolution of the identity and one can show that a unitary image of any possible resolution of the identity is close to the one constructed in this example.

Let us mention several more properties of a resolution of the identity E.

The additivity of E easily implies its *monotonicity*, i.e., $\forall\alpha, \beta \in \mathfrak{R}$,

$$\alpha \subseteq \beta \Rightarrow E(\alpha) \leq E(\beta), \tag{1.10}$$

where the inequality between the operators is understood in the sense of forms (i.e., $(\forall A, B \in \mathcal{L}(H)) : A \leq B \Leftrightarrow (Af, f)_H \leq (Bf, f)_H$ $(\forall f \in H)$; in this connection, see Section 8.5).

Indeed, since $\beta = \alpha \cup (\beta\setminus\alpha)$, we have $E(\beta) = E(\alpha) + E(\beta\setminus\alpha)$ and $E(\beta\setminus\alpha) \geq 0 \Rightarrow E(\beta) \geq E(\alpha)$. □

As in the case of a scalar measure, the following theorem is true:

Theorem 1.2. *Let $(\alpha_n)_{n=1}^{\infty}$, $(\beta_n)_{n=1}^{\infty}$ be decreasing and increasing sequences of sets $\alpha_n, \beta_n \in \mathfrak{R}$: $\alpha_1 \supseteq \alpha_2 \supseteq \ldots$, $\beta_1 \subseteq \beta_2 \subseteq \ldots$. Then, in the sense of strong convergence in H,*

$$\lim_{n\to\infty} E(\alpha_n) = E(\cap_{n=1}^{\infty}\alpha_n) \quad \textit{and} \quad \lim_{n\to\infty} E(\beta_n) = E(\cup_{n=1}^{\infty}\beta_n). \tag{1.11}$$

Proof. We set $\alpha = \cap_{n=1}^{\infty}\alpha_n$ and $\gamma_n = \alpha_n \setminus \alpha$ $(n \in \mathbb{N})$. Then $\gamma_1 \supseteq \gamma_2 \supseteq \ldots$, $\cap_{n=1}^{\infty}\gamma_n = \emptyset$. Consider measure (1.6) for fixed $f \in H$. Then, according to the well-known result for scalar measures (Section 1.6), $\|E(\gamma_n)f\|_H^2 = \rho_{f,f}(\gamma_n) \to 0$, i.e., $E(\gamma_n) \to 0$ strongly. But $E(\gamma_n) = E(\alpha_n) - E(\alpha)$, whence we get the first equality in (1.1).

The second one is proved similarly, one should only set $\beta = \cup_{n=1}^{\infty}\beta_n$ and $\gamma_n = \beta \setminus \beta_n$ $(n \in \mathbb{N})$. □

1.2 Theorem on Extension

We consider another property of a resolution of the identity. As in the case of a scalar measure, in the construction of a resolution of the identity, it is often convenient to define it on a certain algebra of sets, and then apply the theorem on extension similar to Theorem 1.5.2 and proved with its help.

Thus, let R be a space and let $\mathfrak{R}$ be an *algebra* of subsets of R. As above, the set function $\mathfrak{R} \ni \alpha \mapsto E(\alpha) \in \mathcal{L}(H)$ is called a resolution of the identity provided that conditions (a) and (b) are satisfied with an additional requirement that $\cup_{j=1}^{\infty} \alpha_j \in \mathfrak{R}$ in (b). It is clear that Theorem 1.1 and Remark 1.1 remain true in the case where $\mathfrak{R}$ is an algebra.

Theorem 1.3. *Let E be a resolution of the identity on the algebra $\mathfrak{R}$. Then there exists its extension to a resolution of the identity E_σ on its σ-hull $\mathfrak{R}_\sigma$, i.e., the restriction $E_\sigma \restriction \mathfrak{R} = E$. The resolution of the identity E_σ is uniquely defined for given E.*

Proof. We fix $f \in H$ and consider a finite measure $\mathfrak{R} \ni \alpha \mapsto \rho_{f,f}(\alpha) = \big(E(\alpha)f, f\big)_H \geq 0$. According to the standard theory of extensions (Section 1.5), there exists a measure $\mathfrak{R}_\sigma \ni \alpha \mapsto \hat{\rho}_{f,f}(\alpha) \geq 0$ such that $\hat{\rho}_{f,f} \restriction \mathfrak{R} = \rho_{f,f}$. By virtue of the polarization formula, the charge $\mathfrak{R} \ni \alpha \mapsto \rho_{f,g}(\alpha) = \big(E(\alpha)f, g\big)_H \in \mathbb{C}$ can be expressed as a linear combination of four measures of the form $\rho_{h,h}$ $(h \in H)$. Extending each of these measures from $\mathfrak{R}$ to $\mathfrak{R}_\sigma$ and taking the corresponding linear combination, we conclude that the charge $\rho_{f,g}$ is also extended to the charge $\mathfrak{R}_\sigma \ni \alpha \mapsto \hat{\rho}_{f,g}(\alpha) \in \mathbb{C}$.

For fixed $\alpha \in \mathfrak{R}_\sigma$ the mapping

$$H \oplus H \ni \langle f, g \rangle \mapsto \hat{\rho}_{f,g}(\alpha) \in \mathbb{C} \tag{1.12}$$

is bilinear. Indeed, we know that $\mathfrak{R}_\sigma$ coincides with the monotone hull $\mathfrak{R}$ (Theorem 1.7.1) and $\hat{\rho}_{f,f}$ (and, hence, $\hat{\rho}_{f,g}$) is constructed by successive monotone extensions, beginning with the sets $\alpha \in \mathfrak{R}$. Since bilinearity is preserved under monotone extensions, the required property follows from the bilinearity of (1.12) for $\alpha \in \mathfrak{R}$.

Mapping (1.12) is continuous. Indeed, $\hat{\rho}_{f,f}(\alpha) \geq 0$ for $\alpha \in \mathfrak{R}_\sigma$ and, therefore, by virtue of the Cauchy-Buniakowski inequality,

$$|\hat{\rho}_{f,g}(\alpha)|^2 \leq \hat{\rho}_{f,f}(\alpha)\hat{\rho}_{g,g}(\alpha) \leq \hat{\rho}_{f,f}(R)\hat{\rho}_{g,g}(R) = \|f\|_H^2 \cdot \|g\|_H^2.$$

Let $E_\sigma(\alpha)$ denote the bounded operator in H associated with the continuous bilinear form (1.12) (Theorem 8.5.1). Then

$$\hat{\rho}_{f,g}(\alpha) = \Big(E_\sigma(\alpha)f, g\Big)_H \quad (f, g \in H; \alpha \in \mathfrak{R}_\sigma). \tag{1.13}$$

For $\alpha \in \mathfrak{R}$, $E_\sigma(\alpha) = E(\alpha)$ is a projector. Successive monotone extensions $\hat{\rho}_{f,g}$ are extensions from $\alpha \in \mathfrak{R}$ by weak limits of monotone sequences of projectors; therefore, $E_\sigma(\alpha)$ is also a projector for any $\alpha \in \mathfrak{R}_\sigma$ (Section 8.5). Certainly,

$E_\sigma(\emptyset) = 0$ and $E_\sigma(R) = \mathbb{I}$. The set function $\mathfrak{R}_\sigma \ni \alpha \mapsto E_\sigma(\alpha)$ is countably additive in the sense of weak convergence of the corresponding series. According to (1.3), this is a reformulation of the property of countable additivity of the charge $\hat{\rho}_{f,g}(f, g \in H)$. Hence, E_σ is a resolution of the identity defined on $\mathfrak{R}_\sigma$. By construction, it coincides with E on $\mathfrak{R}$. The last statement of the theorem follows from the uniqueness of the extension of the charge from $\mathfrak{R}$ to $\mathfrak{R}_\sigma$. □

Remark 1.2. In the definition of a resolution of the identity (both on a σ-algebra and on an algebra), it is not necessary to require that $E(R) = \mathbb{I}$. For such a *quasiresolution of the identity*, all previous results evidently hold.

Exercises

1.1. Let $\langle R, \mathfrak{R}\rangle$ be a measurable space. Consider an additive operator-valued set function $\mathfrak{R} \ni \alpha \mapsto F(\alpha) \in \mathcal{L}(H)$ that satisfies condition (a) in the definition of a resolution of the identity. Prove that F is a resolution of the identity if and only if one of the following conditions holds:

(i) $(\forall x \in H)(\forall (\alpha_n)_{n=1}^{\infty} \subseteq \mathfrak{R}) : \big(E(\cup_{n=1}^{\infty}\alpha_n)x, x\big)_H \le \sum_{n=1}^{\infty}\big(E(\alpha_n)x, x\big)_H$;

(ii) $\big(\forall (\beta_n)_{n=1}^{\infty} \subseteq \mathfrak{R} : \beta_1 \subseteq \beta_2 \subseteq \ldots\big) : \operatorname{s.lim}_{n\to\infty} E(\beta_n) = E\big(\cup_{n=1}^{\infty}\beta_n\big)$;

(iii) $\big(\forall (\alpha_n)_{n=1}^{\infty} \subseteq \mathfrak{R} : \alpha_1 \supseteq \alpha_2 \supseteq \ldots\big) : E\big(\cap_{n=1}^{\infty}\alpha_n\big) = \operatorname{s.lim}_{n\to\infty} E(\alpha_n)$.

1.2. Let $\langle R, \mathfrak{R}\rangle$ be a measurable space and let $\mathfrak{R} \ni \alpha \mapsto F(\alpha) \in \mathcal{L}(H)$ be an operator-valued set function satisfying the conditions

(i) $(\forall \alpha \in \mathfrak{R}) : F(\alpha) = \big(F(\alpha)\big)^*$;

(ii) $(\forall \alpha, \beta \in \mathfrak{R}) : F(\alpha)F(\beta) = F(\alpha \cap \beta)$;

(iii) $(\forall \alpha_1, \alpha_2 \in \mathfrak{R} : \alpha_1 \cap \alpha_2 = \emptyset) : F(\alpha_1 \cup \alpha_2) = F(\alpha_1) + F(\alpha_2)$.

Prove that

(a) F satisfies condition (a) in the definition of a resolution of the identity;

(b) conditions (i)–(iii) are equivalent to conditions (i), (ii), and (iii)′ $F(\emptyset) = 0$ and $(\forall \alpha \in \mathfrak{R})$: $F(\hat{\alpha}) = \mathbb{I} - F(\alpha)$;

(c) condition (ii) follows from conditions (i), (iii), and (iii)″ $F(\emptyset) = 0$ and $F(R) = \mathbb{I}$.

1.3. Let E be a resolution of the identity on $\langle \mathbb{R}, \mathfrak{B}(\mathbb{R})\rangle$. Prove that the projector-valued function $\mathbb{R} \ni t \mapsto \mathcal{E}(t) = E((-\infty, t))$ satisfies the conditions:

(i) $(\forall t, s \in \mathbb{R}) : \mathcal{E}(t)\mathcal{E}(s) = \mathcal{E}\big(\min\{t, s\}\big)$;

(ii) $(\forall t \in \mathbb{R})$: $\operatorname{s.lim}_{\tau\to t-} \mathcal{E}(\tau) = \mathcal{E}(t)$;

(iii) $\operatorname{s.lim}_{t\to+\infty} \mathcal{E}(t) = \mathbb{I}$, $\operatorname{s.lim}_{t\to-\infty} \mathcal{E}(t) = 0$;

(iv) $(\forall t, s \in \mathbb{R} : t < s) : \mathcal{E}(t) \le \mathcal{E}(s)$.

1.4. Let $\mathcal{E}(t)(t \in \mathbb{R})$ be a projector-valued function satisfying conditions (i)–(iii) in Exercise 1.3. Prove that there exists a unique resolution of the identity E such that $(\forall t \in \mathbb{R}) : \mathcal{E}(t) = E((-\infty, t))$.

Hint. By using the theorem on the construction of a measure for a given nondecreasing function, construct a family of charges $\{\omega_{x,y} | \ x, \ y \in H\}$ for given functions $\left(\mathcal{E}(t)x, y\right)_H$ $(x, y \in H)$. Prove that $H \times H \ni \langle x, y\rangle \mapsto \omega_{x,y}(\alpha) \in \mathbb{C}$ is a bounded bilinear form for fixed $\alpha \in \mathfrak{B}(\mathbb{R})$. By virtue of the Riesz theorem,

$$\big(\forall \alpha \in \mathfrak{B}(\mathbb{R})\big)\big(\exists F(\alpha) \in \mathcal{L}(H)\big)(\forall x, y \in H) : \omega_{x,y}(\alpha) = \big(F(\alpha)x, y\big)_H.$$

Check that $F(\cdot)$ is a resolution of the identity. See also Theorem 6.4.

1.5. Let $H = L_2(R, \mathfrak{R}, \mu)$, $\{\mathfrak{R}_t, t \geq 0\}$ be a family of σ-algebras of the subsets of R such that

(i) $(\forall t, s \in R : t < s) : \mathfrak{R}_t \subseteq \mathfrak{R}_s$;

(ii) $(\forall t > 0) : (\cup_{\tau < t} \mathfrak{R}_\tau)_\sigma = \mathfrak{R}_t$.

We set $\mathcal{E}(t) = M(\cdot|\mathfrak{R}_t)\big(t \in (0, \infty)\big)$ (see Exercises 5.2.7, 5.2.8). Prove that $\mathcal{E}(t)$ satisfies conditions (i)–(iv) in Exercise 1.3.

2 The Construction of Spectral Integrals

To prove relations of the type (0.4), it is first necessary to define spectral integrals, i.e., integrals of complex-valued functions $R \ni \lambda \mapsto F(\lambda) \in \mathbb{C}$ with respect to a resolution of the identity that are operators in H, namely,

$$\int_R F(\lambda)\, dE(\lambda), \tag{2.1}$$

Let us present the required construction. It is similar to the standard construction of the Lebesgue integral with respect to a scalar-valued measure (see Chapter 3) and is given step-by-step, beginning with the simplest classes of functions. Some difficulties appear only in the case of unbounded functions F. Thus, we assume that a measurable space $\langle R, \mathfrak{R}\rangle$ and a resolution of the identity E on R are given. Points of R are denoted by $\lambda, \mu, \ldots$.

2.1 Integrals of Simple Functions

Recall that a function $R \ni \lambda \mapsto F(\lambda) \in \mathbb{C}$ is called a simple function (or a step function) if it is a linear combination of the indicators $\chi_\alpha(\lambda)$ of sets $\alpha \in \mathfrak{R}$. The collection of all simple functions is denoted by $S(R, \mathfrak{R}) = S$; S is an algebra with respect to ordinary summation and multiplication and contains all constants.

It is clear that every simple function can be represented in the form of a linear combination of the indicators of disjoint sets, i.e.,

$$F(\lambda) = \sum_{k=1}^{n} F_k \chi_{\alpha_k}(\lambda), \quad \Big(F_k \in \mathbb{C}; \alpha_k \cap \alpha_j = \emptyset, k \neq j; \lambda \in R\Big). \tag{2.2}$$

By definition, we set

$$\int_R F(\lambda)dE(\lambda) = \int_R \Big(\sum_{k=1}^{n} F_k\chi_{\alpha_k}(\lambda)\Big)dE(\lambda) = \sum_{k=1}^{n} F_k E(\alpha_k) \in \mathcal{L}(H). \quad (2.3)$$

Clearly, in (2.2), one can assume that the sets α_k form a decomposition of the set $R : \cup_{k=1}^{n}\alpha_k = R$. Representation (2.2) of a simple function F is, obviously, not unique because the sets α_k are not defined uniquely for given F (thus, they can be subdivided into smaller sets). Nevertheless, integral (2.3) is defined correctly and does not depend on representation (2.2).

This is a consequence of the finite additivity of E and can be proved just as for the ordinary Lebesgue integral. Namely, it is necessary to take two representations (2.2) that correspond to two distinct decompositions of the space R, i.e., $R = \cup_{k=1}^{n}\alpha_k = \cup_{j=1}^{m}\beta_j$. Then we construct the superposition of these decompositions, i.e., the decomposition $R = \cup_{k=1}^{n} \cup_{j=1}^{m} \alpha_k \cap \beta_j$ and show that the integral constructed according to each original decomposition is equal to the integral that corresponds to the superposition of decompositions. The integral thus constructed possesses a series of simple properties.

(1) *Linearity*, i.e.,

$$\int_R (aF(\lambda) + bG(\lambda))dE(\lambda) = a\int_R F(\lambda)dE(\lambda) + b\int_R G(\lambda)dE(\lambda) \quad (2.4)$$

$$\Big(a, b \in \mathbb{C}; F, G \in S\Big).$$

Indeed, to prove (2.4) it is necessary to write F and G in the form (2.2) for the same decomposition R, by superposing the existing decompositions for F and G. After this procedure, relation (2.4) follows from (2.3). □

(2) The following property is unusual, it is called the *multiplicativity of an integral*

$$\int_R F(\lambda)dE(\lambda) \cdot \int_R G(\lambda)dE(\lambda) = \int_R F(\lambda)G(\lambda)dE(\lambda) \quad (F, G \in S) \quad (2.5)$$

and yields the property of *commutability* of any two integrals.

Indeed, this property is a simple consequence of the orthogonality of the measure E (see (1.5)). To prove it, we construct a superposition of the corresponding decompositions of R and write F and G by using the same decomposition as follows:

$$F(\lambda) = \sum_{j=1}^{n} F_j\chi_{\alpha_j}(\lambda) \quad \text{and} \quad G(\lambda) = \sum_{k=1}^{n} G_k\chi_{\alpha_k}(\lambda).$$

Then, in view of (2.3) and the relations $E(\alpha_j)\,E(\alpha_k) = \delta_{jk}\,E(\alpha_j)$, we get

$$\begin{aligned}\int_R F(\lambda)dE(\lambda)\int_R G(\lambda)dE(\lambda) &= \left(\sum_{j=1}^n F_j E(\alpha_j)\right)\left(\sum_{k=1}^n G_k E(\alpha_k)\right)\\ &= \sum_{j,k=1}^n F_j G_k E(\alpha_j)E(\alpha_k) = \sum_{j=1}^n F_j G_j E(\alpha_j)\\ &= \int_R F(\lambda)G(\lambda)dE(\lambda).\end{aligned}$$ □

(3)

$$\left(\int_R F(\lambda)dE(\lambda)\right)^* = \int_R \overline{F(\lambda)}dE(\lambda) \quad (F \in S). \tag{2.6}$$

Indeed, by virtue of the selfadjointness of $E(\alpha)$, we have

$$\left(\int_R F(\lambda)dE(\lambda)\right)^* = \left(\sum_{k=1}^n F_k E(\alpha_k)\right)^* = \sum_{k=1}^n \bar{F}_k E(\alpha_k) = \int_R \bar{F}(\lambda)\,dE(\lambda).$$ □

(4)

$$\left(\left(\int_R F(\lambda)dE(\lambda)\right)f, g\right)_H = \int_R F(\lambda)d(E(\lambda)f, g)_H \quad (F \in S; f, g \in H) \tag{2.7}$$

(here, $d(E(\lambda)f, g)_H$ means integration with respect to charge (1.7)).

The proof follows directly from (2.3). □

(5)

$$\left\|\left(\int_R F(\lambda)dE(\lambda)\right)f\right\|_H^2 = \int_R |F(\lambda)|^2 d(E(\lambda)f, f)_H \quad (F \in S; f \in H). \tag{2.8}$$

Indeed, by virtue of (2.5)–(2.7), we have

$$\begin{aligned}\left\|\left(\int_R F(\lambda)dE(\lambda)\right)f\right\|_H^2 &= \left(\left(\int_R F(\lambda)dE(\lambda)\right)f, \left(\int_R F(\lambda)dE(\lambda)\right)f\right)_H\\ &= \left(\left(\int_R F(\lambda)dE(\lambda)\right)^*\left(\int_R F(\lambda)dE(\lambda)\right)f, f\right)_H\\ &= \left(\left(\int_R \overline{F(\lambda)}dE(\lambda)\right)\left(\int_R F(\lambda)dE(\lambda)\right)f, f\right)_H\\ &= \left(\left(\int_R |F(\lambda)|^2 dE(\lambda)\right)f, f\right)_H\\ &= \int_R |F(\lambda)|^2 d(E(\lambda)f, f)_H.\end{aligned}$$ □

(6)

$$\left\| \int_R F(\lambda) dE(\lambda) \right\| \leq \sup \{ |F(\lambda)| \quad | \lambda \in R \} \qquad (F \in S). \tag{2.9}$$

Indeed, according to (2.8), we have ($\forall f \in H$):

$$\begin{aligned} \left\| \left(\int_R F(\lambda) dE(\lambda) \right) f \right\|_H^2 &= \int_R |F(\lambda)|^2 \, d(E(\lambda)f, f)_H \\ &\leq \sup \left\{ |F(\lambda)|^2 \; | \lambda \in R \right\} \; (E(R)f, f)_H \\ &= \sup \left\{ |F(\lambda)|^2 \; | \lambda \in R \right\} \|f\|_H^2. \end{aligned}$$

□

2.2 Integrals of Bounded Measurable Functions

Denote the collection of all bounded measurable functions connected with a measurable space $\langle R, \mathfrak{R} \rangle$ by $L_\infty(R, \mathfrak{R}) = L_\infty$. Just as S, this collection is an algebra with respect to the standard algebraic operations.

It has already been proved (see Theorem 2.5.2) that every function $F \in L_\infty$ can be uniformly approximated by a properly chosen sequence $(F_n)_{n=1}^\infty$ of simple functions $F_n \in S$: $\sup\{|F_n(\lambda) - F(\lambda)| | \lambda \in R\} \to 0$ as $n \to \infty$. By definition, we set ($\forall F \in L_\infty$):

$$\mathcal{L}(H) \ni \int_R F(\lambda) dE(\lambda) = \lim_{n \to \infty} \int_R F_n(\lambda) dE(\lambda), \tag{2.10}$$

where the limit is understood in the operator norm.

This definition is correct. First, the limit in (2.10) exists because, in view of (2.4) and (2.9),

$$\begin{aligned} \left\| \int_R F_n(\lambda) dE(\lambda) - \int_R F_m(\lambda) dE(\lambda) \right\| \\ = \left\| \int_R \Big(F_n(\lambda) - F_m(\lambda) \Big) dE(\lambda) \right\| \\ \leq \sup \{ |F_n(\lambda) - F_m(\lambda)| \quad | \lambda \in R \} \xrightarrow[m,n \to \infty]{} 0, \end{aligned} \tag{2.11}$$

and the space of operators $\mathcal{L}(H)$ is complete. Further, limit (2.10) does not depend on the choice of the sequence $\left(F_n\right)_{n=1}^\infty$ that approximates F. If $\left(F_n'\right)_{n=1}^\infty$ is another sequence of this sort, then, by analogy with (2.11), one can show that the corresponding integrals have the same limit.

(7). *The integrals of bounded measurable functions $F, G \in L_\infty$ also possess properties (1)–(6).*

To prove this fact, one should write the relations (2.4)–(2.9) for approximating functions from S and then pass to the limit. It is easy to see that the limit transition is always possible. □

2.3 Integrals of Unbounded Measurable Functions

Consider the collection $L_0(R, \mathfrak{R}, E) = L_0$ of functions of the form $R \ni \lambda \mapsto F(\lambda) \in \mathbb{C} \cup \{\infty\}$ measurable with respect to $\mathfrak{R}$ and almost everywhere finite with respect to an operator-valued measure E. Clearly, the last statement means that

$$E(\{\lambda \in R \mid |F(\lambda)| = \infty\}) = 0. \tag{2.12}$$

Just as S and L_∞, by virtue of the additivity of the measure E, the collection L_0 forms an algebra with the ordinary operations of summation and multiplication of functions (and with standard formal rules for operations with ∞). The definition of integral (2.1) for $F \in L_0$ becomes now more complicated, since this integral is, generally speaking, an unbounded operator with a specific domain. Let us describe this domain.

Lemma 2.1. *Let $F \in L_0$. Then the set*

$$\mathcal{D}_F = \left\{ f \in H \mid \int_R |F(\lambda)|^2 d\Big(E(\lambda) f, f\Big)_H < \infty \right\} \tag{2.13}$$

is linear and everywhere dense in H.

Proof. It is clear that $f \in \mathcal{D}_F \Rightarrow af \in \mathcal{D}_F \quad (a \in \mathbb{C})$. Let us prove the additivity of $\mathcal{D}_F$, i.e., that $f, g \in \mathcal{D}_F \Rightarrow f + g \in \mathcal{D}_F$. Let $\alpha \in \mathfrak{R}$. According to (1.6), we have

$$\begin{aligned}(E(\alpha)(f+g), f+g)_H &= \|E(\alpha)(f+g)\|_H^2 \\ &\le (\|E(\alpha)f\|_H + \|E(\alpha)g\|_H)^2 \\ &\le 2\left(\|E(\alpha)f\|_H^2 + \|E(\alpha)g\|_H^2\right) \\ &= 2\left((E(\alpha)f, f)_H + (E(\alpha)g, g)_H\right).\end{aligned} \tag{2.14}$$

This implies that the existence of the integral in (2.13) for the measures appearing on the right-hand side of (2.14) yields its existence for the measure on the left-hand side of (2.14), i.e., $f + g \in \mathcal{D}_F$. Thus, $\mathcal{D}_F$ is a linear set.

To prove that $\mathcal{D}_F$ is dense in H, we consider the sets $\alpha_n = \{\lambda \in R \mid |F(\lambda)| > n\} \in \mathfrak{R} (n \in \mathbb{N})$. Evidently, $\alpha_1 \supseteq \alpha_2 \supseteq \dots$ and $\alpha = \cap_{n=1}^{\infty} \alpha_n$ consists of the points where $|F(\lambda)| = \infty$. Since the function F is finite almost everywhere with respect to the measure E, we have $E(\alpha) = 0$. It is easy to conclude that

$$\mathcal{R}\Big(E(R \setminus \alpha_n)\Big) \subseteq \mathcal{D}_F \qquad (n \in \mathbb{N}). \tag{2.15}$$

Indeed, according to (1.15), for all $\gamma \in \mathfrak{R}$ and for all $h \in H$, we have

$$\begin{aligned}(E(\gamma)E(R \setminus \alpha_n)h,\, E(R \setminus \alpha_n)h)_H \\ &= \big(E(R \setminus \alpha_n)E(\gamma)E(R \setminus \alpha_n)h, h\big)_H \\ &= \big(E(\gamma \cap (R \setminus \alpha_n))h, h\big)_H \\ &\Rightarrow \int_\gamma d\big(E(\lambda)E(R \setminus \alpha_n)h, E(R \setminus \alpha_n)h\big)_H \\ &= \int_\gamma \chi_{R\setminus\alpha_n}(\lambda) d\big(E(\lambda)h, h\big)_H, \qquad (2.16)\end{aligned}$$

where $\chi_{R\setminus\alpha_n}$ is the indicator of the set $R \setminus \alpha_n$. Therefore, if a vector f in the integral in (2.13) belongs to $\mathcal{R}\big(E(R\setminus\alpha_n)\big)$, then the integrand is a bounded function $|F(\lambda)|^2\chi_{R\setminus\alpha_n}(\lambda)$. Therefore, the integral is finite, i.e., (2.15) is true.

By using (2.15) and Theorem 1.2, one can easily prove that $\mathcal{D}_F$ is dense in H. Indeed,

$$(\forall f \in H) : \mathcal{D}_F \ni E(R \setminus \alpha_n)f \xrightarrow[n\to\infty]{} E(R)f - E(\alpha)f = f. \qquad \square$$

We now proceed to the definition of integral (2.1). As usual, for $F \in L_0$ and $N \geq 0$, we denote by F_N its cutoff function, i.e., the bounded function of the form $F_N(\lambda) = F(\lambda)$ for $\lambda \in \{\lambda \in R \mid |F(\lambda)| \leq N\}$ and $F_N(\lambda) = N$ for all other $\lambda \in R$. By definition, for $f \in \mathcal{D}_F$, we set

$$\mathcal{I}_F f = \int_R F(\lambda)dE(\lambda)f = \lim_{N\to\infty} \int_R F_N(\lambda)dE(\lambda)f \qquad (2.17)$$

in the sense of convergence in H.

It is easy to show that this limit exists. Thus, for any $M, N \geq 0$, according to (2.4) and (2.8) (for the function from L_∞), we obtain

$$\begin{aligned}\left\|\int_R F_M(\lambda)dE(\lambda)f - \int_R F_N(\lambda)dE(\lambda)f\right\|_H^2 \\ &= \left\|\int_R \Big(F_M(\lambda) - F_N(\lambda)\Big)\, dE(\lambda)f\right\|_H^2 \\ &= \int_R |F_M(\lambda) - F_N(\lambda)|^2\, d\Big(E(\lambda)f, f\Big)_H \xrightarrow[M,N\to\infty]{} 0\end{aligned}$$

for $f \in \mathcal{D}_F$.

The last relation is the consequence of the fact that the integral in (2.13) exists.

The integral $\mathcal{I}_F$ (2.17) exists, generally speaking, as an unbounded operator with a dense domain $\mathcal{D}(\mathcal{I}_F) = \mathcal{D}_F$ in H defined by relation (2.13).

Let us describe the properties of the integral $\mathcal{I}_F$. First, note that (6) is meaningless.

(8). *The integral of unbounded functions $F \in L_0$ possesses properties (4) and (5) with $f \in \mathcal{D}(\mathcal{I}_F)$ in (2.7) and (2.8).*

To prove this, it is necessary to write (2.7) and (2.8) for the cutoff functions and pass to the limit as $N \to \infty$. □

2.4 Other Properties of Spectral Integrals

The generalization of properties (1)–(3) appears to be more complicated.

Theorem 2.1. *Let $F \in L_0$. Then the operator $\mathcal{I}_F$ of the form (2.17) with a dense domain $\mathcal{D}(\mathcal{I}_F) = \mathcal{D}_F$ is closed. The equality $(\mathcal{I}_F)^* = \mathcal{I}_{\bar{F}}$, $\mathcal{D}\big((\mathcal{I}_F)^*\big) = \mathcal{D}_F$ holds, i.e., property (3) admits the following generalization:*

$$\left(\int_R F(\lambda)dE(\lambda)\right)^* = \int_R \overline{F(\lambda)}\, dE(\lambda) \qquad (F \in L_0). \tag{2.18}$$

Note that *if $F \in L_0$ is a real-valued function, then the operator $\mathcal{I}_F$ is selfadjoint.*

Proof. Let $f, g \in \mathcal{D}_F$. By virtue of (2.6) (for the functions from L_∞), we have $(\forall N \geq 0)$:

$$\left(\int_R F_N(\lambda)dE(\lambda)f, g\right)_H = \left(f, \int_R \overline{F_N(\lambda)}dE(\lambda)g\right)_H.$$

Passing here to the limit as $N \to \infty$, which is possible in view of the equality $\mathcal{D}_{\bar{F}} = \mathcal{D}_F$, we obtain $(\mathcal{I}_F f, g)_H = (f, \mathcal{I}_{\bar{F}} g)_H$. But this means that $\mathcal{I}_{\bar{F}} \subseteq (\mathcal{I}_F)^*$.

Let us prove the opposite inclusion. We have

$$(\mathcal{I}_F f, g)_H = (f, g^*)_H \ \big(f \in \mathcal{D}(\mathcal{I}_F) = \mathcal{D}_F, g \in \mathcal{D}\big((\mathcal{I}_F)^*\big)\big) \text{ and } g^* = (\mathcal{I}_F)^* g \tag{2.19}$$

We introduce the same notation as in the proof of Lemma 2.1 and substitute $f = E(R \setminus \alpha_n)h$ $(h \in H)$ in (2.18). This is possible by virtue of (2.15). Then, by analogy with (2.16), in view of (1.5), we get $E(\gamma)E(R \setminus \alpha_n)h = E(\gamma \cap (R \setminus \alpha_n))h$ and, consequently,

$$\mathcal{I}_F f = \left(\int_R F(\lambda)dE(\lambda)\right)\left(E(R \setminus \alpha_n)h\right) = \int_R F(\lambda)\chi_{R\setminus\alpha_n}(\lambda)dE(\lambda)h. \tag{2.20}$$

By inserting this expression in (2.19) and taking the boundedness of the function $F(\lambda)\chi_{R\setminus\alpha_n}(\lambda)$ and (2.6) into account, we obtain

$$\begin{aligned}\left(\int_R F(\lambda)\chi_{R\setminus\alpha_n}(\lambda)dE(\lambda)h, g\right)_H &= \left(h, \int_R \overline{F(\lambda)}\chi_{R\setminus\alpha_n}(\lambda)dE(\lambda)g\right)_H \\ &= (E(R \setminus \alpha_n)h, g^*)_H \\ &= (h, E(R \setminus \alpha_n)g^*)_H,\end{aligned}$$

whence, in view of the fact that $h \in H$ is arbitrary,

$$\int_R \overline{F(\lambda)}\chi_{R\setminus\alpha_n}(\lambda)dE(\lambda)g = E(R \setminus \alpha_n)g^*.$$

According to (2.8) (for the functions from L_0), it follows from this equality that

$$\int_R |F(\lambda)\chi_{R\setminus\alpha_n}(\lambda)|^2 d\Big(E(\lambda)g, g\Big)_H = \|E(R\setminus\alpha_n)g^*\|_H^2 \le \|g^*\|_H^2 \quad (n \in \mathbb{N}).$$

By passing here to the limit as $n \to \infty$, we conclude that $g \in \mathcal{D}_F = \mathcal{D}(\mathcal{I}_{\bar{F}})$. Hence, $(\mathcal{I}_F)^* \subseteq \mathcal{I}_{\bar{F}}$, i.e., $(\mathcal{I}_F)^* = \mathcal{I}_{\bar{F}}$.

The closeness of the operator $\mathcal{I}_F$ follows from the equality $\mathcal{I}_F = (\mathcal{I}_{\bar{F}})^*$. □

Theorem 2.2. *Let F,$G \in L_0$ and $a, b \in \mathbb{C}$. The following equalities, which generalize properties (1) and (2), are true:*

$$\int (aF(\lambda) + bG(\lambda))\, dE(\lambda) = \Big(a \int_R F(\lambda)dE(\lambda) + b \int_R G(\lambda)dE(\lambda)\Big)^{\sim}; \qquad (2.21)$$

$$\int_R F(\lambda)G(\lambda)dE(\lambda) = \Big(\int_R F(\lambda)dE(\lambda) \int_R G(\lambda)dE(\lambda)\Big)^{\sim}. \qquad (2.22)$$

Proof. Let us show that (2.21) is true, i.e., that $\mathcal{I}_{F+G} = (\mathcal{I}_F + \mathcal{I}_G)^{\sim}$ (the relation $\mathcal{I}_{aF} = a\mathcal{I}_F$ is obvious). Let

$$f \in \mathcal{D}(\mathcal{I}_F + \mathcal{I}_G) = \mathcal{D}(\mathcal{I}_F) \cap \mathcal{D}(\mathcal{I}_G) = \mathcal{D}_F \cap \mathcal{D}_G.$$

The estimate

$$|F(\lambda) + G(\lambda)|^2 \le 2(|F(\lambda)|^2 + |G(\lambda)|^2)$$

implies that, in this case, $f \in \mathcal{D}_{F+G} = \mathcal{D}(\mathcal{I}_{F+G})$. By virtue of (2.4) (for bounded functions), we have $\mathcal{I}_{F_N+G_N}f = \mathcal{I}_{F_N}f + \mathcal{I}_{G_N}f$ for $N \ge 0$. Passing here to the limit as $N \to \infty$, we obtain $\mathcal{I}_{F+G}f = \mathcal{I}_F f + \mathcal{I}_G f$ ($\mathcal{I}_{F_N+G_N}f \to \mathcal{I}_{F+G}f$ because $f \in \mathcal{D}_{F+G}$). Thus, $\mathcal{I}_F + \mathcal{I}_G \subseteq \mathcal{I}_{F+G}$.

To prove the inclusion $\mathcal{I}_{F+G} \subseteq (\mathcal{I}_F + \mathcal{I}_G)^{\sim}$, we denote the sets α_n constructed in the proof of Lemma 2.1 for F,G, and $F+G$ by α_n, β_n, and γ_n, respectively. We set $\delta_n = \alpha_n \cup \beta_n \cup \gamma_n$ ($n \in \mathbb{N}$), $\delta_1 \supseteq \delta_2 \supseteq \dots$, $E\big(\bigcap_{n=1}^{\infty} \delta_n\big) = 0$.

Let $f \in \mathcal{D}(\mathcal{I}_{F+G}) = \mathcal{D}_{F+G}$. Then $f_n = E(R \setminus \delta_n)f \in \mathcal{D}(\mathcal{I}_{F+G}) \cap \mathcal{D}(\mathcal{I}_F) \cap \mathcal{D}(\mathcal{I}_G)$ and $f_n \to f$ and $\mathcal{I}_{F+G}f_n \to \mathcal{I}_{F+G}f$ as $n \to \infty$ in H. The last relation follows from the fact that (see (2.20), (2.8) for $F \in L_0$ and (8))

$$\mathcal{I}_{F+G}f_n = \int_R (F(\lambda) + G(\lambda))\, \chi_{R\setminus\delta_n}(\lambda)dE(\lambda)f, \qquad (2.23)$$

$$\|\mathcal{I}_{F+G}f_n - \mathcal{I}_{F+G}f\|_H^2 = \int_R |F(\lambda) + G(\lambda)|^2\, \chi_{\delta_n}(\lambda)\, d(E(\lambda)f, f)_H \to 0, \quad n \to \infty.$$

In addition, there exists $\lim_{n\to\infty}(\mathcal{I}_F + \mathcal{I}_G) f_n = \lim_{n\to\infty}\mathcal{I}_{F+G} f_n$. This follows from the existence of $\lim_{n\to\infty}\mathcal{I}_F f_n$ and $\lim_{n\to\infty}\mathcal{I}_G f_n$ by virtue of the argument similar to (2.23). This indicates that $f \in \mathcal{D}\big((\mathcal{I}_F+\mathcal{I}_G)^\sim\big)$ and $\mathcal{I}_{F+G}f = (\mathcal{I}_F+\mathcal{I}_G)^\sim f$. The required inclusion is proved. We have $\mathcal{I}_{F+G} = (\mathcal{I}_F+\mathcal{I}_G)^\sim$.

Let us now prove equality (2.22). First, we note that the equality

$$(E(\gamma)\mathcal{I}_G f, \mathcal{I}_G f)_H = \int_R |G(\lambda)|^2 \chi_\gamma(\lambda) d\Big(E(\lambda)f, f\Big)_H \quad (f \in \mathcal{D}(\mathcal{I}_G)) \tag{2.24}$$

holds for all $\gamma \in \mathfrak{R}$.

Indeed, for $G \in L_\infty$, according to (3) and (2) (for bounded functions), we obtain

$$\big(E(\gamma)\mathcal{I}_G f, \mathcal{I}_G f\big)_H = \big(\mathcal{I}_{\overline{G}}\mathcal{I}_{\chi_\gamma}\mathcal{I}_G f, f\big)_H = \big(\mathcal{I}_{|G|^2\chi_\gamma} f, f\big)_H,$$

i.e., relation (2.24). In the general case of $G \in L_0$, we write (2.24) for the cutoff function G_N and then pass to the limit as $N \to \infty$. This limit transition is possible both on the left-hand side (since $\mathcal{I}_{G_N} f \to \mathcal{I}_G f$ by virtue of (2.17)), and on the right-hand side (by virtue of the estimate $|G_N(\lambda)|^2 \le |G(\lambda)|^2$ and inclusion $f \in \mathcal{D}(\mathcal{I}_G)$).

By using (2.24), one can easily show that

$$\mathcal{D}(\mathcal{I}_F\mathcal{I}_G) = \mathcal{D}(\mathcal{I}_{FG}) \cap \mathcal{D}(\mathcal{I}_G). \tag{2.25}$$

Indeed, in view of (2.24), we have

$$\begin{aligned}
\mathcal{D}(\mathcal{I}_F\mathcal{I}_G) &= \{f \in \mathcal{D}(\mathcal{I}_G) | \mathcal{I}_G f \in \mathcal{D}(\mathcal{I}_F)\} \\
&= \Big\{f \in \mathcal{D}(\mathcal{I}_G) | \int_R |F(\lambda)|^2 d\Big(E(\lambda)\mathcal{I}_G f, \mathcal{I}_G f\Big)_H < \infty\Big\} \\
&= \Big\{f \in \mathcal{D}(\mathcal{I}_G) \mid \int_R |F(\lambda)|^2 |G(\lambda)|^2 d\Big(E(\lambda)f, f\Big)_H < \infty\Big\} \\
&= \mathcal{D}(\mathcal{I}_{FG}) \cap \mathcal{D}(\mathcal{I}_G).
\end{aligned}$$

It follows from (2.25) that $\mathcal{I}_F\mathcal{I}_G \subseteq \mathcal{I}_{FG}$. Thus, let $f \in \mathcal{D}(\mathcal{I}_F\mathcal{I}_G) \subseteq \mathcal{D}(\mathcal{I}_{FG}) \cap \mathcal{D}(\mathcal{I}_G)$.

For the cutoff functions, one can clearly write $\mathcal{I}_{F_N}\mathcal{I}_{G_M} f = \mathcal{I}_{F_N G_M} f$. Let us pass in this equality to the limit as $M \to \infty$. Since $\mathcal{I}_{G_M} f \to \mathcal{I}_G f$ and $\mathcal{I}_{F_N}$ is continuous, on the left-hand side, we obtain $\mathcal{I}_{F_N}\mathcal{I}_G f$. On the right-hand side, we get $\mathcal{I}_{F_N G} f : f \in \mathcal{D}(\mathcal{I}_{F_N G}) = \mathcal{D}_{F_N G}$ because F_N is bounded. Then it is necessary to pass to the limit as $M \to \infty$ under the integral sign in the expression $\|\mathcal{I}_{F_N G_M} f - \mathcal{I}_{F_N G} f\|_H^2$ (written as an integral according to (5) and (8)). Thus, $\mathcal{I}_{F_N}\mathcal{I}_G f = \mathcal{I}_{F_N G} f$ $(N \ge 0)$.

Let us pass to the limit as $N \to \infty$ in the last equality. This is possible, since $f \in \mathcal{D}(\mathcal{I}_{FG})$. As a result, we conclude that $\mathcal{I}_F\mathcal{I}_G f = \mathcal{I}_{FG} f$ on f under consideration. The inclusion $\mathcal{I}_F\mathcal{I}_G \subseteq \mathcal{I}_{FG}$ is proved.

The inclusion $\mathcal{I}_{FG} \subseteq (\mathcal{I}_F\mathcal{I}_G)^\sim$ is proved just as the inclusion $\mathcal{I}_{F+G} \subseteq (\mathcal{I}_F + \mathcal{I}_G)^\sim$ in the first part of the theorem. Namely, we denote the sets α_n constructed in the proof of Lemma 2.1 for F, G, and FG by α_n, β_n, and γ_n, respectively. We set $\delta_n = \alpha_n \cup \beta_n \cup \gamma_n$ $(n \in \mathbb{N})$; $\delta_1 \supseteq \delta_2 \supseteq \dots,$, $E(\bigcap_{n=1}^{\infty} \delta_n) = 0$. Further, for $f \in \mathcal{D}(\mathcal{I}_{FG}) = \mathcal{D}_{FG}$, we construct a sequence $f_n = E(R\setminus\delta_n)f \in \mathcal{D}(\mathcal{I}_{FG}) \cap \mathcal{D}(\mathcal{I}_F) \cap \mathcal{D}(\mathcal{I}_G)$. It is easy to show that this sequence gives the required approximation. □

REMARK 2.1. If one of the functions (F or G) in Theorem 2.2 is bounded, then it is not necessary to take the closure on the right-hand sides of equalities (2.21) and (2.22).

Indeed, assume, for example, that $G \in L_\infty$. Then the operator $\mathcal{I}_G$ is bounded, $\mathcal{D}(\mathcal{I}_F + \mathcal{I}_G) = \mathcal{D}(\mathcal{I}_F)$, and $\mathcal{D}(\mathcal{I}_F\mathcal{I}_G) = \mathcal{D}(\mathcal{I}_{FG})$. As a result, just in the first parts of the proofs of relations (2.21) and (2.22), we obtain not the inclusions $\mathcal{I}_F + \mathcal{I}_G \subseteq \mathcal{I}_{F+G}$ and $\mathcal{I}_F\mathcal{I}_G \subseteq \mathcal{I}_{FG}$ but the equalities of the operators. □

As when integrating over a scalar measure, we suppose that $(\forall \alpha \in \mathfrak{R}, F \in L_0)$

$$\int_\alpha F(\lambda)dE(\lambda) = \int_R F(\lambda)\chi_\alpha(\lambda)dE(\lambda). \tag{2.26}$$

Integrals of the form (2.26) are often encountered in what follows. The domains of these integrals are given by relation (2.13), where R is replaced by α. We also stress that integrals (2.17) and (2.13) remain unchanged if the integrand changes on a set of E-measure zero.

We now introduce an important definition that generalizes the notion of a bounded normal operator (see Section 8.5).

An operator A acting on the Hilbert space H is called normal if it is densely defined and commutes with the adjoint operator, i.e.,

$$A\,A^* = A^*\,A. \tag{2.27}$$

It is clear that selfadjoint and unitary operators are normal. The following important assertion is true:

For all $F \in L_0$, operator (2.17) is normal.

Indeed, according to Theorem 2.1, we have $B_1 = (\mathcal{I}_F)^*\mathcal{I}_F = \mathcal{I}_{\bar{F}}\mathcal{I}_F$ and, according to Theorem 12.7.3, this operator is closed. Therefore, equality (2.22) now gives $B_1 = \tilde{B}_1 = \mathcal{I}_{|F|^2}$. Furthermore, in view of the remark after the proof of Theorem 12.7.3, we get $B_2 = \mathcal{I}_F(\mathcal{I}_F)^* = \mathcal{I}_{|F|^2}$. Thus, $B_1 = B_2$, i.e., the operator $\mathcal{I}_F$ is normal. □

We stress that if $F, G \in L_\infty$ (or, more generally, $F \in L_\infty$ and $G \in L_0$), then the operators $\mathcal{I}_F$ and $\mathcal{I}_G$ commute (see (2) and Remark 2.1). If $F, G \in L_0$, then commutation is, in a certain sense, conventional, i.e., $(\mathcal{I}_F\mathcal{I}_G)^\sim = (\mathcal{I}_G\mathcal{I}_F)^\sim$. This follows from (2.22).

3 Image of a Resolution of the Identity. Change of Variables in Spectral Integrals. Product of Resolutions of the Identity

In this section, we clarify how to perform a change of variables in spectral integrals and construct the Cartesian product of resolutions of the identity. These constructions are similar to those realized in the case of scalar measures (Sections 4.2 and 5.4).

3.1 Image of a Resolution of the Identity

Let $\langle R, \mathfrak{R}\rangle$ be a measurable space, R' be another space of points $\lambda', \mu', \ldots$, and $R \ni \lambda \to \varphi(\lambda) \in R'$ be a fixed one-to-one mapping of R into R'. Without loss of generality, we can assume that $\varphi(R) = R'$. Given $\mathfrak{R}$ and φ, we can define on R' (in a standard way) the σ-algebra $\mathfrak{R}'$ that consists of all sets $\alpha' \subseteq R'$ whose preimage $\varphi^{-1}(\alpha')$ lies in $\mathfrak{R}$. In Section 5.4, it was shown that $\mathfrak{R}'$ is indeed a σ-algebra. Thus, we have constructed the measurable space $\langle R', \mathfrak{R}'\rangle$.

Let E be a resolution of the identity given on $\langle R, \mathfrak{R}\rangle$. We now describe the standard method for constructing the image of E, i.e., the resolution of the identity E' in $\langle R', \mathfrak{R}'\rangle$. Thus, we set

$$\mathfrak{R}' \ni \alpha' \mapsto E'(\alpha') = E\Big(\varphi^{-1}(\alpha')\Big). \tag{3.1}$$

It is not difficult to show that the operator-valued measure (3.1) is a resolution of the identity, i.e., satisfies the requirements (a) and (b) in Section 1. In fact, (a) is absolutely obvious while (b) follows from the equality

$$\begin{aligned} E'\left(\bigcup_{j=1}^{\infty} \alpha'_j\right) &= E\left(\varphi^{-1}\left(\bigcup_{j=1}^{\infty} \alpha'_j\right)\right) = E\left(\bigcup_{j=1}^{\infty} \varphi^{-1}(\alpha'_j)\right) \\ &= \sum_{j=1}^{\infty} E(\varphi^{-1}(\alpha'_j)) = \sum_{j=1}^{\infty} E'(\alpha'_j) \\ &(\alpha'_j \in \mathfrak{R}', \quad \alpha'_j \cap \alpha'_k = \emptyset, \quad j \neq k) \end{aligned} \tag{3.2}$$

(here, we have used the facts that $\varphi^{-1}(\cup_{j=1}^{\infty} \alpha'_j) = \cup_{j=1}^{\infty} \varphi^{-1}(\alpha'_j)$, where the sets $\varphi^{-1}(\alpha'_j)$ are mutually disjoint (see Section 5.4).

Let $R' \ni \lambda' \mapsto F'(\lambda') \in \mathbb{C} \cup \{\infty\}$ be a function defined in R'. We define a function in R by setting $R \ni \lambda \mapsto F'(\varphi(\lambda)) = (F' \circ \varphi)(\lambda) \in \mathbb{C} \cup \{\infty\}$, i.e., by taking the superposition of two mappings φ and F'. For any Borel set $\delta \subseteq \mathbb{C}$, the inclusion $(F')^{-1}(\delta) \in \mathfrak{R}'$ is equivalent to the fact that $(F' \circ \varphi)^{-1}(\delta) \in \mathfrak{R}$ (since $\varphi^{-1}((F')^{-1}(z)) = (F' \circ \varphi)^{-1}(z)(z \in \mathbb{C})$). Therefore, the function F' is measurable with respect to $\mathfrak{R}'$ if and only if $F' \circ \varphi$ is measurable with respect to $\mathfrak{R}$.

The following theorem gives the rule of the change of variables in spectral integrals:

Theorem 3.1. *Let $L_0(R', \mathfrak{R}', E')$ be the collection of complex-valued functions defined on R', measurable with respect to $\mathfrak{R}'$, and finite E'-almost everywhere. If $F' \in L_0(R', \mathfrak{R}', E')$, then $F' \circ \varphi \in L_0(R, \mathfrak{R}, E)$ and the following formulas of the change of variables hold:*

$$\int_R F'(\varphi(\lambda))dE(\lambda) = \int_{R'} F'(\lambda')dE'(\lambda') = \mathcal{I}_{F'}; \tag{3.3}$$

$$\begin{aligned} \mathcal{D}(\mathcal{I}_{F'}) &= \left\{ f \in H \;\Big|\; \int_R |F'(\varphi(\lambda))|^2 d\Big(E(\lambda)f, f\Big)_H < \infty \right\} \\ &= \left\{ f \in H \;\Big|\; \int_{R'} |F'(\lambda')|^2 d\Big(E'(\lambda')f, f\Big)_H < \infty \right\}. \end{aligned} \tag{3.4}$$

Proof. Let $\alpha' = \{\lambda' \in R' \mid |F'(\lambda')| = \infty\}$, then $\varphi^{-1}(\alpha') = \{\lambda \in R \mid |F'(\varphi(\lambda))| = \infty\}$. Therefore, the equality $E(\varphi^{-1}(\alpha')) = E'(\alpha') = 0$ means that the the function $F' \circ \varphi$ is finite E-almost everywhere. The first statement of the theorem is proved.

For simple functions $F' \in S(R', \mathfrak{R}')$, equality (3.3) immediately follows from the definition of integral (2.3) and the resolution of the identity E' (3.1), namely, for a partition of R' into disjoint sets $\alpha'_k \in \mathfrak{R}'$, we have

$$\begin{aligned} \int_{R'} F'(\lambda)dE'(\lambda') &= \int_{R'} \left(\sum_{k=1}^{n} F'_k \chi_{\alpha'_k}(\lambda') \right) dE'(\lambda') \\ &= \sum_{k=1}^{n} F'_k E'(\alpha'_k) = \sum_{k=1}^{n} F'_k E\Big(\varphi^{-1}(\alpha'_k)\Big) \\ &= \int_R \left(\sum_{k=1}^{n} F'_k \chi_{\varphi^{-1}(\alpha_k)}(\lambda) \right) dE(\lambda) = \int_R F'(\varphi(\lambda))dE(\lambda). \end{aligned}$$

Here, we have used the fact that $\chi_{\alpha'}(\varphi(\lambda)) = \chi_{\varphi^{-1}(\alpha')}(\lambda)$ for $\alpha' \subseteq R'$.

Equality (3.3) admits an obvious extension to the functions F' from the class $L_\infty(R', \mathfrak{R}')$ by the limit transition in the uniform approximation of $F' \in L_\infty(R', \mathfrak{R}')$ by $F'_n \in S(R', \mathfrak{R}')$.

In the case where $F' \in L_0(R', \mathfrak{R}', E')$, we use (2.17) and take the relation $F'_N(\varphi(\lambda)) = (F' \circ \varphi)_N(\lambda)$ and the fact that the integrals in (3.4) are equal into account. This equality is guaranteed by the standard change of variables in the integral with respect to a scalar measure. □

3.2 Product of Resolutions of the Identity

Here, we describe how to construct the direct product of resolutions of the identity.

Let $\langle R_1, \mathfrak{R}_1\rangle, \langle R_2, \mathfrak{R}_2\rangle$ be two measurable spaces with resolutions of the identity E_1 and E_2, respectively. The values of these resolutions of the identity are projectors in the same Hilbert space H. Suppose that E_1 and E_2 commute, i.e., the operators $E_1(\alpha_1)$ and $E_2(\alpha_2)$ commute for all $\alpha_1 \in \mathfrak{R}_1$ and $\alpha_2 \in \mathfrak{R}_2$.

As in the case of scalar measures, one can try to construct the direct product E of resolutions of the identity E_1 and E_2 on the space $R = R_1 \times R_2$. More precisely, denote by $\mathfrak{R}$ the direct product $\mathfrak{R}_1 \times \mathfrak{R}_2$ of the σ-algebras $\mathfrak{R}_1$ and $\mathfrak{R}_2$ composed of all subsets of $R_1 \times R_2$ that belong to the σ-span of all possible rectangles $\alpha_1 \times \alpha_2$ $(\alpha_1 \in \mathfrak{R}_1, \alpha_2 \in \mathfrak{R}_2)$ (see Section 4.1). It is necessary to construct a resolution of the identity E in the measurable space $\langle R, \mathfrak{R}\rangle$ such that

$$E(\alpha_1 \times \alpha_2) = E_1(\alpha_1)E_2(\alpha_2) \quad (\alpha_1 \in \mathfrak{R}_1,\ \alpha_2 \in \mathfrak{R}_2) \tag{3.5}$$

(i.e., the measure E of a rectangle is the product of the corresponding measures of its sides). Note that, in view of the fact that $E_1(\alpha_1)$ and $E_2(\alpha_2)$ commute, operator (3.5) is a projector.

This E is called the direct product of E_1 and E_2 and denoted by $E = E_1 \times E_2$.

We arrive at the quite unexpected result that, unlike in the scalar case, the direct product E does not always exist. Nevertheless, in "proper" cases, it exists.

Thus, let R be a complete metric separable space and let $\mathfrak{R} = \mathfrak{B}(R)$ be the σ-algebra of its Borel subsets.

The resolution of the identity defined on $\mathfrak{B}(R)$ is called the Borel resolution of the identity.

Theorem 3.2. *Let E_1 and E_2 be two commuting Borel resolutions of the identity in the spaces R_1 and R_2, respectively. Then condition (3.5) determines a unique resolution of the identity $E = E_1 \times E_2$ defined on $\mathfrak{R} = \mathfrak{B}\,(R_1 \times R_2)$.*

Before proving this theorem, we recall that every scalar finite measure $\mathfrak{B}(R) \ni \alpha \mapsto \mu(\alpha) \geq 0$ is automatically regular (Section 6.8), i.e.,

$$\mu(\alpha) = \inf\,\{\mu(o) \mid o \supseteq \alpha,\ o \text{ is open}\} \quad (\alpha \in \mathfrak{B}(R)). \tag{3.6}$$

By passing to the complements, we easily conclude that (3.6) is equivalent to the relation

$$\mu(\alpha) = \sup\,\{\mu(\varphi) \mid \varphi \subseteq \alpha,\ \varphi \text{ is closed}\} \quad (\alpha \in \mathfrak{B}(R)). \tag{3.7}$$

Moreover, supremum in (3.7) can be taken only over compact sets $\varphi \subseteq \alpha$.

Proof. Denote by $\mathfrak{R}'$ the algebra spanned by the collection of all rectangles $\alpha_1 \times \alpha_2$ $\big(\alpha_1 \in \mathfrak{B}(R_1), \alpha_2 \in \mathfrak{B}(R_2)\big)$. The function $E(\alpha_1 \times \alpha_2) = E_1(\alpha_1)E_2(\alpha_2)$ is uniquely extended by additivity to a finite additive operator-valued function $\mathfrak{R}' \ni \alpha \mapsto E(\alpha)$. This can be proved in a standard way just as in the case of a scalar measure (Section 4.2). Each value of $E(\alpha)$ $(\alpha \in \mathfrak{R}')$ is a projector since α is representable as a union of finitely many disjoint rectangles $\alpha_1 \times \alpha_2$ and, therefore, $E(\alpha)$ is equal to a finite sum of projectors $E(\alpha_1 \times \alpha_2)$ which are mutually orthogonal by virtue of the orthogonality relation for E_1 and E_2. Thus, $\mathfrak{R}' \ni \alpha \mapsto E(\alpha)$ is a projector-valued finitely additive set function on the algebra $\mathfrak{R}', E(\emptyset) = 0$, and $E(R_1 \times R_2) = \mathbb{I}$. Let us prove that this function is countably additive.

Note that, for every rectangle $\alpha_1 \times \alpha_2$ $\big(\alpha_2 \in \mathfrak{B}(R_1), \alpha_2 \in \mathfrak{B}(R_2)\big)$, $f \in H$ and $\varepsilon > 0$, one can find two rectangles $o_1 \times o_2$ and $\varphi_1 \times \varphi_2$, where $o_1 \supseteq \alpha_1$ and $o_2 \supseteq \alpha_2$ are open and $\varphi_1 \subseteq \alpha_1$ and $\varphi_2 \subseteq \alpha_2$ are compact sets, such that

$$\begin{aligned} \big(E(o_1 \times o_2)f, f\big)_H - \big(E(\alpha_1 \times \alpha_2)f, f\big)_H &< \varepsilon, \\ \big(E(\alpha_1 \times \alpha_2)f, f\big)_H - \big(E(\varphi_1 \times \varphi_2)f, f\big)_H &< \varepsilon. \end{aligned} \tag{3.8}$$

Let us establish, for example, the first relation. In view of the fact that scalar measures on $\mathfrak{B}(R_1)$ and $\mathfrak{B}(R_2)$ are regular, we select, for given $\delta > 0$, an open set $o_1 \supseteq \alpha_1$ and $o_2 \supseteq \alpha_2$ such that $\big(E_1(o_1 \setminus \alpha_1)f, f\big)_H < \delta$ and $\big(E_2(o_2 \setminus \alpha_2)f, f\big)_H < \delta$. Since

$$(o_1 \times o_2) \setminus (\alpha_1 \times \alpha_2) = \big((o_1 \setminus \alpha_1) \times o_2\big) \cup \big(\alpha_1 \times (o_2 \setminus \alpha_2)\big),$$

we have

$$\begin{aligned} (E(o_1 \times o_2)f, f)_H &- (E(\alpha_1 \times \alpha_2)f, f)_H \\ &= \big(E\big((o_1 \setminus \alpha_1) \times o_2\big)f, f\big)_H + \big(E\big(\alpha_1 \times (o_2 \setminus \alpha_2)\big)f, f\big)_H \\ &= (E_1(o_1 \setminus \alpha_1)E_2(o_2)f, f)_H + (E_1(\alpha_1)E_2(o_2 \setminus \alpha_2)f, f)_H \\ &\le \|f\|_H(\|E_1(o_1 \setminus \alpha_1)f\|_H + \|E_2(o_2 \setminus \alpha_2)f\|_H) \\ &= \|f\|_H \Big(\big(E_1(o_1 \setminus \alpha_1)f, f\big)_H^{1/2} + \big(E_2(o_2 \setminus \alpha_2)f, f\big)_H^{1/2}\Big) \\ &< 2\|f\|_H \delta^{1/2}. \end{aligned}$$

By choosing sufficiently small $\delta > 0$, we arrive at the required result.

Let $\alpha_1, \alpha_2, \ldots \in \mathfrak{R}'$ be mutually disjoint and such that $\alpha = \cup_{k=1}^{\infty} \alpha_k \in \mathfrak{R}'$. To prove that E is countably additive, it suffices to show that

$$(E(\alpha)f, f)_H \le \sum_{k=1}^{\infty} (E(\alpha_k)f, f)_H \tag{3.9}$$

for any $f \in H$ (the inverse inequality follows from the finite additivity and monotonicity of the function $\mathfrak{R}' \ni \beta \mapsto (E(\beta)f, f)_H \ge 0$; therefore, (3.9) means that these expressions are equal). The proof of (3.9) is similar to those presented in Sections 1.9 and 1.11 and can be described as follows.

Since α is equal to the union of finitely many disjoint rectangles, we can apply to each the second inequality in (3.8) and find, for given $\varepsilon > 0$, a compact set $\mathfrak{R}' \ni \varphi \subseteq \alpha$ such that $\big(E(\alpha)f, f\big)_H - \big(E(\varphi)f, f\big)_H < \varepsilon$. Similarly, by using the first inequality in (3.8), for every $k \in \mathbb{N}$, one can find an open set $\mathfrak{R}' \ni o_k \supseteq \alpha_k$ such that $\big(E(o_k)f, f\big)_H - \big(E(\alpha_k)f, f\big)_H < \varepsilon/2^k$. The family $(o_k)_{k=1}^{\infty}$ covers φ. Since φ is compact, there exists $n \in \mathbb{N}$ such that $\cup_{k=1}^{n} o_k \supseteq \varphi$. This, the monotonicity and finite semiadditivity of the scalar measure, and the last two inequalities yield

$$(E(\alpha)f, f)_H < (E(\varphi)f, f)_H + \varepsilon \le \left(E\left(\cup_{k=1}^{n} o_k\right)f, f\right)_H + \varepsilon$$
$$\le \sum_{k=1}^{\infty} (E(o_k)f, f)_H + \varepsilon < \sum_{k=1}^{\infty} (E(\alpha_k)f, f)_H + 2\varepsilon.$$

Passing here to the limit as $\varepsilon \to 0$, we obtain (3.9). The countable additivity of E is proved.

Thus, we have constructed the resolution of the identity E on the algebra $\mathfrak{R}'$. By Theorem 1.3, it can be extended to the resolution of the identity E_σ on $(\mathfrak{R}')_\sigma = \mathfrak{R}$. This resolution of the identity is clearly the required one. The equality $\mathfrak{R} = \mathfrak{B}(R_1 \times R_2)$ follows from the definition of the topology in $R_1 \times R_2$. □

Certainly, this assertion remains true for an arbitrary finite number of resolutions of the identity and admits a generalization for the case of infinitely many resolutions of the identity.

Exercises

3.1. Let E_1 and E_2 be the resolutions of the identity on $\langle \mathbb{R}, \mathfrak{B}(\mathbb{R})$ considered in Examples 1.1 and 1.2, respectively. Find the images of these resolutions of the identity for the mappings (a) $\varphi_1(\lambda) = \lambda^3$; (b) $\varphi_2(\lambda) = \lambda^2$.

3.2. Let $E_1, \dots, E_n$ be pairwise commuting Borel resolutions of the identity in the spaces $R_1, \dots, R_n$, respectively. Prove that there exists a unique resolution of the identity E on $\langle \times_{k=1}^{n} R_k, \mathfrak{B}(\times_{k=1}^{n} R_k)\rangle$ such that $(\forall \alpha_k \in \mathfrak{B}(R_k), k = 1, \dots n)$:

$$E(\alpha_1 \times \dots \times \alpha_n) = E_1(\alpha_1) \dots E_n(\alpha_n) \stackrel{\text{def}}{=} (\times_{k=1}^{n} E_k)(\times_{k=1}^{n} \alpha_k).$$

3.3. Let $\big(E_n\big)_{n=1}^{\infty}$ be a sequence of pairwise commuting resolutions of the identity on $\langle \mathbb{R}, \mathfrak{B}(\mathbb{R})\rangle$. Consider a collection $\mathfrak{S}$ of sets from $\mathbb{R}^\infty = \mathbb{R} \times \mathbb{R} \times \dots$ of the form $\{\alpha \times \mathbb{R}^\infty | \ \alpha \in \mathfrak{B}(\mathbb{R}^n), n \in \mathbb{N}\}$ (these sets are called cylindrical). On $\langle \mathbb{R}^\infty, \mathfrak{S}\rangle$, we define a projector-valued set function E, by setting $E(\alpha \times \mathbb{R}^\infty) = \big(\times_{k=1}^{n} E_k\big)(\alpha)$ on $\alpha \times \mathbb{R}^\infty$, where $\alpha \in \mathfrak{B}(\mathbb{R}^n)$. Prove that

(a) $\mathfrak{S}$ is an algebra of sets;

(b) E is the resolution of the identity on $\langle \mathbb{R}^\infty, \mathfrak{S}\rangle$.

The extension of E to $\mathfrak{S}_\sigma$ is called the product $\times_{k=1}^{\infty} E_k$ of the infinitely many resolutions of the identity $E_1, E_2, \dots$.

4 Spectral Decomposition of Bounded Selfadjoint Operators

It seems reasonable to construct spectral decompositions first for the case of bounded selfadjoint operators where they are quite simple and natural. There are many proofs of the spectral theorem. Here, we present one of the most instructive ideas, namely, the "descent to the spectrum" in the theory of analytic functions of operators (Section 10.4). It appears to be useful to perform starting steps of this procedure for a general bounded operator with real spectrum.

4.1 The Spectral Theorem

Consider a bounded operator A acting on a Hilbert space H. Assume that its spectrum $S(A)$ is a (closed) set in a finite interval $(a, b) \subset \mathbb{R}$. Denote its resolvent by $R_z = (A - z\mathbb{I})^{-1}$ $\left(z \in \mathbb{C} \setminus S(A)\right)$.

Let $F(z)$ be an analytic function defined in a certain (complex) neighbourhood of the interval $[a, b]$. The collection of all functions of this sort is denoted by $\mathcal{A}([a, b])$. For any closed contour γ that encloses the spectrum $S(A)$ and lies in the domain of the function F, one can write the integral

$$F(A) = -\frac{1}{2\pi i} \oint_\gamma F(z) R_z dz. \tag{4.1}$$

As we know (Section 10.4), this integral is independent of the choice of the contour γ of the indicated type. The mapping

$$\mathcal{A}([a, b]) \ni F \mapsto F(A) \in \mathcal{L}(H) \tag{4.2}$$

is a homomorphism of the algebra of functions $\mathcal{A}([a, b])$ to the algebra $\mathcal{L}(H)$ of bounded operators in H that transforms the function $F(z) \equiv 1$ to the identity operator and the function $F(z) = z$ to the operator A.

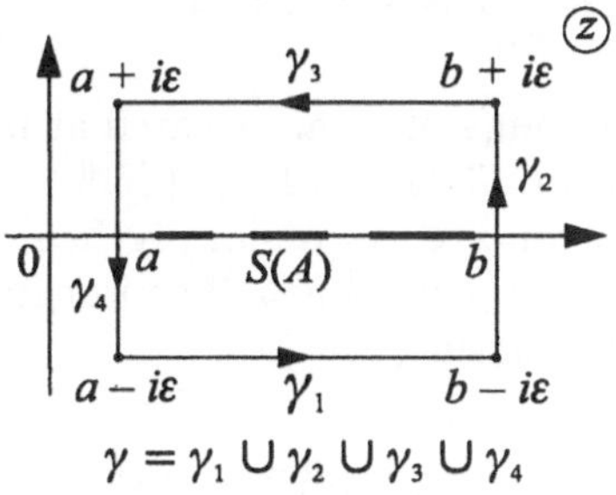

Fig. 9

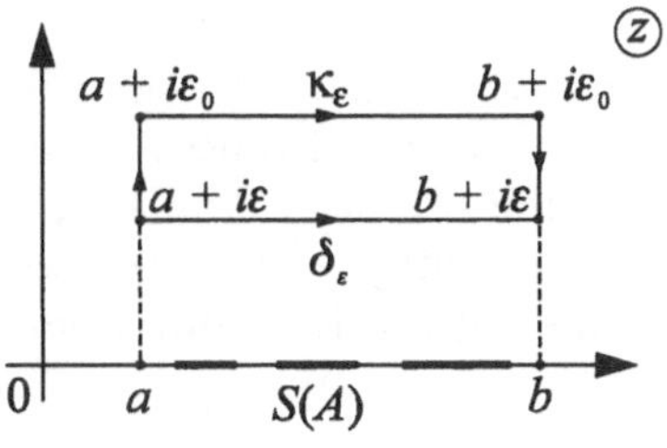

Fig. 10

Recall that $\mathcal{A}([a,b])$ is an algebra with respect to the standard algebraic operations

$$(\forall F, G \in \mathcal{A}([a,b])) : (F+G)(z) = F(z) + G(z), (FG)(z) = F(z)G(z),$$

where the functions $F+G$ and FG are defined on the intersection of the domains of the functions F and G (note that $\mathcal{A}([a,b])$ contains constants).

Thus, the indicated assertion means that the following equalities hold:

$$(F+G)(A) = F(A) + G(A), \quad (FG)(A) = F(A)G(A),$$

$$1(A) = \mathbb{I}, \quad z(A) = A \tag{4.3}$$

Lemma 4.1. *The following formula, where the limit is understood in the operator norm, holds:*

$$\begin{aligned} F(A) &= \lim_{\varepsilon \to +0} \frac{1}{2\pi i} \int_a^b F(\lambda)(R_{\lambda+i\varepsilon} - R_{\lambda-i\varepsilon})d\lambda \\ &= \lim_{\varepsilon \to +0} \frac{\varepsilon}{\pi} \int_a^b F(\lambda) R_{\lambda+i\varepsilon} R_{\lambda-i\varepsilon} d\lambda \quad (F \in \mathcal{A}([a,b])) . \end{aligned} \tag{4.4}$$

Proof. The contour γ in relation (4.1) is chosen as indicated in Fig. 9 with a sufficiently small $\varepsilon > 0$. Then this contour is split into the arcs $\gamma_1, \dots, \gamma_4$ as shown in the figure. Let I_k be integral (4.1) over γ_k. We have

$$\begin{aligned} F(A) &= -\frac{1}{2\pi i} \oint_\gamma F(z) R_z dz = I_1 + I_2 + I_3 + I_4 \\ &= -\frac{1}{2\pi i} \int_a^b F(\lambda - i\varepsilon) R_{\lambda - i\varepsilon} d\lambda + \frac{1}{2\pi i} \int_a^b F(\lambda + i\varepsilon) R_{\lambda+i\varepsilon} d\lambda + I_2 + I_4 \\ &= \frac{1}{2\pi} \int_a^b (F(\lambda + i\varepsilon) R_{\lambda+i\varepsilon} - F(\lambda - i\varepsilon) R_{\lambda - i\varepsilon}) d\lambda + I_2 + I_4 . \end{aligned}$$

Let us estimate the integral I_2. Since the points of γ_2 are located at positive (independent of ε) distances from the spectrum $S(A)$, we have $\|R_z\| \le c_1$ for $z \in \gamma_2$ uniformly in ε. Similarly, $|F(z)| \le c_2$ for $z \in \gamma_2$ uniformly in ε, where $\varepsilon > 0$ is sufficiently small. It is now clear that $\|I_2\| \le \varepsilon \pi^{-1} c_1 c_2$. The integral I_4 is estimated similarly. It follows from these estimates that

$$F(A) = \lim_{\varepsilon \to +0} \frac{1}{2\pi i} \int_a^b (F(\lambda + i\varepsilon) R_{\lambda - i\varepsilon} - F(\lambda - i\varepsilon) R_{\lambda - i\varepsilon}) d\lambda . \tag{4.5}$$

in the operator norm.

To pass from representation (4.5) to the left equality in (4.4), it is necessary to prove that

$$\varphi(\varepsilon) = \left\| \int_a^b (F(\lambda + i\varepsilon) - F(\lambda))\, R_{\lambda+i\varepsilon} d\lambda \right\| \xrightarrow[\varepsilon\to 0]{} 0, \tag{4.6}$$

and a similar relation with $\lambda + i\varepsilon$ replaced by $\lambda - i\varepsilon$. Here, we prove (4.6); the second relation is established analogously.

Denote the horizontal interval that connects the points $a + i\varepsilon$ and $b + i\varepsilon$ by δ_ε and let κ_ε be a path that connects the same points and passes through $a + i\varepsilon_0$ and $b + i\varepsilon_0$ (Fig. 10); $\kappa_0 = \kappa_\varepsilon$ for $\varepsilon = 0$. Since the function $F(z) - F(z - i\varepsilon)$ ($\varepsilon \in (0, \varepsilon_0]$, where $\varepsilon_0 > 0$, is a sufficiently small fixed number) is analytic in z and R_z is analytic outside the spectrum $S(A)$, we obtain the required result

$$\begin{aligned}\int_a^b (F(\lambda + i\varepsilon) - F(\lambda))\, R_{\lambda+i\varepsilon} d\lambda &= \int_{\delta_\varepsilon} (F(z) - F(z - i\varepsilon))\, R_z dz \\ &= \int_{\kappa_\varepsilon} (F(z) - F(z - i\varepsilon))\, R_z dz;\end{aligned}$$

$$\begin{aligned}\varphi(\varepsilon) &= \left\| \int_{\kappa_\varepsilon} (F(z) - F(z - i\varepsilon))\, R_z dz \right\| \\ &\leq \int_{\kappa_\varepsilon} |F(z) - F(z - i\varepsilon)|\, \|R_z\| ds \\ &\leq \int_{\kappa_0} |F(z) - F(z - i\varepsilon)|\, \|R_z\| ds \\ &\leq c \max\{|F(z) - F(z - i\varepsilon)| \mid z \in \kappa_0\} \xrightarrow[\varepsilon\to 0]{} 0.\end{aligned}$$

Here, we have used the fact that the path κ_0 lies at the positive distance from $S(A)$ and, therefore, $\|R_z\|$ ($z \in \kappa_0$) is bounded.

The right equality in (4.4) directly follows from the Hilbert formula $R_z - R_\zeta = (z - \zeta) R_z R_\zeta$, where z and ζ are regular points (Section 8.8). □

Relation (4.4) demonstrates that the function $F(A)$ of the operator A is calculated according to the jump of the resolvent on the spectrum. Depending on the character of the resolvent and, in particular, on its behaviour near the spectrum, the limit in (4.4) may exist for broader classes W of functions $F(\lambda)$ than the functions analytic in a neighbourhood of the spectrum. Therefore, this limit can be taken as a definition of the function $F(A)$ of A for $F \in W$. If W contains the indicators $\chi_\alpha(\lambda)$ of the sets $\alpha \subseteq \mathbb{R}$, then the operator-valued set function $\alpha \mapsto \chi_\alpha(A) = E(\alpha)$ is a function of the resolution-of-the-identity type.

Just in this way, we now construct the resolution of the identity for a self-adjoint operator. For operators of other types with a certain behaviour of the resolvent near the spectrum, W can be chosen as a set of sufficiently smooth functions. This also gives certain spectral-type representations but we do not consider them here.

Theorem 4.1. *Let A be a bounded selfadjoint operator. Then one can define a resolution of the identity E on the σ-algebra $\mathfrak{B}(\mathbb{R})$ of Borel subsets of the real axis such that the following spectral representation is true:*

$$A = \int_{\mathbb{R}} \lambda \, dE(\lambda). \tag{4.7}$$

The proof of (4.7) leads to a measure E concentrated on a finite interval that covers the spectrum of A. Therefore, the limits of integration in (4.7) are , in fact, finite. Furthermore, it is proved in what follows that $\mathbb{R}$ in (4.7) can be replaced by the spectrum $S(A)$ of the operator A.

Thus, let the operator A be selfadjoint. Then

$$R_z^* = \left((A - z\mathbb{I})^{-1}\right)^* = \left(A^* - \bar{z}\mathbb{I}\right)^{-1} = R_{\bar{z}}$$

and relation (4.4) can be rewritten in the following form:

$$F(A) = \lim_{\varepsilon \to +0} \frac{\varepsilon}{\pi} \int_a^b F(\lambda) R_{\lambda+i\varepsilon}^* R_{\lambda+i\varepsilon} d\lambda \quad \left(F \in \mathcal{A}([a,b])\right). \tag{4.8}$$

Lemma 4.2. *Assume that $F \in \mathcal{A}([a,b])$ takes real values on the real axis (the algebra of such F is denoted by $\mathcal{A}_{\mathrm{Re}}([a,b])$). The operator $F(A)$ is selfadjoint and admits the estimate*

$$\begin{gathered} |(F(A)f,g)_H| \le \max\{|F(\lambda)| \mid \lambda \in [a,b]\} \, \|f\|_H \|g\|_H \\ \left(f, g \in H; F \in \mathcal{A}_{\mathrm{Re}}([a,b])\right). \end{gathered} \tag{4.9}$$

Proof. Since $F(\lambda)$ is real-valued, the integrand in (4.8) is a selfadjoint operator. Therefore, the integral itself and $F(A)$ are also selfadjoint operators.

In view of the fact that $\left(R_{\lambda+i\varepsilon}^* R_{\lambda+i\varepsilon} f, f\right)_H \ge 0$, we have

$$\left| \frac{\varepsilon}{\pi} \int_a^b F(\lambda) \left(R_{\lambda+i\varepsilon}^* R_{\lambda+i\varepsilon} f, f\right)_H d\lambda \right| \\ \le \max\{|F(\lambda)| \mid \lambda \in [a,b]\} \frac{\varepsilon}{\pi} \int_a^b 1 \cdot \left(R_{\lambda+i\varepsilon}^* R_{\lambda+i\varepsilon} f, f\right)_H d\lambda$$

for any $f \in H$. By passing to the limit as $\varepsilon \to 0$, we conclude that (see (4.3))

$$\begin{aligned} |(F(A)f,f)_H| &\le \max\{|F(\lambda)| \mid \lambda \in [a,b]\} \, (1(A)f,f)_H \\ &= \max\{|F(\lambda)| \mid \lambda \in [a,b]\} \, \|f\|_H^2. \end{aligned} \tag{4.10}$$

However, for a bounded selfadjoint operator B, the estimate $\left|(Bf,f)_H\right| \le c\|f\|_H^2$ $(f \in H)$ implies that $\left|(Bf,g)_H\right| \le c\|f\|_H \|g\|_H (f, g \in H)$ (Section 10.1). Therefore, (4.10) yields (4.9). □

Proof of Theorem 4.1. (1). Consider the space $C_{\mathrm{Re}}([a,b]) = C_{\mathrm{Re}}$ of continuous real-valued functions φ defined on $[a,b]$ with the standard norm $\|\varphi\|_C = \max\{|\varphi(\lambda)| \mid \lambda \in [a,b]\}$. Recall (see Section 7.5) that a linear continuous complex-valued functional l on C_{Re} has the form

$$l(\varphi) = \int_a^b \varphi(\lambda) d\omega(\lambda) \ (\varphi \in C_{\mathrm{Re}}), \tag{4.11}$$

where ω is a (complex-valued) charge uniquely defined on the σ-algebra of Borel subsets of $[a,b]$ for a given l, namely, $\mathfrak{B}([a,b]) \ni \alpha \mapsto \omega(\alpha) \in \mathbb{C}$. In this case,

$$\|l\| = V(\omega;[a,b])\,; (\forall \alpha \in \mathfrak{B}([a,b])) : |\omega(\alpha)| \le V(\omega;[a,b]) = \|l\|. \tag{4.12}$$

We fix $f, g \in H$ and consider a linear functional in F

$$C_{\mathrm{Re}} \supset \mathcal{A}_{\mathrm{Re}}([a,b]) \ni F \mapsto l_{f,g}(F) = (F(A)f, g)_H \in \mathbb{C} \tag{4.13}$$

defined on a set dense in C_{Re} (its denseness follows from the classical Weierstrass theorem, since $\mathcal{A}_{\mathrm{Re}}([a,b])$ contains polynomials with real coefficients). This functional is continuous in $\mathcal{A}_{\mathrm{Re}}([a,b])$ by virtue of (4.9) and $\|l_{f,g}\| \le \|f\|_H \|g\|_H$. We extend this function by continuity to the whole C_{Re} preserving the notation $l_{f,g}$. Relations (4.11) and (4.12) imply the following representation with a certain charge $\omega_{f,g}$ (Section 7.5):

$$l_{f,g}(\varphi) = \int_a^b \varphi(\lambda) d\omega_{f,g}(\lambda) \qquad (\varphi \in C_{\mathrm{Re}});$$

$$(\forall \alpha \in \mathfrak{B}([a,b])) : |\omega_{f,g}(\alpha)| \le \|l_{f,g}\| \le \|f\|_H \|g\|_H$$

$$l_{f,g}(F) = (F(A)f, g)_H \qquad (f, g \in H; F \in \mathcal{A}_{\mathrm{Re}}([a,b])). \tag{4.14}$$

We now fix $\alpha \in \mathfrak{B}([a,b])$ and consider the mapping

$$H \oplus H \ni \langle f, g \rangle \mapsto \omega_{f,g}(\alpha) \in \mathbb{C}. \tag{4.15}$$

It is easy to see that this mapping is a bilinear form. Indeed, let $a_1, a_2 \in \mathbb{C}$ and $f_1, f_2, g \in H$. Then

$$\begin{aligned}
\int_a^b F(\lambda) d\omega_{a_1 f_1 + a_2 f_2, g}(\lambda) &= l_{a_1 f_1 + a_2 f_2, g}(F) \\
&= (F(A)(a_1 f_1 + a_2 f_2), g)_H \\
&= a_1 (F(A) f_1, g)_H + a_2 (F(A) f_2, g)_H \\
&= a_1 l_{f_1, g}(F) + a_2 l_{f_2, g}(F) \\
&= \int_a^b F(\lambda) d\left(a_1 \omega_{f_1, g}(\lambda) + a_2 \omega_{f_2, g}(\lambda)\right).
\end{aligned}$$

Due to the denseness of $\mathcal{A}_{\mathrm{Re}}([a,b])$ in C_{Re}, this enables us to conclude that $a_1\omega_{f_1,g} + a_2\omega_{f_2,g} = \omega_{a_1 f_1 + a_2 f_2, g}$, i.e., (4.15) is linear in the first variable. Its antilinearity in the second variable is proved similarly.

The bilinear form (4.15) is continuous (this follows from estimate (4.14)). Thus, by the theorem on representation of bilinear forms (Section 8.5), there exists an operator $E(\alpha) \in \mathcal{L}(H)$ such that

$$\omega_{f,g}(\alpha) = (E(\alpha)f, g)_H \; (\alpha \in \mathfrak{B}([a,b]); f, g \in H). \tag{4.16}$$

Let us show that the set function $\mathfrak{B}([a,b]) \ni \alpha \mapsto E(\alpha) \in \mathcal{L}(H)$ is a resolution of the identity on the measurable space $\langle [a,b], \mathfrak{B}([a,b]) \rangle$. We use the following equality implied by (4.16) and (4.14):

$$(F(A)f, g)_H = \int_a^b F(\lambda)\, d\,(E(\lambda)f, g)_H \quad (F \in \mathcal{A}_{\mathrm{Re}}([a,b]); f, g \in H). \tag{4.17}$$

First, we note that, for any $\alpha \in \mathfrak{B}([a,b])$, the operator $E(\alpha)$ is selfadjoint. Indeed, by virtue of the selfadjointness of $F(A)$, we have

$$\begin{aligned} \int_a^b F(\lambda)\, d\,(E(\lambda)f, g)_H &= (F(A)f, g)_H \\ &= \overline{(F(A)g, f)_H} = \int_a^b F(\lambda)\, \overline{d\,(E(\lambda)g, f)_H} \\ &(F \in \mathcal{A}_{\mathrm{Re}}([a,b]); f, g \in H). \end{aligned} \tag{4.18}$$

Since $\mathcal{A}_{\mathrm{Re}}([a,b])$ is dense in C_{Re}, we conclude that $(E(\alpha)f, g)_H = \overline{(E(\alpha)g, f)}_H$, i.e., $(E(\alpha))^* = E(\alpha)$.

(2). Let us show that if $\alpha \in \mathfrak{B}([a,b])$ is a union of finitely many disjoint half intervals of the form $[c,d) \subset [a,b]$, then $E(\alpha)$ is a projector. In fact, for the indicator $\chi_\alpha(\lambda)$ of a set with the indicated structure, one can easily construct a uniformly bounded sequence $(F_n)_{n=1}^\infty$ of functions $F_n \in \mathcal{A}_{\mathrm{Re}}([a,b])$ such that $(\forall \lambda \in [a,b]) : F_n(\lambda) \to \chi_\alpha(\lambda)$ as $n \to \infty$ (first, we find a sequence of continuous functions satisfying this requirement, and then, for each continuous function of this sort, construct a sufficiently close approximation by an analytic function in the metric of C).

If we write relation (4.17) for $F = F_n$ and pass to the limit under the integral sign (this is obviously possible), then we obtain

$$\begin{aligned} \lim_{n\to\infty} (F_n(A)f, g)_H &= \lim_{n\to\infty} \int_a^b F_n(\lambda)\, d\,(E(\lambda)f, g)_H \\ &= \int_a^b \chi_\alpha(\lambda)\, d\,(E(\lambda)f, g)_H = (E(\alpha)f, g)_H, \end{aligned}$$

i.e., $E(\alpha)$ is the weak limit of the operators $F_n(A)$ as $n \to \infty$.

Further, by using the selfadjointness of $F_n(A)$, the second equality in (4.3) and relation (4.18), we obtain, $\forall n, m \in \mathbb{N}$ and $\forall f, g \in H$,

$$(F_m(A)f, F_n(A)g)_H = (F_n(A)F_m(A)f, g)_H$$
$$= ((F_nF_m)(A)f, g)_H = \int_a^b F_n(\lambda)F_m(\lambda)d\,(E(\lambda)f, g)_H \, .$$

On the left-hand and right-hand sides of the equality, we pass to the limit as $n \to \infty$ and then as $m \to \infty$. As a result, we get

$$\left(E(\alpha)f, E(\alpha)g\right)_H = \left(E(\alpha)f, g\right)_H \quad (f, g \in H).$$

In view of the selfadjointness of $E(\alpha)$ established above, this equality means that $E(\alpha)$ is a projector. Statement (2) is proved.

(3). The unions of half intervals considered above form an algebra of subsets of the space $[a, b)$; it is denoted by $\mathfrak{R}$. The projector-valued set function $\mathfrak{R} \ni \alpha \mapsto E(\alpha)$ satisfies the condition of countable additivity (with the weak convergence of the series); this follows from the countable additivity of the charge $\omega_{f,g}$ on $\mathfrak{B}([a, b]) \supset \mathfrak{R}$ and formula (4.16). Therefore, this function is a quasiresolution of the identity (in the sense indicated at the end of Section 1; in fact, E is an ordinary resolution of the identity: it is not difficult to show that $E(\{b\}) = 0$ and, therefore, $E([a, b)) = E([a, b]) = 1(A) = \mathbb{I}$, see (4.17)).

By virtue of Theorem 1.3, E can be extended from $\mathfrak{R}$ to a quasiresolution of the identity E_σ on the σ-span $\mathfrak{R}_\sigma = \mathfrak{B}([a, b))$. Thus, for any $f, g \in H$, there are two charges $\omega_{f,g}(\alpha) = (E(\alpha)f, g)_H$ and $\rho_{f,g}(\alpha) = (E_\sigma(\alpha)f, g)_H$ on the σ-algebra $\mathfrak{B}([a, b))$. They coincide in $\mathfrak{R}$ and, therefore, by virtue of the uniqueness of the extension of a scalar measure, they coincide on $\mathfrak{R}_\sigma$, i.e., $E(\alpha) = E_\sigma(\alpha)$ and, hence, E_α is a projector $\forall \alpha \in \mathfrak{B}([a, b))$. Since $E([a, b]) = \mathbb{I}$, we conclude that $E(\alpha)$ is a projector for $\alpha \in \mathfrak{B}([a, b])$. Thus, E defined by relation (4.16) is a resolution of the identity in the space $\langle [a, b], \mathfrak{B}([a, b]) \rangle$.

By setting $F(\lambda) = \lambda$ in relation (4.17), in view of (4.3), we obtain

$$A = \int_a^b \lambda dE(\lambda). \tag{4.19}$$

The resolution of the identity thus constructed can be extended from $\langle [a, b], \mathfrak{B}([a, b]) \rangle$ to $\langle \mathbb{R}, \mathfrak{B}(\mathbb{R}) \rangle$ by setting it equal to zero outside $[a, b]$. More precisely, we assume that it is equal to $E\,(\alpha \cap [a, b])$ on $\alpha \in \mathfrak{B}(\mathbb{R})$. As a result, relation (4.19) turns into (4.7) and we get the statement of Theorem 4.1. □

4.2 Functions of Operators and Their Spectrum

REMARK 4.1. Analytic functions of the form (4.1) of selfadjoint operators $A \in \mathcal{L}(H)$ admit a representation as a spectral integral

$$F(A) = \int_{\mathbb{R}} F(\lambda) dE(\lambda) \quad (F \in \mathcal{A}([a,b])) \tag{4.20}$$

(in (4.20), it is assumed that $S(A) \subset (a,b)$; therefore, E is concentrated on (a,b)).

Indeed, for $F \in \mathcal{A}_{\mathrm{Re}}([a,b])$, relation (4.20) is equivalent to (4.17). For a general $F \in \mathcal{A}([a,b])$, one can indicate the following decomposition valid in a certain complex neighbourhood of (a,b):

$$\begin{aligned} F(z) &= F_1(z) + iF_2(z), \\ F_1(z) &= \frac{1}{2}\left(F(z) + \overline{F(\bar{z})}\right) \\ F_2(z) &= \frac{1}{2i}\left(F(z) - \overline{F(\bar{z})}\right). \end{aligned} \tag{4.21}$$

We have $F_1, F_2 \in \mathcal{A}_{\mathrm{Re}}([a,b])$. If we now write representation (4.20) for these functions and use (4.21), then we obtain the general formula (4.20). □

The spectral representation (4.7) enables us to generalize the definition of a function of an operator A from the class of analytic functions to the class $L_0(\mathbb{R}, \mathfrak{B}(\mathbb{R}), E)$ of functions of the form $\mathbb{R} \ni \lambda \mapsto F(\lambda) \in \mathbb{C} \cup \{\infty\}$ measurable with respect to $\mathfrak{B}(\mathbb{R})$ and almost everywhere finite with respect to E. Namely, we set

$$\begin{gathered} L_0(\mathbb{R}, \mathfrak{B}(\mathbb{R}), E) \ni F \mapsto F(A) = \int_{\mathbb{R}} F(\lambda) dE(\lambda), \\ \mathcal{D}(F(A)) = \left\{ f \in H \middle| \int_{\mathbb{R}} |F(\lambda)|^2 d\Big(E(\lambda)f, f\Big)_H < \infty \right\}. \end{gathered} \tag{4.22}$$

For $F \in \mathcal{A}([a,b])$, relation (4.22) turns into the old definition (4.1). In the general case, according to Section 3, (4.22) are closed normal operators. Moreover, operators (4.22) satisfy equalities (2.21) and (2.22) that generalize the first two equalities in (4.3). The reader, clearly, has already understood and we stress once again that the second relation in (4.3) (multiplicativity) implies the orthogonality (1.5) for E which, in turn, ensures multiplicativity (2.22) for a broader class of functions than the class of analytic functions.

As in the case of a scalar measure (see Section 11.1, Remark 1.4), the support of a general resolution of the identity E on the measurable space $\langle \mathbb{R}, \mathfrak{B}(\mathbb{R}) \rangle$ is understood as the intersection of all closed sets $\varphi \subseteq \mathbb{R}$ of full measure E, i.e., such that $E(\varphi) = \mathbb{1}$. As earlier, this support is denoted by supp$E$. It is closed and, hence, belongs to $\mathfrak{B}(\mathbb{R})$. It is easy to see that

$$E(\mathrm{supp}E) = \mathbb{1}. \tag{4.23}$$

Indeed, let $\text{supp}E = \bigcap_{\xi\in\Xi}\varphi_\xi$, where $\varphi_\xi \subseteq \mathbb{R}$ are all possible closed sets of full measure. By passing to the complements, we conclude that $\mathbb{R}\setminus\text{supp}E = \bigcup_{\xi\in\Xi}(\mathbb{R}\setminus\varphi_\xi) = o$. It is possible to indicate countably many sets $\mathbb{R}\setminus\varphi_\xi$ whose union is equal to o; this follows from the fact that every open set $\mathbb{R}\setminus\varphi_\xi$ on the axis can be obtained as the union of rational open intervals (i.e., intervals with rational ends), and the set of these intervals is countable. Thus, $\mathbb{R}\setminus\text{supp}R = \bigcup_{n=1}^{\infty}(\mathbb{R}\setminus\varphi_{\xi_n})$. But

$$E(\mathbb{R}\setminus\varphi_{\xi_n}) = E(\mathbb{R}) - E(\varphi_{\xi_n}) = 0.$$

The countable additivity of E now implies that $E(\mathbb{R}\setminus\text{supp}E) = 0$, i.e., E (supp E) $= \mathbb{1}$. □

Also note that *if o is an open set on the axis $\mathbb{R}$ such that $o\cap\text{supp}E \neq \emptyset$, then $E(o) \neq 0$.*

Indeed, otherwise, the closed set $\varphi = \mathbb{R}\setminus o$ must be one of the sets that appear in the intersection in the definition of supp E. But this set cannot contain supp E, and we arrive at a contradiction. □

It is clear from the reasonings presented above that *the integrals over $\mathbb{R}$ in relations (4.7), (4.20), and (4.22) can be replaced by the integrals over* supp$E$.

Theorem 4.2. *The spectrum of a selfadjoint operator $A \in \mathcal{L}(H)$ coincides with the support of its resolution of the identity, namely $S(A) = \text{supp}E$. Thus, the integrals over $\mathbb{R}$ in relations (4.7), (4.20), and (4.22) can be replaced by the integrals over $S(A)$.*

Proof. Let $z \in \mathbb{C}\setminus\text{supp}E$. We prove that this point is regular for A, i.e., that $S(A) \subseteq$ supp E. Thus, the function $\text{supp}E \ni \lambda \mapsto (\lambda - z)^{-1} \in \mathbb{C}$ is bounded. Therefore, the spectral integral $B = \int_{\text{supp}E}(\lambda - z)^{-1}dE(\lambda)$ is a bounded operator which, by virtue of (4.7) and (2.5) (for bounded functions), satisfies the equality

$$\begin{aligned}(A - z\mathbb{1})B &= \int_{\text{supp}E}(\lambda - z)dE(\lambda)\int_{\text{supp}E}(\lambda - z)^{-1}dE(\lambda)\\ &= \int_{\text{supp}E}(\lambda - z)(\lambda - z)^{-1}dE(\lambda) = \mathbb{1}.\end{aligned}$$

In other words, $\exists(A - z\mathbb{1})^{-1} = B$.

Let us prove that $\text{supp}E \subseteq S(A)$. Assume the contrary. In this case, one can find a point $\lambda_0 \in \text{supp}E$ regular for A. At the point λ_0, the inequality

$$\|(A - \lambda_0\mathbb{1})f\|_H \geq c_{\lambda_0}\|f\|_H \qquad (f \in H)$$

should hold with a certain constant $c_{\lambda_0} > 0$. Therefore, to arrive at the contradiction, one must prove that there exists a sequence $(f_n)_{n=1}^{\infty}$ of vectors $f_n \in H$ such that

$$\|f_n\|_H = 1, \|(A - \lambda_0\mathbb{1})f_n\|_H \xrightarrow[n\to\infty]{} 0. \tag{4.24}$$

According to the already proved second property of supp$E$, $0 \neq E((\lambda_0 - n^{-1}, \lambda_0+n^{-1}))$ for any $n \in \mathbb{N}$. Let $f_n \in \mathcal{R}(E((\lambda_0-n^{-1}, \lambda_0+n^{-1})))$ and $\|f_n\|_H = 1$. By using the formula (2.8) (for bounded functions), we obtain

$$\|(A - \lambda_0 \mathbb{I})f_n\|_H^2 = \int_{\mathbb{R}} |\lambda - \lambda_0|^2 d\,(E(\lambda)f_n, f_n)_H$$
$$= \int_{\lambda_0-\frac{1}{n}}^{\lambda_0+\frac{1}{n}} |\lambda - \lambda_0|^2 d\,(E(\lambda)f_n, f_n)_H \to 0, n \to \infty.$$

Here, we have used the fact that $f_n = E\,((\lambda_0 - n^{-1}, \lambda_0 + n^{-1}))f_n$ and, by virtue of the orthogonality of E, the limits of the spectral integral can be cut off.

Thus, (4.24) is established and this proves that supp$E \subseteq S(A)$. □

Recall that formula (4.1) is true for every function $F(z)$ analytic in the complex neighbourhood of the spectrum (but not of the interval $(a, b) \supseteq S(A)$) (Section 10.4). It is not difficult to show that the operator $F(A)$ thus defined coincides with the operator defined by (4.20), where $\mathbb{R}$ is replaced by $S(A)$.

REMARK 4.2. The resolution of the identity E that appears in (4.7) is uniquely defined for a given selfadjoint operator $A \in \mathcal{L}(H)$. It is called the resolution of the identity of the operator A.

Indeed, consider two representations of A by different resolutions of the identity E_1 and E_2. According to Theorem 4.2, integration in (4.7) can be carried out over $S(A)$. By taking the integer powers of A and using (2.5) (for bounded functions), we obtain

$$\begin{aligned}&\int_{S(A)} \lambda^n dE_1(\lambda) = \int_{S(A)} \lambda^n dE_2(\lambda)\\ &\quad\Rightarrow \int_{S(A)} \lambda^n d\,((E_1(\lambda)f, g)_H - (E_2(\lambda)f, g)_H) = 0 \quad (f, g \in H)\end{aligned} \tag{4.25}$$

for any $n \in \mathbb{Z}_+$. Then we consider linear combinations and conclude that the last equality holds for every polynomial $P(\lambda)$ (instead of λ^n). Finally, by passing to uniform limits, we prove this equality for an arbitrary continuous function $F(\lambda)$ (recall that $S(A)$ is bounded). But this yields that the charge with respect to which we integrate is equal to zero, i.e.,

$$(E_1(\alpha)f, g)_H - (E_2(\alpha)f, g)_H = 0 \quad (\forall \alpha \in \mathfrak{B}(\mathbb{R}), \forall f, g \in H) \Rightarrow E_1 = E_2. \quad \square$$

To conclude this section, we note that the notion of the support of a resolution of the identity can also be introduced in the general case of a measurable space $\langle R, \mathfrak{R}\rangle$ provided that R is topological and $\mathfrak{R}$ contains closed sets. Assume that supp$E$ is equal to the intersection of all closed $\varphi \in \mathfrak{R}$ of full measure. In the general case, for the support defined in this way, one may observe various pathologies (for example, $\operatorname{supp} E = \emptyset$ in the case where $\mathfrak{R}$ contains sufficiently many closed sets of full measure). At the same time, the following statement is true:

REMARK 4.3. If R is a topological space with a countable basis of neighbourhoods and $\mathfrak{R}$ is the σ-algebra $\mathfrak{B}(R)$ of its Borel subsets, then the properties of the support indicated above and, in particular, equality (4.23), remain true.

Indeed, this fact can be proved by the same argument as above if we replace the intervals with rational ends by the neighbourhoods of a countable basis. □

Exercises

4.1. Find resolutions of the identity for the following operators in $L_2((a,b))$:

(a) $(Af)(x) = xf(x)$;

(b) $(Af)(x) = a(x)f(x)$, where $a = \bar{a} \in C([a,b])$.

4.2. Let $A = A^* \in \mathcal{L}(H)$, let E be its resolution of the identity, and let $\alpha \in \mathfrak{B}(\mathbb{R})$. We set $H_\alpha = E(\alpha)H$. Prove that

(a) H_α is an invariant subspace for A;

(b) $S(A \restriction H_\alpha) \subseteq \tilde{\alpha}$.

4.3. Let $A = A^* \in \mathcal{L}(H)$ and let E be its resolution of the identity. We set $|A| = \sqrt{A^2}$, $A_+ = 1/2\big(|A| + A\big)$ and $A_- = 1/2\big(|A| - A\big)$. Find the resolutions of the identity of the operators $|A|$, A_+, and A_-.

4.4. Let $A = A^* \in \mathcal{L}(H)$ and let E be its resolution of the identity. Prove that $S(A) = \big\{\lambda \in \mathbb{R} \mid (\forall \varepsilon > 0) : E((\lambda - \varepsilon, \lambda + \varepsilon)) \neq 0\big\}$.

4.5. Let $A = A^* \in \mathcal{L}(H)$ and let E be its resolution of the identity. Prove that

$$\frac{1}{2}\left(E((c,d)) + E([c,d])\right) = \text{s.} \lim_{\varepsilon \to \infty} \frac{\varepsilon}{\pi} \int_c^d \left((c-\lambda)^2 + \varepsilon^2\right)^{-1} dE(\lambda)$$

for any $c, d \in \mathbb{R}$ such that $c < d$ (cf. (6.17)).

5 Spectral Decompositions for Unitary and Bounded Normal Operators

5.1 Spectral Theorem for Unitary Operators

For a unitary operator, a spectral decomposition is constructed as in Section 4 with the axis replaced by the unit circle. In this case, the corresponding construction is less descriptive but, in a certain sense, more general. Thus, in the next section, we describe how to construct a spectral decomposition for an arbitrary selfadjoint (even unbounded) operator from a given spectral decomposition of a unitary operator.

Let U be a unitary operator in H and let $S(U)$ be its spectrum. It is well known that $S(U)$ is a closed set on the unit circle $\mathbb{T} = \{z \in \mathbb{C} \mid |z| = 1\}$. Denote by $\mathcal{A}(\mathbb{T})$ the collection of all functions analytic in a certain neighbourhood of $\mathbb{T}$. Just as $\mathcal{A}([a,b])$, the class $\mathcal{A}(\mathbb{T})$ is an algebra with respect to the standard algebraic operations.

Relations (4.1)–(4.3) remain true with A replaced by U, $\mathcal{A}([a,b])$ by $\mathcal{A}(\mathbb{T})$, and γ by a contour that encloses the unit circle and is bypassed in the required direction. For example, $\gamma = \gamma_1 \cup \gamma_2$, where $\gamma_1 (\gamma_2)$ is a circle whose radius is greater than one (less than one) which is bypassed counterclockwise (clockwise) (Section 10.4).

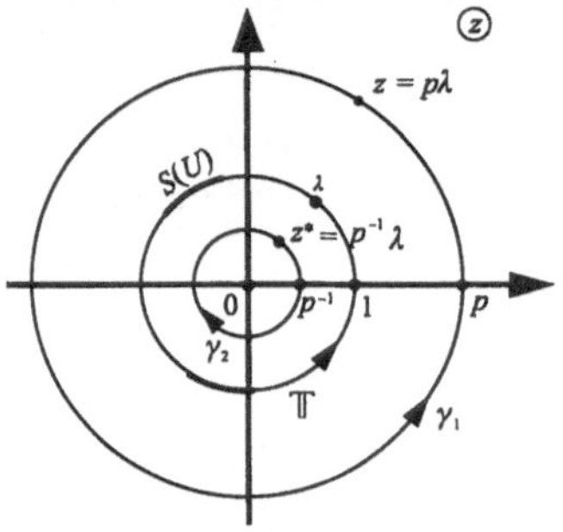

Fig. 11

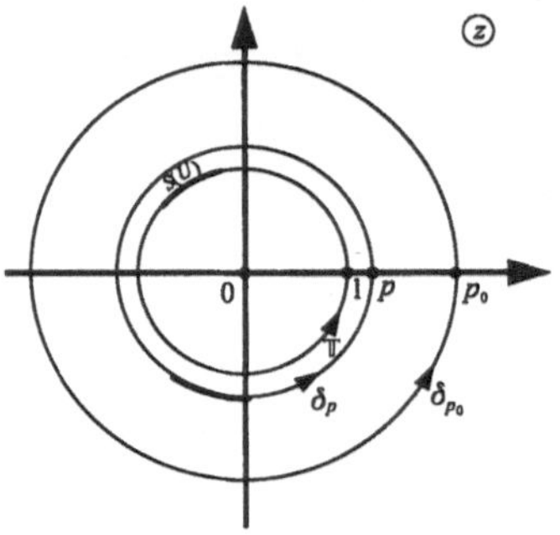

Fig. 12

The role of the transition $\mathbb{C} \ni z \mapsto \bar{z} \in \mathbb{C}$ is now played by the reflection with respect to the unit circle: $\mathbb{C} \setminus \{0\} \ni z \mapsto z^* = \bar{z}^{-1} \in \mathbb{C} \setminus \{0\}$. Figures 9 and 10 are replaced here by Figures 11 and 12.

In Fig. 11, γ_1 and γ_2 are the circles centered at the origin whose radii are $p > 1$ and $p^{-1} < 1$, respectively. Every point $\lambda \in \mathbb{T}$, is associated with the points $z = p\lambda \in \gamma_1$ and $z^* = p^{-1}\lambda \in \gamma_2$ (cf. $\lambda \mapsto \lambda - i\epsilon$, $\lambda + i\epsilon$ in Fig. 9). For $F \in \mathcal{A}(\mathbb{T})$ and sufficiently small p^{-1}, we can write (γ_1 is bypassed in counterclockwise direction):

$$\begin{aligned} F(U) &= -\frac{1}{2\pi i} \oint_{\gamma} F(z) R_z dz = \frac{1}{2\pi i} \left(\oint_{\gamma_2} F(z) R_z dz - \oint_{\gamma_1} F(z) R_z dz \right) \\ &= \frac{1}{2\pi i} \oint_{\mathbb{T}} \left(p^{-1} F(p^{-1}\lambda) R_{p^{-1}\lambda} - pF(p\lambda) R_{p\lambda} \right) d\lambda. \end{aligned} \tag{5.1}$$

Relation (5.1) is similar to (4.5). By analogy with (4.6), we have

$$\left\| \oint_{\mathbb{T}} (pF(p\lambda) - F(\lambda))\, R_{p\lambda} d\lambda \right\| \xrightarrow[p \to 1]{} 0,$$

$$\left\| \oint_{\mathbb{T}} (p^{-1}F(p^{-1}\lambda) - F(\lambda))\, R_{p^{-1}\lambda} d\lambda \right\| \xrightarrow[p \to 1]{} 0. \tag{5.2}$$

Let us prove the first relation. Assume that $p_0 > 1$ is close to one and fixed and that $p \in (1, p_0]$. The function $F_1(z) = pF(z) - F(p^{-1}z)$ is analytic in a domain that contains a ring with the boundary $\mathbb{T} \cup \delta_{p_0}$ (see Fig.12); R_z is analytic outside $S(U)$. The integral under the sign of the norm is equal to $\oint_{\delta_p} \left(pF(z) - F(p^{-1}z)\right) R_z p^{-1} dz$. Due to the indicated analyticity, it is equal to the same integral carried out over δ_{p_0}. By passing in the latter to the limit as $p \to 1$, we arrive at the first relation in (5.2).

Relations (5.1) and (5.2) and the Hilbert identity yield the following formula, similar to (4.4) (convergence is understood in the sense of the operator norm):

$$F(U) = \lim_{p\to 1} \frac{1}{2\pi i} \oint_{\mathbb{T}} F(\lambda)(R_{p^{-1}\lambda} - R_{p\lambda})d\lambda$$

$$= \lim_{p\to 1} \frac{p^{-1} - p}{2\pi i} \oint_{\mathbb{T}} \lambda F(\lambda) R_{p^{-1}\lambda} R_{p\lambda} d\lambda \quad (F \in \mathcal{A}(\mathbb{T}))\,. \tag{5.3}$$

Let us prove an analogue of Lemma 4.2. The role of the equality $R_z^* = R_{\bar{z}}$ for a selfadjoint operator A is now played by the formula

$$R_z^* = -z^* R_{z^*} U \quad \left(z^* = \bar{z}^{-1};\ z \in \mathbb{C} \setminus \{0\}\right), \tag{5.4}$$

obtained by simple calculation. Indeed,

$$R_z^* = \left((U - z\mathbb{I})^{-1}\right)^* = \left(U^* - \bar{z}\mathbb{I}\right)^{-1} = \left(-\bar{z}U^{-1}(U - \bar{z}^{-1}\mathbb{I})\right)^{-1} = -\bar{z}^{-1} R_{\bar{z}^{-1}} U.$$

It follows from (5.4) that $R_{z^*} = -\overline{z}U^{-1}R_z^*$. By substituting this expression in (5.3), we get an analogue of relation (4.8) $(\lambda = e^{i\varphi}, \varphi \in [0, 2\pi))$

$$\begin{aligned} F(U) &= \lim_{p\to 1} \frac{p^{-1} - p}{2\pi i} \oint_{\mathbb{T}} \lambda F(\lambda)(-p\overline{\lambda}U^{-1}) R_{p\lambda}^*\, R_{p\lambda}\, d\lambda \\ &= \lim_{p\to 1} \frac{p^2 - 1}{2\pi i} \int_0^{2\pi} F(e^{i\varphi}) U^{-1} R_{pe^{i\varphi}}^*\, R_{pe^{i\varphi}}\, ie^{i\varphi}\, d\varphi \\ &= U^{-1} \lim_{p\to 1} \frac{p^2 - 1}{2\pi} \int_0^{2\pi} e^{i\varphi}\, F\left(e^{i\varphi}\right)\, R_{pe^{i\varphi}}^*\, R_{pe^{i\varphi}}\, d\varphi. \end{aligned}$$

Multiplying this equality by U and replacing the function $zF(z)$ by $F(z)$, we arrive at the required formula

$$F(U) = \lim_{p\to 1} \frac{p^2 - 1}{2\pi} \int_0^{2\pi} F\left(e^{i\varphi}\right)\, R_{pe^{i\varphi}}^*\, R_{pe^{i\varphi}}\, d\varphi \qquad (F \in \mathcal{A}(\mathbb{T})). \tag{5.5}$$

(Note that an arbitrary function $F \in \mathcal{A}(\mathbb{T})$ can be represented in the form $F(z) = zF_1(z)$, where $F_1 \in \mathcal{A}(\mathbb{T})$; therefore, this procedure is correct.)

Lemma 5.1. *Let $A_{\mathrm{Re}}(\mathbb{T})$ be the class of functions from $A(\mathbb{T})$ that take real values in $\mathbb{T}$. Then the operator $F(U)$ is selfadjoint and satisfies the estimate*

$$\left|(F(U)f,g)_H\right| \le \max\{|F(\lambda)| \mid \lambda \in \mathbb{T}\}\, \|f\|_H\, \|g\|_H \qquad (5.6)$$
$$(F \in \mathcal{A}_{\mathrm{Re}}(\mathbb{T}); f,g \in H).$$

Proof. The proof is analogous to the proof of Lemma 4.2. Thus, it follows from (5.5) and the fact that $F(e^{i\varphi})$ is real that the operator $F(U)$ is selfadjoint. Further, (5.6) follows from a similar estimate for $g = f$, which is a consequence of the inequality

$$\left|\frac{p^2-1}{2\pi}\int_0^{2\pi} F(e^{i\varphi})\,(R^*_{pe^{i\varphi}}\, R_{pe^{i\varphi}}\, f, f)_H\, d\varphi\right|$$
$$\le \max\{|F(\lambda)| \mid \lambda \in \mathbb{T}\}\,\frac{p^2-1}{2\pi}\int_0^{2\pi} 1\cdot\left(R^*_{pe^{i\varphi}}\, R_{pe^{i\varphi}}\, f, f\right)_H d\varphi \quad (f \in H).$$
□

The following theorem (similar to Theorems 4.1 and 4.2 and Remarks 4.1 and 4.2) is true:

Theorem 5.1. *Let U be a unitary operator. Then, on the σ-algebra $\mathfrak{B}(\mathbb{T})$ of the Borel subsets of the unit circle $\mathbb{T} = \{z \in \mathbb{C} \mid |z| = 1\}$, the resolution of the identity E of the operator U is defined such that the following representation is true:*

$$U = \int_{\mathbb{T}} \lambda dE(\lambda) = \int_0^{2\pi} e^{i\varphi} dE(e^{i\varphi}). \qquad (5.7)$$

In (5.7), $\mathbb{T}$ can be replaced by the spectrum $S(U)$ of the operator U. For every function $F(z)$ analytic in the neighbourhood of $S(U)$, the equality

$$F(U) = \int_{S(U)} F(\lambda)\, dE(\lambda) \qquad (5.8)$$

holds, where the operator $F(U)$ is constructed as a contour integral of the type (4.1). The resolution of the identity E from (5.7) is defined uniquely.

Proof. The proof repeats the proofs in Section 4 and is based on estimate (5.6). Thus, for fixed $f, g \in H$, we consider a linear functional

$$C_{\mathrm{Re}}(\mathbb{T}) \supset \mathcal{A}_{\mathrm{Re}}(\mathbb{T}) \ni F \mapsto l_{f,g}(F) = (F(U)f,g)_H \in \mathbb{C},$$

where $C_{\mathrm{Re}}(\mathbb{T}) = C_{\mathrm{Re}}$ is a space of real-valued continuous functions on $\mathbb{T}$ with the uniform norm. This functional is continuous in C_{Re} by virtue of estimate (5.6). At the same time, $\mathcal{A}_{\mathrm{Re}}(\mathbb{T})$ is dense in C_{Re} by virtue of the trigonometrical version of the Weierstrass theorem. Therefore, this functional can be extended by continuity to the whole C_{Re} and admits a representation of the form (4.14) with a uniquely defined charge $\omega_{f,g}$ (given on $\mathfrak{B}(\mathbb{T})$), namely,

$$l_{f,g}(\varphi) = \int_{\mathbb{T}} \varphi(\lambda) d\omega_{f,g}(\lambda) \quad (\varphi \in C_{\mathrm{Re}});$$
$$(\forall \alpha \in \mathfrak{B}(\mathbb{T})) : \ |\omega_{f,g}(\alpha)| \leq \|l_{f,g}\| \leq \|f\|_H \|g\|_H,$$
$$l_{f,g}(F) = \big(F(U)f, g\big)_H \quad (f, g \in H; F \in \mathcal{A}_{\mathrm{Re}}(\mathbb{T})) . \tag{5.9}$$

By analogy with Section 4, one can deduce from (5.9) that, for fixed α, $\omega_{f,g}(\alpha)$ is a continuous bilinear form with respect to $f, g \in H$. Therefore, it admits a representation $\omega_{f,g}(\alpha) = \big(E(\alpha)f, g\big)_H$ with a certain operator $E(\alpha) \in \mathcal{L}(H)$. For α that coincides with a union of finitely many half intervals in the circle $\mathbb{T}$, in view of (4.3), we can prove that $E(\alpha)$ is a projector. Then, by using Theorem 1.3, we establish the same property for arbitrary $\alpha \in \mathfrak{B}(\mathbb{T})$. This implies that $\mathfrak{B}(\mathbb{T}) \ni \alpha \mapsto E(\alpha)$ is a resolution of the identity.

The formula

$$F(U) = \int_{\mathbb{T}} F(\lambda) \, dE(\lambda) \tag{5.10}$$

for $F \in \mathcal{A}_{\mathrm{Re}}(\mathbb{T})$ is a consequence of the last equality in (5.9). Further, for $F \in \mathcal{A}(\mathbb{T})$, it can be derived from the decomposition

$$F(z) = F_1(z) + iF_2(z); F_1(z) = \frac{1}{2}\big(F(z) + \overline{F(z^*)}\big),$$
$$F_2(z) = \frac{1}{2i}\big(F(z) - \overline{F(z^*)}\big) \in \mathcal{A}_{\mathrm{Re}}\,(\mathbb{T}).$$

In particular, for $F(z) = z$, relation (5.7) follows from (5.10).

We now introduce supp$E$ (see Remark 4.3). The exact repetition of the proof of Theorem 4.2 yields the equality supp$E = S(U)$. This enables us to replace $\mathbb{T}$ in (5.10) by $S(U)$ and then pass to a similar formula for the functions $F(z)$ analytic only in a neighbourhood of $S(U)$.

The proof of the fact that E is uniquely defined for given U by relation (5.7) is similar to Remark 4.2. One should only take $n \in \mathbb{Z}$ and use the fact that linear combinations of the functions λ^n ($n \in \mathbb{Z}$) are dense in $C(\mathbb{T})$. □

As in the case of selfadjoint operators, representation (5.7) makes it possible to generalize the notion of a function of an operator U for arbitrary functions from the class $L_0\big(\mathbb{T}, \mathfrak{B}(\mathbb{T}), E\big)$ of E-almost everywhere finite functions defined in $\mathbb{T}$ and measurable with respect to $\mathfrak{B}(\mathbb{T})$. This generalization is given by the formula

$$L_0\,(\mathbb{T}, (\mathbb{T}), E) \ni F \mapsto F(U) = \int_{\mathbb{T}} F(\lambda) dE(\lambda),$$
$$\mathcal{D}\,(F(U)) = \left\{ f \in H | \int_{\mathbb{T}} |F(\lambda)|^2 d(E(\lambda)f, f)_H < \infty \right\}. \tag{5.11}$$

The properties of this mapping were described in Section 2.

5.2 Spectral Theorem for Normal Operators

Let us construct the spectral decomposition for a bounded normal operator A. The corresponding resolution of the identity will be constructed as a direct product of the resolutions of the identity of two commuting selfadjoint operators associated with A. First, we prove the following theorem:

Theorem 5.2. *Let A_1 and A_2 be two selfadjoint bounded operators. In order that their resolutions of the identity be commuting, it is necessary and sufficient that their resolvents $R_{z_1}(A_1)$ and $R_{z_2}(A_2)$ commute for some fixed z_1 and z_2 regular for the operators A_1 and A_2, respectively.*

Proof. Let E_1 and E_2 be the resolutions of the identity of the operators A_1 and A_2, respectively. If E_1 and E_2 commute, then the representation $R_{z_j}(A_j) = \int_{\mathbb{R}}(\lambda - z_j)^{-1}dE_j(\lambda)$ $(j = 1,2)$ implies the commutability of resolvents. Indeed, this can be proved by using definition (2.10) of the integral of a bounded function as a uniform limit.

Conversely, assume that $R_{z_1}(A_1)$ and $R_{z_2}(A_2)$ commute. First, we prove that $R_{\zeta_1}(A_1)$ and $R_{\zeta_2}(A_2)$ commute for every ζ_1 and ζ_2 regular for the operators A_1 and A_2, respectively.

Indeed, let B be an operator acting on H and let $z, \zeta \in \mathbb{C}$ be two regular points of this operator. Then the operator $\left(\mathbb{1} - (z-\zeta)R_\zeta(B)\right)^{-1}$ exists and

$$\left(\mathbb{1} - (z-\zeta)R_\zeta(B)\right)^{-1} = \mathbb{1} + (z-\zeta)\, R_z(B), \tag{5.12}$$

This can be proved by simple verification with the use of the Hilbert identity. According to this identity and (5.12), we obtain

$$R_{\zeta_1}(A_1) - R_{z_1}(A_1) = (\zeta_1 - z_1)R_{z_1}(A_1)R_{\zeta_1}(A_1),$$

$$R_{\zeta_1}(A_1)\left(\mathbb{1} - (\zeta_1 - z_1)R_{z_1}(A_1)\right) = R_{z_1}(A_1),$$

$$R_{\zeta_1}(A_1) = \left(\mathbb{1} - (\zeta_1 - z_1)R_{z_1}(A_1)\right)^{-1} R_{z_1}(A_1).$$

It follows from the last equality and the commutability of $R_{z_1}(A_1)$ and $R_{z_2}(A_2)$ that $R_{\zeta_1}(A_1)$ and $R_{z_2}(A_2)$ are commuting. By analogy, this implies that $R_{\zeta_1}(A_1)$ and $R_{\zeta_2}(A_2)$ also commute.

Let (a, b) be a sufficiently large interval that contains the spectra $S(A_1)$ and $S(A_2)$. The commutability of resolvents just established, relation (4.1) written for A_1 and A_2, and the definition of the contour integral as the limit of the corresponding integral sums with respect to the operator norm imply that $F_1(A_1)$ and $F_2(A_2)$ commute for any $F_1, F_2 \in \mathcal{A}([a,b])$.

Consider sets α_1 and α_2; each of them is a union of finitely many disjoint half intervals $[c,d) \subset [a,b)$; such sets form an algebra $\mathfrak{R}$ on $[a,b)$. As in step (2) of the proof of Theorem 4.1, we construct the sequences $(F_{1,n})_{n=1}^{\infty}$ and $(F_{2,m})_{m=1}^{\infty}$ of functions from $\mathcal{A}_{\text{Re}}([a,b])$ such that $F_{1,n}(A_1) \xrightarrow[n\to\infty]{} E_1(\alpha_1)$ and $F_{2,m}(A_2) \xrightarrow[m\to\infty]{} E_2(\alpha_2)$ weakly in H. For any $f, g \in H$, we have

$$(F_{1,n}(A_1)f, F_{2,m}(A_2)g)_H = (F_{2,m}(A_2)F_{1,n}(A_1)f, g)_H$$
$$= (F_{1,n}(A_1)F_{2,m}(A_2)f, g)_H = (F_{2,m}(A_2)f, F_{1,n}(A_1)g)_H.$$

By passing in the equality obtained to the limit as $m \to \infty$ and then as $n \to \infty$, we obtain

$$\big(E_1(\alpha_1)f, E_2(\alpha_2)g\big)_H = \big(E_2(\alpha_2)f, E_1(\alpha_1)g\big)_H.$$

This means that $E_1(\alpha_1)$ and $E_2(\alpha_2)$ commute for any $\alpha_1,\ \alpha_2 \in \mathfrak{R}$. For arbitrary $\alpha_1,\ \alpha_2 \in \mathfrak{B}\big([a,b)\big)$, this commutability is established in a simple way by using the uniqueness of the extension of a scalar measure. Thus, we fix $\alpha_2 \in \mathfrak{R}$ and $f, g \in H$. In $\mathfrak{B}\big([a,b)\big) = \mathfrak{R}_\sigma$, we have two charges with respect to α_1, namely,

$$\omega(\alpha_1) = (E_1(\alpha_1)f, E_2(\alpha_2)g)_H \qquad \text{and} \qquad \nu(\alpha_1) = \big(E_2(\alpha_2)f, E_1(\alpha_1)g\big)_H$$

These charges coincide in $\mathfrak{R}$. Therefore, they coincide in $\mathfrak{R}_\sigma$. Further, we fix $\alpha_1 \in \mathfrak{B}\big([a,b)\big)$ and consider charges with respect to α_2. As a result, we obtain

$$\big(E_1(\alpha_1)f, E_2(\alpha_2)g\big)_H = \big(E_2(\alpha_2)f, E_1(\alpha_1)g\big)_H,$$

i.e., the required commutability of E_1 and E_2. □

Note that there exists another version of the final part of the proof of this theorem (see Theorem 6.3 in the Section 6).

REMARK 5.1. In order that the resolutions of the identity of bounded selfadjoint operators A_1 and A_2 commute, it is necessary and sufficient that these operators be commuting.

Indeed, if E_1 and E_2 commute, then, just as above, the representation $A_j = \int_{\mathbb{R}} \lambda dE_j(\lambda)$ $(j = 1,2)$ implies that A_1 and A_2 are commuting. Conversely, it follows from the commutability of A_1 and A_2 and the formula

$$R_z(A_j) = -\sum_{n=0}^{\infty} z^{-n-1}A_j^n \quad (j = 1,2)$$

valid for sufficiently large $|z|$ that $R_z(A_1)$ and $R_z(A_2)$ commute for such z, and the problem is reduced to Theorem 5.2. □

Consider a bounded normal operator in H, i.e., an operator $A \in \mathcal{L}(H)$ such that $A^*A = AA^*$.

Theorem 5.3. *Let A be a bounded normal operator. Then, on the σ-algebra $\mathfrak{B}(\mathbb{C})$ of Borel subsets of the complex plane, one can construct the resolution of the identity E of the operator A satisfying the spectral following representation:*

$$A = \int_{\mathbb{C}} \lambda dE(\lambda). \tag{5.13}$$

In (5.13), $\mathbb{C}$ *can be replaced by the spectrum* $S(A)$ *of the operator* A*. Any function* $F(z)$ *analytic in a neighbourhood of* $S(A)$ *satisfies the equality*

$$F(A) = \int_{S(A)} F(\lambda) dE(\lambda), \tag{5.14}$$

where the operator $F(A)$ *is constructed as a contour integral of the type (4.1) (Section 10.4). The resolution of the identity* E *from (5.13) is determined uniquely.*

Proof. Consider the standard decomposition of A into bounded selfadjoint operators, namely,

$$A = A_1 + iA_2 \quad \left(A_1 = \mathrm{Re} A = \frac{1}{2}(A + A^*), \quad A_2 = \mathrm{Im} A = \frac{1}{2i}(A - A^*)\ \right). \tag{5.15}$$

Since A is normal, the operators A_1 and A_2 commute, but then, according to Remark 5.1, their resolutions of the identity E_1 and E_2 commute, as well.

According to Theorem 3.2, we shall construct the direct product $E = E_1 \times E_2$. For this resolution of the identity, we can write

$$A_j = \int_{\mathbb{C}} \lambda_j dE(\lambda) \quad (j = 1, 2;\ \mathbb{C} \ni \lambda = \lambda_1 + i\lambda_2, \lambda_1, \lambda_2 \in \mathbb{R})\,. \tag{5.16}$$

Indeed, let $j = 1$. Since E_1 and E_2 are concentrated on finite intervals (a_1, b_1) and $(a_2, b_2), E$ is concentrated on $(a_1, b_1) \times (a_2, b_2)$, i.e., in (5.16), we, in fact, integrate a bounded function. By using the definition (2.10) of a spectral integral and approximating the function $F(\lambda) = \lambda_1$ by functions $F_n(\lambda)$ of the variable λ_1, we conclude that the integral in (5.16) is equal to

$$\int_{\mathbb{R}} \lambda_1 dE_1(\lambda_1) = A_1. \tag{5.17}$$

In other words, we have transformed (5.16) as follows:

$$\int_{\mathbb{C}} \lambda_1 dE(\lambda) = \int_{\mathbb{R}} \int_{\mathbb{R}} \lambda_1 dE_1(\lambda_1) E_2(\lambda_2) = \int_{\mathbb{R}} \lambda_1 dE_1(\lambda_1) E_2(\mathbb{R}) = A_1, \tag{5.18}$$

i.e., the first integration was carried out with respect to the variable λ_2 that does not appear in the integrand and then we have used the fact that $E_2(\mathbb{R}) = \mathbb{I}$.

By adding equality (5.16) with $j = 1$ to the same equality multiplied by i for $j = 2$, we arrive at (5.13).

As before, by using Remark 4.3 we introduce $\mathrm{supp} A = S(A)$. This enables us to replace $\mathbb{C}$ by $S(A)$ in (5.13).

To prove equality (5.14), it suffices to show that the representation

$$R_z = (A - z\mathbb{I})^{-1} = \int_{S(A)} (\lambda - z)^{-1} dE(\lambda) \tag{5.19}$$

holds for $z \notin S(A)$. Indeed, it is necessary to substitute (5.19) in (4.1), to change the order of integration (this possibility is easily substantiated), and to make use of the Cauchy formula for $F(z)$. Representation (5.19), in turn, follows from (5.13) and (2.5) for bounded functions.

Finally, let us prove that the definition of E in (5.13) for given A is unique. Let L be some other resolution of the identity in $\mathfrak{B}(\mathbb{C})$ that satisfies the equality

$$A = \int_{\mathbb{C}} \lambda dL(\lambda). \tag{5.20}$$

As above, by using Remark 4.3 and Theorem 4.2, we can prove that $\mathbb{C}$ can be replaced by $S(A)$ in (5.20) and, hence, the integration in this equality is carried out over a bounded set, i.e., the integrand can be regarded as a bounded function.

According to (2.4) and (2.6) for bounded functions and (5.15), it follows from (5.20) that

$$A_1 = \int_{\mathbb{C}} \lambda_1 dL(\lambda) = \int_{\mathbb{R}} \lambda_1 dL_1(\lambda_1). \tag{5.21}$$

Here, L_1 is "the projection of the resolution of the identity L onto the axis λ", i.e., the image of L under the mapping $R = \mathbb{C} \ni \lambda = \lambda_1 + i\lambda_2 \mapsto \lambda_1 \in \mathbb{R} = R'$ (see Section 3). The second equality in (5.21) is, in fact, relation (3.3) for the special case where $F'(\lambda_1) = \lambda_1$. It follows from (3.1) that

$$\mathfrak{B}(\mathbb{R}) \ni \alpha_1 \mapsto L_1(\alpha_1) = L(\alpha_1 \times \mathbb{R}). \tag{5.22}$$

The equalities similar to (5.21) and (5.22) also hold for A_2. In this case, $\mathfrak{B}(\mathbb{R}) \ni \alpha_2 \mapsto L_2(\alpha_2) = L(\mathbb{R} \times \alpha_2)$. By virtue of the orthogonality of L, this equality and (5.22) imply that

$$L_1(\alpha_1)L_2(\alpha_2) = L(\alpha_1 \times \mathbb{R})L(\mathbb{R} \times \alpha_2) = L\left((\alpha_1 \times \mathbb{R}) \cap (\mathbb{R} \times \alpha_2)\right) = L(\alpha_1 \times \alpha_2)$$

for any $\alpha_1, \alpha_2 \in \mathfrak{B}(\mathbb{R})$.

Thus, L is defined for given L_1 and L_2 by formula (3.5) and, in view of the fact that Theorem 3.2 guarantees the uniqueness of the definition of the direct product, to prove the required uniqueness, it suffices to show that $L_1 = E_1$ and $L_2 = E_2$.

At the same time, the operator A_1 is uniquely determined for given A and admits two spectral representations (5.17) and (5.21). Since this operator is selfadjoint, according to Remark 4.2, we have $L_1 = E_1$. Similarly, $L_2 = E_2$. □

Clearly, for the spectral representation (5.13), one can make the same conclusions from the considerations in Section 2 as have already been made for selfadjoint and unitary operators.

Also note that Theorem 5.3 covers Theorem 5.1. Indeed, for a unitary operator U, we have $U^* = U^{-1}$ and, therefore, it is normal. Representation (5.13) for U, where $\mathbb{C}$ is replaced by $S(U) \subseteq \mathbb{T}$, turns into representation (5.7).

The case of a unitary operator is also instructive in the following sense: The proof of Theorem 5.3 may create the wrong impression that $S(A) = S(A_1) \times S(A_2)$ (as in the case of supports of multiplied scalar measures). But we have just seen that, in fact, one can only guarantee the validity of the inclusion

$$S(A) \subseteq S(A_1) \times S(A_2); \tag{5.23}$$

for unitary $A = U$ it is strict (the equality in (5.23) is impossible, since the relation $E_1(\alpha_1)E_2(\alpha_2) = 0$ does not mean that at least one of the factors is equal to zero).

Exercises

5.1. Let $u \in C(\mathbb{R})$ be such that $(\forall t \in \mathbb{R})$: $|u(t)| = 1$. Find the resolution of the identity of the operator of multiplication by the function u acting on $L_2(\mathbb{R})$.

5.2. Let $a \in C(\mathbb{R})$ be bounded. Find the resolution of the identity of the operator of multiplication by the function a acting on $L_2(\mathbb{R})$.

5.3. Let $A_1 = A_1^*$ and $A_2 = A_2^* \in \mathcal{L}(H)$. Prove that the resolutions of the identity A_1 and A_2 commute if and only if one of the following conditions is satisfied:

(i) $(\forall t, s \in \mathbb{R}) : \left[e^{itA_1}, e^{isA_2}\right] = 0;$

(ii) the Cayley transforms of the operators A_1 and A_2 commute.

5.4. Let $A = A^* \in \mathcal{L}(H)$, $|A|$, A_+, and A_- be the operators defined in Exercise 4.3. Prove that $|A|$ (A_+ and A_-, respectively) is the least nonnegative operator $B \in \mathcal{L}(H)$ which commutes with A and satisfies the inequalities $B \geq A$, and $B \geq -A$ ($B \geq A$ and $B \geq -A$, respectively).

6 Spectral Decompositions of Unbounded Operators

6.1 Selfadjoint Operators

In this section, results similar to those presented in Sections 4 and 5 are obtained for unbounded operators. Let us first consider a selfadjoint operator A. To prove the required results we use the information on the mappings of spaces that can be found in Section 3. Our idea is as follows: To construct the resolution of the identity E of the operator A, we first find a bounded (selfadjoint or unitary) operator for which the resolution of the identity has already been constructed; its proper image would give the required E.

Theorem 6.1. *Let A be an arbitrary selfadjoint operator. Then the resolution of the identity E of the operator A is defined on the σ-algebra $\mathfrak{B}(\mathbb{R})$ of Borel subsets of the axis and the following spectral representation is true:*

$$A = \int_{\mathbb{R}} \lambda dE(\lambda), \quad \mathcal{D}(A) = \left\{ f \in H \,\middle|\, \int_{\mathbb{R}} \lambda^2 d(E(\lambda)f, f)_H < \infty \right\}. \tag{6.1}$$

In (6.1), $\mathbb{R}$ can be replaced by the spectrum $S(A)$ of the operator A. The resolution of the identity E from (6.1) is defined uniquely.

Proof. We shall give two proofs of representation (6.1): on the basis of Theorem 4.1 and on the basis of Theorem 5.1. Unfortunately, in the first case, one should additionally require that A must have a real regular point. In view of the notation introduced in Section 3, throughout the proof of Theorem 6.1, it is convenient to denote A, E, and λ from its formulation by A', E', and λ', respectively.

(1). Assume that a given selfadjoint operator A' possesses a regular point $\lambda'_0 \in \mathbb{R}$. Let us choose a sufficiently small interval $(a', b') \ni \lambda'_0$ that consists of regular points of A'. Denote $R' = (-\infty, a'] \cup [b', +\infty)$ and $R = [a, b]$, where $a = (a' - \lambda'_0)^{-1} < 0$ and $b = (b' - \lambda'_0)^{-1} > 0$. In other words, R is the image of R' under the mapping

$$R' \ni \lambda' \mapsto \lambda = \left(\lambda' - \lambda'_0\right)^{-1} \in R. \tag{6.2}$$

Consider the mapping

$$R \ni \lambda \mapsto \lambda' = \varphi(\lambda) = \lambda^{-1} + \lambda'_0 \in R', \tag{6.3}$$

which is inverse to (6.2).

According to (6.2), we introduce the operator

$$A = \left(A' - \lambda'_0 \mathbb{I}\right)^{-1}, \tag{6.4}$$

i.e., the resolvent of the operator A' at the point $\lambda'_0 \in \mathbb{R}$. It is easy to show (e.g., by using (5.12)) that its spectrum lies in the interval (a, b). The operator A is bounded and selfadjoint. Let E be its resolution of the identity on $\mathfrak{R} = \mathfrak{B}(R)$, which exists by virtue of Theorem 4.1. Denote by E' the image of E under mapping (6.3) given by relation (3.1) on the σ-algebra $\mathfrak{R}'$, which obviously coincides with $\mathfrak{B}(R')$.

It is easy to show that E' is the required resolution of the identity of the operator A', i.e., that (6.1) holds. Indeed, on the basis of the formula of the change of variable (3.3) for the function $R' \ni \lambda' \mapsto F'(\lambda') = \lambda' \in R$, by virtue of (6.4), we have

$$\int_{R'} \lambda' dE'(\lambda') = \int_R (\lambda^{-1} + \lambda'_0) dE(\lambda) = A^{-1} + \lambda'_0 \mathbb{I} = A'.$$

Relation (3.4) gives the domain $\mathcal{D}(A')$ indicated in (6.1).

(2). Let A' be an arbitrary selfadjoint operator. We set $R' = \mathbb{R} \cup \{\infty\}$ and $R = \mathbb{T}$ (the unit circle). Instead of (6.2), we consider the linear-fractional (one-to-one) mapping of the form (12.6.3) with $z = i$, namely,

$$R' \ni \lambda' \mapsto \lambda = \frac{\lambda' + i}{\lambda' - i} \in R. \tag{6.5}$$

As φ, we take the mapping inverse to (6.5), i.e.,

$$R \ni \lambda \mapsto \lambda' = \varphi(\lambda) = i\frac{\lambda + 1}{\lambda - 1} \in R'. \tag{6.6}$$

According to (6.5), we introduce the operator

$$U = (A' + i\mathbb{1})(A' - i\mathbb{1})^{-1}, \tag{6.7}$$

i.e., the Cayley transform of the form (12.6.6) with $z = i$. By virtue of assertion (iii) in Section 12.6, operator (6.7) is unitary. Let E be its resolution of the identity on $\mathfrak{R} = \mathfrak{B}(\mathbb{T})$, which exists according to Theorem 5.1. Denote by E' the image of E under mapping (6.6) defined by relation (3.1) on R'. The sets of this σ-algebra coincide with the sets from $\mathfrak{B}(\mathbb{R})$ and their unions with the point ∞. But $E'(\{\infty\}) = E(\{1\})$ and $E(\{1\}) = 0$, since, otherwise, the operator U would possess eigenvectors that correspond to the eigenvalue one, which is impossible because $\mathcal{D}(A)$ is dense (see (vii), Section 12.6). Thus, $E'(\{\infty\}) = 0$ and, therefore, E' can be regarded as a resolution of the identity defined on $\mathfrak{B}(\mathbb{R})$.

The resolution of the identity E' constructed as a result is the resolution of the identity of the operator A'. Indeed, as above, by virtue of relations (3.3) and (3.4), for the function $R' \ni \lambda' \mapsto F(\lambda') = \lambda' \in \mathbb{R}$, we have

$$\int_{\mathbb{R}} \lambda' dE'(\lambda') = \int_{R'} \lambda' dE'(\lambda') = \int_{\mathbb{T}} i\frac{\lambda + 1}{\lambda - 1} dE(\lambda) = i\,(U + \mathbb{1})\,(U - \mathbb{1})^{-1} = A'.$$

Here, we have used relation (12.6.13) for the inverse Cayley transformation. Relation (6.1) is proved.

(3). Let us establish the remaining assertion of Theorem 6.1. As in Theorems 4.1, 5.1, and 5.3, one can now replace $\mathbb{R}$ by $S(A)$; this is a consequence of Remark 4.3 and the proof of Theorem 4.2.

The uniqueness of the procedure of determination of E' for given A' from (6.1) can be proved, e.g., as follows:

Let L' be another resolution of the identity defined on $\mathfrak{B}(\mathbb{R})$ that satisfies relation (6.1) with E' replaced by L'. In the notation of (2), we now consider the mapping φ^{-1} of the form (6.5) that maps R' onto R. According to the construction described in Section 3, it generates a resolution of the identity L on R. By repeating the reasoning of (2) in the inverse order, one can easily show that L is the resolution of the identity for unitary operator U. In view of the uniqueness of the definition of resolution of the identity for unitary operator, we get $L = E \Rightarrow L' = E'$. □

REMARK 6.1. Let us mention a useful fact that enables one to prove *the uniqueness of the procedure of determination of E for given A from (6.1).* It is based on the following well-known result from analysis:

Let $\mathfrak{B}(\mathbb{R}) \ni \alpha \mapsto \omega(\alpha) \in \mathbb{C}$ be a charge on the axis representable as a linear combination of finite measures. The function

$$\varphi(z) = \int_{\mathbb{R}} (\lambda - z)^{-1} d\omega(\lambda) \qquad (z \in \mathbb{C} \setminus \mathbb{R}), \tag{6.8}$$

analytic outside supp ω is called the Hilbert transform of the charge ω. Note that the charge ω is uniquely determined for given φ. Moreover, for every finite open interval $\delta \subset \mathbb{R}$, we have

$$\frac{1}{2}\left(\omega(\delta)+\omega(\tilde{\delta})\right) = \lim_{\varepsilon\to+0}\frac{1}{2\pi i}\int_{\delta+i\varepsilon}(\varphi(z)-\varphi(\overline{z}))\,dz \tag{6.9}$$

(as easily follows from the arguments presented in Section 1.17, if one knows the left-hand side in (6.9) for any δ, then a charge is determined uniquely; see also the proof of Theorem 6.3).

Note that relation (6.9) will be proved if we substitute the expression for φ from (6.8) into its right-hand side, change the order of integration, and pass to the limit under the sign of the Lebesgue-Stieltjes integral.

Uniqueness is proved as follows: Let E and L be two resolutions of the identity that appear in representation (6.1) for a certain selfadjoint operator A. Then its resolvent satisfies the equality

$$R_z = \left(A - z\mathbb{1}\right)^{-1} = \int_{\mathbb{R}}(\lambda - z)^{-1}dE(\lambda) \qquad (z \in \mathbb{C}\setminus\mathbb{R}); \tag{6.10}$$

and the same equality with E replaced by L, as follows from the properties of the spectral integrals of bounded functions (more precisely, from (2.5)). By subtracting these equalities and passing to the scalar products, we obtain $(\forall f, g \in H)$:

$$0 = \int_{\mathbb{R}}(\lambda - z)^{-1}\, d\omega_{f,g}(\lambda),$$
$$\mathfrak{B}(\mathbb{R}) \ni \alpha \mapsto \omega_{f,g}(\alpha) = \left((E(\alpha) - L(\alpha))f, g\right)_H \in \mathbb{C},$$

whence $E = L$ because φ uniquely determines ω in (6.8). □

As in Sections 4 and 5, by using spectral integrals, one can construct the theory of functions of operators for unbounded selfadjoint operators A. Thus, for $F \in L_0(\mathbb{R}, \mathfrak{B}(\mathbb{R}), E)$, where E is the resolution of the identity that corresponds to A, we construct the operator

$$F(A) = \int_{\mathbb{R}} F(\lambda)dE(\lambda),$$
$$\mathcal{D}(F(A)) = \left\{f \in H \mid \int_{\mathbb{R}} |F(\lambda)|^2 d(E(\lambda)f, f)_H < \infty\right\}. \tag{6.11}$$

The properties of the correspondence $L_0(\mathbb{R}, \mathfrak{B}(\mathbb{R}), R) \ni F \mapsto F(A)$ were described in Section 2.

Let us mention some important functions of a selfadjoint operator A (operators (6.14)–(6.16) are, generally speaking, unbounded).

(i) *A resolution of the identity:* For the indicator χ_α of a set $\alpha \in \mathfrak{B}(\mathbb{R})$, we have

$$E(\alpha) = \chi_\alpha(A) = \int_{\mathbb{R}} \chi_\alpha(\lambda)dE(\lambda). \tag{6.12}$$

(ii) *A resolvent:* Let $z \notin S(A)$. Then the following representation is true:

$$R_z = \left(A - z\mathbb{I}\right)^{-1} = \int_{S(A)} \frac{1}{\lambda - z} dE(\lambda). \tag{6.13}$$

(iii) *The square root of a nonnegative operator A:*

$$\sqrt{A} = \int_{S(A)} \sqrt{\lambda} dE(\lambda) = \int_0^\infty \sqrt{\lambda} dE(\lambda) \geq 0; \quad (\sqrt{A})^2 = A. \tag{6.14}$$

(iv) *The absolute value of an operator:*

$$|A| = \int_{S(A)} |\lambda| dE(\lambda) = \int_{\mathbb{R}} |\lambda| dE(\lambda) \geq 0. \tag{6.15}$$

(v) *The exponent:* For any $z \in \mathbb{C}$,

$$\exp(zA) = e^{zA} = \int_{S(A)} e^{zA} dE(\lambda). \tag{6.16}$$

6.2 Stone's Formula

In what follows, we dwell upon three more problems connected with the resolution of the identity E of a selfadjoint operator A. Thus, by using Remark 6.1 and relation (6.9), we obtain a useful expression for E in terms of the resolvent of A.

Theorem 6.2. *Let E be a resolution of the identity of a selfadjoint operator and let R_z be its resolvent. Then, the formula*

$$\frac{1}{2}\left(E(\delta) + E(\tilde{\delta})\right) = \lim_{\varepsilon \to +\infty} \frac{1}{2\pi i} \int_{\delta + i\varepsilon} (R_z - R_{\bar{z}}) dz \tag{6.17}$$

holds in the sense of strong convergence for every open finite interval $\delta \subset \mathbb{R}$.

Proof. The validity of (6.17) in the sense of weak convergence follows directly from (6.9). Indeed, we fix $f, g \in H$, then $\varphi(z) = (R_z f, g)_H$ is the Hilbert transform of the charge $\omega(\alpha) = \left(E(\alpha) f, g\right)_H$ (this follows from (6.13)) and we can apply (6.9).

To prove strong convergence, it suffices to show that $\left\|\int_{\delta+i\varepsilon}(R_z - R_{\bar{z}}) dz f\right\|_H$ is bounded in $\varepsilon > 0$ for fixed $\delta = (a, b)$ for any $f \in H$. Let $z = x + iy$. According to (6.13), we have

$$\begin{aligned}
&\left\| \int_{\delta+i\varepsilon} (R_z - R_{\bar{z}}) dz f \right\|_H^2 \\
&= \left\| \int_{\delta+i\varepsilon} \left(\int_{\mathbb{R}} ((\lambda - z)^{-1} - (\lambda - \bar{z})^{-1}) dE(\lambda) \right) dz f \right\|_H^2 \\
&= \left\| \int_{\mathbb{R}} \left(\int_{\delta} ((\lambda - x - i\varepsilon)^{-1} - (\lambda - x + i\varepsilon)^{-1}) dx \right) dE(\lambda) f \right\|_H^2 \\
&= 4 \left\| \int_{\mathbb{R}} \left(\int_a^b \frac{\varepsilon}{(\lambda - x)^2 + \varepsilon^2} dx \right) dE(\lambda) f \right\|_H^2 = \psi(\varepsilon).
\end{aligned}$$

It is easy to compute that the internal integral is equal to

$$\arctan(\varepsilon^{-1}(b-\lambda)) - \arctan(\varepsilon^{-1}(a-\lambda)) = \chi(\lambda,\varepsilon).$$

Therefore, by virtue of (2.9) for bounded functions, for any $\varepsilon > 0$, we have

$$\psi(\varepsilon) = 4\left\|\int_{\mathbb{R}} \chi(\lambda,\varepsilon)dE(\lambda)f\right\|_H^2 \leq 4\sup\left\{|\chi(\lambda,\varepsilon)|^2 \,|\lambda \in \mathbb{R}\right\} \|f\|_H^2 \leq c\|f\|_H^2. \quad \square$$

6.3 Commuting Operators

Theorem 6.3. *For general selfadjoint operators A_1 and A_2, Theorem 5.2 holds in the same formulation.*

Proof. As can be seen from the proof of Theorem 5.2, it remains true for unbounded A_1 and A_2 up to the place where the commutability of $R_{\zeta_1}(A_1)$ and $R_{\zeta_2}(A_2)$ is established. In the case under consideration, the proof can be completed by using relation (6.17), which implies that the operators $B_j(\delta_j) = E_j(\delta_j) + E_j(\tilde{\delta_j})$ $(\forall \delta_j = (a_j, b_j))$ with different $j = 1, 2$ are commuting.

Let $c_1 \in \mathbb{R}$ and let $\delta_{11} \supseteq \delta_{12} \supseteq \ldots$ be a sequence of intervals contracting to $c_1 \in \delta_{1n}$. Then $\{c_1\} = \cap_{n=1}^{\infty}\delta_{1n} = \cap_{n=1}^{\infty}\tilde{\delta}_{1n}$ and, by Theorem 1.2, $B_1(\delta_{1n}) \to 2E_1(\{c_1\})$ in the sense of strong convergence as $n \to \infty$. In addition, for any $n \in \mathbb{N}$ and all δ_2, $B_1(\delta_{1n})$ commutes with $B_2(\delta_2)$. Therefore, $E_1(\{c_1\})$ also commute with $B_2(\delta_2)$. But

$$2E_1([a_1, b_1)) = B_1((a_1, b_1)) + E_1(\{a_1\}) - E_1(\{b_1\}),$$

and, consequently, $E_1([a_1, b_1))$ commutes with $B_2(\delta_2)$, as well; here, δ_2 is an arbitrary open interval and $[a_1, b_1)$ is an arbitrary half open interval.

As a result of an analogous procedure for $j = 2$, we conclude that $E_1([a_1, b_1))$ and $E_2([a_2, b_2))$ commute for any $[a_1, b_1), [a_2, b_2) \subset \mathbb{R}$. For arbitrary $\alpha_1, \alpha_2 \in \mathfrak{B}(\mathbb{R})$, the commutability of $E_1(\alpha_1)$ and $E_2(\alpha_2)$ is proved by using Theorem 1.3 with the help of the procedure applied at the end of the proof of Theorem 5.2. $\square$

REMARK 6.2. It is useful to note that (as in Remark 5.1 and on the basis of the same arguments) if one of the operators in Theorem 6.3 A_1 or A_2 is bounded, then, in the formulation of the theorem, one can require the commutability of this operator with the resolvent of the other instead of the commutability of their resolvents.

6.4 The Function E_λ

As in the case of scalar measures, for a resolution of the identity defined on Borel subsets of the real axis, we often introduce (instead of an operator-valued measure) a nondecreasing operator-valued function; this function is also called a resolution of the identity (function). Let us give the corresponding definition.

The operator-valued function $\mathbb{R} \ni \lambda \mapsto E_\lambda$ *whose values are projectors in a fixed Hilbert space* H *is called a resolution of the identity if it satisfies the following conditions:*

(a) monotonicity: $(\forall\lambda,\ \mu \in \mathbb{R},\ \lambda < \mu) \Rightarrow E_\lambda \leq E_\mu$.

(b) completeness: $\lim_{\lambda\to-\infty} E_\lambda = 0$ *and* $\lim_{\lambda\to+\infty} E_\lambda = \mathbb{I}$ *in the sense of strong convergence.*

(c) left-continuity: $\lim_{\lambda\to\mu-0} E_\lambda = E_\mu$ *in the sense of strong convergence.*

Theorem 6.4. *Let* E *be a resolution of the identity (measure) given on* $\mathfrak{B}(\mathbb{R})$*. Then*

$$\mathbb{R} \ni \lambda \mapsto E_\lambda = E\big((-\infty, \lambda)\big) \tag{6.18}$$

is a resolution of the identity (function). Conversely, for a given resolution of the identity E_λ*, one can construct a resolution of the identity* E *on* $\mathfrak{B}(\mathbb{R})$ *such that* E_λ *and* E *satisfy (6.18).*

Proof. Assume that a resolution of the identity E is given on $\mathfrak{B}(\mathbb{R})$ and construct E_λ in accordance with (6.18). Property (a) holds by virtue of (1.10). To prove that $\lim_{\lambda\to-\infty} E(\lambda) = 0$, it suffices to show that if $\lambda_1 > \lambda_2 > \ldots$ and $\lambda_n \to -\infty$ as $n \to \infty$, then $E_{\lambda_n} \to 0$. By virtue of Theorem 1.2, we have

$$\lim_{n\to\infty} E_{\lambda_n} = \lim_{n\to\infty} E((-\infty, \lambda_n)) = E\left(\bigcap_{n=1}^{\infty} (-\infty, \lambda_n)\right) = E(\emptyset) = 0.$$

The fact that $\lim_{\lambda\to+\infty} E_\lambda = \mathbb{I}$ can be established similarly.

To prove (c), it suffices to check that if $\lambda_1 < \lambda_2 < \ldots < \mu$ and $\lim_{n\to\infty} \lambda_n = \mu$, then $E_\mu - E_{\lambda_n} \to 0$. But $E_\mu - E_{\lambda_n} = E\left([\lambda_n, \mu)\right) \to 0$ by virtue of Theorem 1.2, since $\cap_{n=1}^{\infty} [\lambda_n, \mu) = \emptyset$.

The converse statement of the theorem is proved by using Theorem 1.3. Denote by $\mathfrak{R}$ the algebra of sets α each of which is a union of finitely many half intervals of the form $[\lambda, \mu)$. On each interval of this sort, we set $E\left([\lambda, \mu)\right) = E_\mu - E_\lambda$. Since E_μ and E_λ are projectors and $E_\lambda \leq E_\mu$, it is not difficult to show that $E\big([\lambda, \mu)\big)$ is also a projector (note that this inequality implies the inclusion $\mathrm{Ker} E_\mu \subseteq \mathrm{Ker} E_\lambda$ and, hence, $\mathcal{R}(E_\lambda) \subseteq \mathcal{R}(E_\mu)$).

The set function obtained as a result is additive in $\mathfrak{R}$. To prove that it is countably additive, we fix $f \in H$ and consider a scalar measure $\mathfrak{R} \ni \alpha \mapsto \rho_{f,f}(\alpha) = \big(E(\alpha)f, f\big)_H \geq 0$. It is constructed for a given nondecreasing bounded function

$$\mathbb{R} \ni \lambda \mapsto \varphi_{f,f}(\lambda) = (E_\lambda f, f)_H \geq 0$$

(in accordance with the construction of E) by the standard procedure described in Section 1.14. By virtue of Theorem 1.14.1, $\rho_{f,f}$ is absolutely additive on $\mathfrak{R}$, and this, due to the arbitrariness of f, yields the absolute additivity of E. Finally, according to Theorem 1.3, we extend E to a resolution of the identity E_σ on $\mathfrak{R}_\sigma = \mathfrak{B}(\mathbb{R})$, and E_σ is just the required resolution of the identity. □

As in the theory of Stieltjes integrals, E is usually identified with the corresponding E_λ.

6.5 The Case of Normal Operators

Recall that a closed densely defined operator A is called *normal* if $A^*A = AA^*$.

Theorem 6.5. *Let A be an arbitrary normal operator. Then, on the σ-algebra $\mathfrak{B}(\mathbb{C})$ of Borel subsets of the complex plane, one can define a resolution of the identity E of the operator A such that the spectral representation*

$$A = \int_{\mathbb{C}} \lambda dE(\lambda), \quad \mathcal{D}(A) = \left\{ f \in H \middle| \int_{\mathbb{C}} |\lambda|^2 d(E(\lambda)f, f)_H < \infty \right\} \tag{6.19}$$

is true.

In (6.19), $\mathbb{C}$ can be replaced by the spectrum $S(A)$ of the operator A. The resolution of the identity E from (6.19) is determined uniquely.

Proof. As in the case of Theorem 6.1, we give two different proofs of representation (6.19). The first proof is very simple. It is based on the reduction to the case of bounded normal operator and is similar to the first proof of Theorem 6.1. To realize this version of the proof, one must additionally assume that the inverse operator A^{-1} exists and is bounded (or, more generally, that the operator A possesses a regular point).

The second proof works in the general case but is much more complicated.

I. In the first proof of the theorem, it is convenient to denote, as above, A, E, and λ by A', E', and λ', respectively. In this case, we assume that $\lambda' = 0$ is a regular point of the operator A'. Let $\varepsilon' > 0$ be so small that the circle $\{\lambda' \in \mathbb{C} \mid |\lambda'| < \varepsilon'\}$ consists of regular points of A'. Denote $R' = \{\lambda' \in \mathbb{C} \mid |\lambda'| \geq \varepsilon\}$, $R = \{\lambda \in \mathbb{C} \mid |\lambda| \leq \varepsilon^{-1}\}$. The space R is the image of R' under the mapping $R' \ni \lambda' \mapsto \lambda = \lambda'^{-1} \in R$. Consider the inverse mapping

$$R \ni \lambda \mapsto \lambda' = \varphi(\lambda) = \lambda^{-1} \in R'. \tag{6.20}$$

Consider the bounded operator $A = A'^{-1}$; its spectrum is located in R. This operator is normal. Indeed, by virtue of (3.8) and (12.3.13)

$$\begin{aligned} A^*A = (A'^{-1})^* A'^{-1} = (A'^*)^{-1} A'^{-1} = (A'A'^*)^{-1} \\ = (A'^*A')^{-1} = A'^{-1}(A'^*)^{-1} = A'^{-1}(A'^{-1})^* = AA^*. \end{aligned}$$

Let E be a resolution of the identity of the operator A, which exists according to Theorem 5.3; it may be regarded as given in $\mathfrak{R} = \mathfrak{B}(R)$. Denote by E' the image of E under mapping (6.20); E' is defined by relation (3.1) in $\mathfrak{R}' = \mathfrak{B}(R')$. As above, E' is a resolution of the identity of A'. In fact, by virtue of (3.3) for the function $R' \ni \lambda' \mapsto F(\lambda') = \lambda' \in \mathbb{C}$, we obtain the required formula

$$\int_{R'} \lambda' dE'(\lambda') = \int_R \lambda^{-1} dE(\lambda) = A^{-1} = A';$$

$$\mathcal{D}(A') = \left\{ f \in H \;\middle|\; \int_{R'} |\lambda'|^2 d(E'(\lambda')f, f)_H < \infty \right\}.$$

II. Thus, let A be a general normal operator. Consider the two operators

$$B = \mathbb{I} + A^*A, \quad \text{and} \quad C = A\big(\mathbb{I} + A^*A\big)^{-1}. \tag{6.21}$$

According to Theorem 12.7.2, the operator A^*A is selfadjoint and nonnegative; therefore, B is selfadjoint and invertible. The operator C is well-defined because $\mathcal{R}((\mathbb{I} + A^*A)^{-1}) = \mathcal{D}(A^*A) \subseteq \mathcal{D}(A)$.

Let us establish several auxiliary facts (they become evident if we assume that the operator A admits representation (6.19); one must only take into account the properties of spectral integrals presented in Section 2).

Lemma 6.1. *The operator C is bounded. The relations*

$$C = A\big(\mathbb{I} + A^*A\big)^{-1} = \Big(\big(\mathbb{I} + A^*A\big)^{-1} A\Big)^{\sim}, \tag{6.22}$$

$$C^* = \Big(\big(\mathbb{I} + A^*A\big)^{-1} A^*\Big)^{\sim} = A^*\big(\mathbb{I} + A^*A\big)^{-1}. \tag{6.23}$$

are true.

Proof. The boundedness of C means that there exists $c > 0$ such that, for any $f \in H$, the inequality $\left\|A\big(\mathbb{I} + A^*A\big)^{-1} f\right\|_H \le c\, \|f\|_H$ is satisfied. After the change $\big(\mathbb{I} + A^*A\big)^{-1} f = g$, this inequality turns into $\|Ag\|_H \le c \left\|\big(\mathbb{I} + A^*A\big)g\right\|_H$ $\quad (g \in \mathcal{D}(A^*A))$. We have

$$\begin{aligned} \|Ag\|_H^2 &= (Ag, Ag)_H = (A^*Ag, g)_H \le ((\mathbb{I} + A^*A)g, g)_H \\ &\le \|(\mathbb{I} + A^*A)g\|_H \, \|g\|_H \le \left\|\big(\mathbb{I} + A^*A\big)g\right\|_H^2 \end{aligned}$$

as required.

Let us prove (6.22). It suffices to show that $A\big(\mathbb{I}+A^*A\big)^{-1} f = \big(\mathbb{I}+A^*A\big)^{-1} Af$ for $f \in \mathcal{D}(A)$. The change $\big(\mathbb{I} + A^*A\big)^{-1} f = g$ transforms this equality into the equivalent relation

$$Ag = \big(\mathbb{I} + A^*A\big)^{-1} A\big(\mathbb{I} + A^*A\big)g \quad (g \in \mathcal{D}(A(\mathbb{I} + A^*A))). \tag{6.24}$$

At the same time,

$$A(\mathbb{I} + A^*A)g = Ag + AA^*Ag = Ag + A^*A^2g = (\mathbb{I} + A^*A)Ag$$

but this means that (6.24) is true.

Let us prove the first equality in (6.23). We have $(\forall f \in H)(\forall g \in \mathcal{D}(A^*))$:

$$(Cf, g)_H = \left(A(\mathbb{I} + A^*A)^{-1}f, g\right)_H = \left((\mathbb{I} + A^*A)^{-1}f, A^*g\right)_H$$
$$= \left(f, (\mathbb{I} + A^*A)^{-1}A^*g\right)_H$$

and this relation gives the required result. The second equality in (6.23) is obtained just as in the case of (6.22), with A replaced by A^* (it follows from Theorem 12.3.3 that the normality of A implies the normality of A^*). □

Lemma 6.2. *The operator C is normal.*

Proof. It is necessary to prove that $C^*C = CC^*$ or, equivalently, that

$$(Cf, Cg)_H = (CC^*f, g)_H = (C^*Cf, g)_H = (C^*f, C^*g)_H \quad (f, g \in H). \quad (6.25)$$

By using the first equality in (6.22) and the second in (6.23), we can rewrite (6.25) in the form

$$(A(\mathbb{I} + A^*A)^{-1}f, \quad A(\mathbb{I} + A^*A)^{-1}g)_H$$
$$= (A^*(\mathbb{I} + A^*A)^{-1}f, A^*(\mathbb{I} + A^*A)^{-1}g)_H \qquad (f, g \in H). \quad (6.26)$$

The vectors $(\mathbb{I} + A^*A)^{-1}f = f_1$ and $(\mathbb{I} + A^*A)^{-1}g = g_1$ run through the whole $\mathcal{D}(A^*A)$ if f and g run through H. Therefore, (6.26) is equivalent to the equality $(Af_1, Ag_1)_H = (A^*f_1, A^*g_1)(f_1, g_1 \in \mathcal{D}(A^*A))$, which holds because the operator A is normal. □

Lemma 6.3. *The resolutions of the identity of the selfadjoint operator B and bounded normal operator C commute.*

Proof. Since the resolution of the identity of the operator C is constructed as the direct product of resolutions of the identity of the bounded selfadjoint operators $\operatorname{Re} C$ and $\operatorname{Im} C$ (see Theorem 5.3), it suffices to prove that the resolutions of the identity of the operators B and $\operatorname{Re} C$ (and B and $\operatorname{Im} C$) are commuting. According to Theorem 6.3 and Remark 6.2, for this purpose, it suffices to verify the commutability of B^{-1} and $\operatorname{Re} C$ and of B^{-1} and $\operatorname{Im} C$ or, which is the same, of B^{-1} and C and of B^{-1} and C^*.

Let us show that B^{-1} and C commute. By using (6.22), we obtain

$$B^{-1}C - CB^{-1} = (\mathbb{I} + A^*A)^{-1}A(\mathbb{I} + A^*A)^{-1} - \left((\mathbb{I} + A^*A)^{-1}A\right)^{\sim}(\mathbb{I} + A^*A)^{-1}. \quad (6.27)$$

But the last expression is indeed equal to zero because $\mathcal{R}((\mathbb{I} + A^*A)^{-1}) = \mathcal{D}(A^*A) \subseteq \mathcal{D}(A)$ and, therefore, the wave in the second term in (6.27) can be omitted. Similarly, by using (6.23), one can easily prove that B^{-1} and C^* commute. □

We now return to the proof of Theorem 6.5. According to Theorem 6.1, the selfadjoint positive operator B admits a representation

$$B = \int_1^\infty t dE_B(t), \tag{6.28}$$

where E_B is the resolution of the identity of B. Since $B \geq \mathbb{I}$, we have $S(B) \subseteq [\mathbb{I}, \infty)$. Therefore, the integral in (6.28) has just these limits of integration. According to Theorem 5.3, the normal bounded operator C admits the following representation:

$$C = \int_{\mathbb{C}} z dE_C(z), \tag{6.29}$$

where E_C is the resolution of the identity of C. Since C is bounded, supp E_C is also bounded or, more precisely, supp$E_C \subseteq \{z \in \mathbb{C} \mid |z| \leq \|C\|\} = B_{\|C\|}(0)$.

The commutability of E_B and E_C was proved in Lemma 6.3. Let us construct, in accordance with Theorem 3.2, their direct product $E = E_B \times E_C$, which is the resolution of the identity defined on Borel subsets of the space $R = [1, \infty) \times \mathbb{C}$, whose points are denoted by $\lambda = (t, z)$; supp $E \subseteq [1, \infty) \times B_{\|C\|}(0)$. In terms of E, relations (6.28) and (6.29) can be rewritten in the form (see (5.18))

$$B = \int_R t dE(\lambda), \qquad C = \int_R z dE(\lambda), \tag{6.30}$$

where, in the first case, we integrate the function $R \ni \lambda = (t, z) \mapsto t \in [1, \infty)$, and, in the second case, the function $R \ni \lambda = (t, z) \mapsto z \in \mathbb{C}$, which can be regarded as bounded (if we replace R by supp E in (6.30)).

According to (6.21), $A = CB$. Moreover, the product of the spectral integrals (6.30) is equal to the integral of the product of integrands (see Section 2). Thus, we obtain the following spectral representation for A:

$$A = \int_R tz dE(\lambda), \quad \mathcal{D}(A) = \left\{ f \in H \middle| \int_R t^2 |z|^2 d(E(\lambda) f, f)_H < \infty \right\}. \tag{6.31}$$

Relation (6.31) can easily be transformed into (6.19) by applying the mapping $R \ni \lambda = (t, z) \mapsto \lambda' = tz \in \mathbb{C} = R'$. Let E' be the image of E under this mapping; E' is the resolution of the identity on $\mathfrak{B}(R') = \mathfrak{B}(\mathbb{C})$. By virtue of (3.3), for the function $R' = \mathbb{C} \ni \lambda' \mapsto \lambda' \in \mathbb{C}$, we have

$$\int_{\mathbb{C}} \lambda' dE'(\lambda') = \int_R tz dE(\lambda) = A$$

$$\mathcal{D}(A) = \left\{ f \in H \;\middle|\; \int_{\mathbb{C}} |\lambda'|^2 d(E'(\lambda') f, f)_H < \infty \right\}.$$

Representation (6.19) is thus proved in the general case.

III. The transition from $\mathbb{C}$ to $S(A)$ in (6.19) is justified in exactly the same way as in Theorems 4.2, 5.3, and 6.1 by using Remark 4.3 and the proof of Theorem 4.2.

IV. The uniqueness of the definition of E by (6.19) is proved by using a modified (somewhat more complicated) version of the reasoning from the proof of Theorem 5.3, and we leave this proof for the reader. In addition, instead of relations (5.15), one must use their version for the unbounded case that are given below (see (6.34)). □

We consider some properties of normal operators, which immediately follow from representation (6.19) and the properties of spectral integrals established in Section 2.

If A is normal, then A^* is also normal and $\mathcal{D}(A^*) = \mathcal{D}(A)$, $\|A^*f\|_H = \|Af\|_H$ ($f \in \mathcal{D}(A)$). If A has the form (6.19), then

$$A^* = \int_{\mathbb{C}} \overline{\lambda} dE(\lambda). \tag{6.32}$$

The absolute value $|A|$ of a normal operator defined by the formula $|A| = \sqrt{A^*A}$ has the following spectral representation:

$$|A| = \int_{\mathbb{C}} |\lambda| dE(\lambda),$$

$$\mathcal{D}(|A|) = \left\{ f \in H \middle| \int_{\mathbb{C}} |\lambda|^2 d(E(\lambda)f, f)_H < \infty \right\} = \mathcal{D}(A) = \mathcal{D}(A^*). \tag{6.33}$$

Given A of the form (6.19), we introduce the operators $\operatorname{Re} A$ and $\operatorname{Im} A$ by the spectral integrals

$$\operatorname{Re} A = \int_{\mathbb{C}} \operatorname{Re} \lambda \, dE(\lambda), \quad \text{and} \quad \operatorname{Im} A = \int_{\mathbb{C}} \operatorname{Im} \lambda \, dE(\lambda) \tag{6.34}$$

with proper domains of definition. This enables us to rewrite relations (5.15) in the form

$$\operatorname{Re} A = \frac{1}{2}(A + A^*)^{\sim},$$

$$\operatorname{Im} A = \frac{1}{2i}(A - A^*)^{\sim},$$

$$A = (\operatorname{Re} A + i\operatorname{Im} A)^{\sim}. \tag{6.35}$$

The operators Re A and Im A are uniquely defined for given A. Conversely, the operator A is uniquely determined in terms of Re A and Im A.

REMARK 6.3. By using Theorem 6.1, one can prove the following result, which is called the spectral theorem in terms of the operators of multiplication (see, e.g., [ReS1, pp. 287–288]). Let us formulate this theorem.

Let A be a selfadjoint operator in H and let $\mathcal{D}(A)$ be its domain of definition. It is stated that there exist a measurable space $\langle R, \mathfrak{R}\rangle$ with a finite measure μ, a unitary operator $U : H \to L_2(R, \mathfrak{R}, d\mu)$, and a real-valued μ-almost everywhere finite function φ in R such that

(i) $f \in \mathcal{D}(A) \Leftrightarrow \varphi(\cdot)(Uf)(\cdot) \in L_2(R, \mathfrak{R}, d\mu)$;

(ii) if $g \in U(\mathcal{D}(A))$, then $(UAU^{-1}g)(\cdot) = \varphi(\cdot)g(\cdot)$.

This result can be readily generalized to the case of normal operators A.

Exercises

6.1. Let $a \in C(\mathbb{R})$. Consider the operator of multiplication by a, namely,

$$L_2(\mathbb{R}) \supseteq \mathcal{D}(A) = \{f \in L_2(\mathbb{R}) | af \in L_2(\mathbb{R})\} \ni f \mapsto Af = af \in L_2(\mathbb{R}).$$

Prove that A is a normal operator (or selfadjoint if $a = \overline{a}$) and find its resolution of the identity.

6.2. Find the resolution of the identity of the operator

$$L_2(\mathbb{R}) \supset \mathcal{D}(A) = \{f \in L_2(\mathbb{R}) \mid \exists\, f' \in L_2(\mathbb{R})\} \ni f \mapsto Af = if' \in L_2(\mathbb{R}).$$

6.3. Find the resolution of the identity of the selfadjoint extension of the operator

$$L_2(\mathbb{R}) \supset \mathcal{D}(A) = C_0^\infty(\mathbb{R}) \ni f \mapsto Af = f'' \in L_2(\mathbb{R}).$$

6.4. Operators A and B acting on H are called metrically equal (notation: $A \overset{\text{m}}{=} B$) if $\mathcal{D}(A) = \mathcal{D}(B)$ and $(\forall f \in \mathcal{D}(A))$: $\|Af\| = \|Bf\|$. Prove that metrically equal operators are simultaneously (a) closed, (b) closable and, furthermore, $\tilde{A} \overset{\text{m}}{=} \tilde{B}$.

6.5. Prove that metrically equal nonnegative operators are equal.

6.6. Prove that, for any operator A with $\mathcal{D}(A)^\sim = H$,

(a) $(\exists! B \geq 0, B \subseteq B^*)$: $B \overset{\text{m}}{=} A$;

(b) $(\exists! C \geq 0, C \subseteq C^*)$: $C \overset{\text{m}}{=} A^*$.

Show that $B = \sqrt{A^*A}$ and $C = \sqrt{AA^*}$.

6.7. Let $A \overset{\text{m}}{=} B$. Prove that there is a unitary operator U in H such that $A = UB$.

6.8. Prove that A is a normal operator if and only if $A^* \overset{\text{m}}{=} A$.

7 Spectral Representation of One-Parameter Unitary Groups and Operator Differential Equations

7.1 Stone's Theorem

Let A be a selfadjoint operator acting on a Hilbert space H and let E be its resolution of the identity. For a given function $\mathbb{R} \times \mathbb{R} \ni (\lambda, t) \mapsto e^{it\lambda} \in \mathbb{C}$, we construct the operator-valued function

$$\mathbb{R} \ni t \mapsto U(t) = \int_{\mathbb{R}} e^{it\lambda} dE(\lambda) = e^{itA} \in \mathcal{L}(H) \tag{7.1}$$

according to (6.16).

It follows from the properties of spectral integrals (Section 2) that the operator $U(t)$ is unitary for all $t \in \mathbb{R}$ and satisfies the equality

$$U(t+s) = U(t)U(s) \qquad (t, s \in \mathbb{R}). \tag{7.2}$$

The function defined by (7.1) is strongly continuous, i.e., $(\forall f \in H)(\forall t \in \mathbb{R})$: $U(s)f \to U(t)f$ as $s \to t$. To prove this fact, we use relation (2.8) and obtain

$$\begin{aligned}\|U(s)f - U(t)f\|_H^2 &= \left\| \int_{\mathbb{R}} (e^{is\lambda} - e^{it\lambda}) dE(\lambda) f \right\|_H^2 \\ &= \int_{\mathbb{R}} \left| e^{is\lambda} - e^{it\lambda} \right|^2 d\big(E(\lambda)f, f\big)_H \xrightarrow[s \to t]{} 0.\end{aligned}$$

Moreover, $U(t)$ *is strongly continuously differentiable,* i.e., for all $f \in \mathcal{D}(A)$ and $t \in \mathbb{R}$, the strong derivative

$$U'(t)f = \lim_{h \to 0} \frac{1}{h} \left(U(t+h) - U(t) \right) f = iU(t)Af \tag{7.3}$$

exists and is a continuous vector function.

Indeed, by virtue of relations (7.1), (6.16), and (2.8) we obtain

$$\begin{aligned}&\left\| iU(t)Af - \frac{1}{h}(U(t+h) - U(t))f \right\|_H^2 \\ &\qquad = \left\| \int_{\mathbb{R}} \left(ie^{it\lambda}\lambda - \frac{1}{h}\left(e^{i(t+h)\lambda} - e^{it\lambda}\right) \right) dE(\lambda) f \right\|_H^2 \\ &\qquad = \int_{\mathbb{R}} \left| i\lambda - \frac{1}{h}(e^{ih\lambda} - 1) \right|^2 d(E(\lambda)f, f)_H.\end{aligned}$$

The expression under the integral sign on the right-hand side of this equality tends to zero as $h \to 0$ for all $\lambda \in \mathbb{R}$ and is uniformly bounded in h by the function $c\lambda^2$. For $f \in \mathcal{D}(A)$, by virtue of the Lebesgue theorem (Section 3.6), we can pass to the limit under the integral sign and get (7.3). The continuity of the derivative follows from the inclusion $f \in \mathcal{D}(A)$. □

The following interpretation of relations (7.1) and (7.3) proves to be useful: Consider an operator differential equation

$$u'(t) = iAu(t) \qquad (t \in \mathbb{R}), \tag{7.4}$$

where $\mathbb{R} \ni t \mapsto u(t) \in H$ is the required solution. The function u is assumed to be strongly continuously differentiable and $u(t) \in \mathcal{D}(A)$ for all t. Such solutions are called strong. The strong solution of equation (7.4) satisfying the initial condition $u(0) = u_0 \in \mathcal{D}(A)$, i.e., the solution of the corresponding Cauchy problem exists and is given by the formula

$$u(t) = U(t)u_0 \qquad (t \in \mathbb{R}). \tag{7.5}$$

It is not difficult to show that this Cauchy problem is uniquely solvable. In more details, we dwell upon these problems in Section 8.

The indicated formulas also admit the following interpretation:

We say that a function $\mathbb{R} \ni t \mapsto U(t)$ whose values are unitary operators in H satisfying relation (7.2) defines a one-parameter unitary group (or, in other words, a unitary representation of the group $\mathbb{R}$).

Thus, (7.1) is an example of a one-parameter unitary group which is, in addition, strongly continuous. The following theorem demonstrates that relation (7.1) gives the general form of these groups.

Theorem 7.1 (Stone). *A strongly continuous one-parameter unitary group $U(t)$ $(t \in \mathbb{R})$ always admits representation (7.1) with a certain resolution of the identity E uniquely determined for a given group. The corresponding operator A is called the infinitesimal generator of this group.*

Proof. Let us construct a linear set $\mathcal{D} \subseteq H$ dense in H and such that the vector function

$$\mathbb{R} \ni t \mapsto U(t)f \in H \tag{7.6}$$

is strongly continuously differentiable for all $f \in \mathcal{D}$ (the construction presented below is a particular case of the construction of the so-called Gårding domain).

For $F \in C_0^\infty(\mathbb{R})$ and $g \in H$, we consider a vector of the form

$$g_F = \int_{\mathbb{R}} F(s)U(s)g\,ds \in H, \tag{7.7}$$

where the integral is understood as the limit of Riemann integral sums in H. By virtue of the continuity of $F(\cdot)$ and $U(\cdot)f$ and finiteness of $F(\cdot)$, the standard arguments enable us to conclude that integral (7.7) exists and possesses natural properties of Riemann integrals. Thus, the set $\mathcal{D}$ can be chosen as the collection of linear combinations of vectors (7.7) with $F \in C_0^\infty(\mathbb{R})$ and $g \in H$.

The set $\mathcal{D}$ is dense in H. To prove this, we consider a vector $H \ni h \perp \mathcal{D}$. Multiplying (7.7) scalarly by this vector, we obtain $\left(\forall F \in C_0^\infty(\mathbb{R})\right)$:

$$0 = (g_F, h)_H = \int_{\mathbb{R}} F(s)\big(U(s)g, h\big)_H ds.$$

In view of the arbitrariness of F in this equality and the continuity of the function $\mathbb{R} \ni s \mapsto \big(U(s)g, h\big)_H \in \mathbb{C}$ (which follows from our assumptions), we conclude that $(\forall s \in \mathbb{R})\colon (U(s)g, h)_H = 0$ $(g \in H)$. In particular, by setting $s = 0$ and using the equality $U(0) = \mathbb{1}$ (a consequence of (7.2)), we obtain $(g, h)_H = 0$ $(\forall g \in H) \Rightarrow h = 0$. Therefore, $\mathcal{D}$ is dense.

To prove that function (7.6) is strongly continuously differentiable, it suffices to consider f of the form (7.7). By using (7.2) and changing variables in the integral, we get $(\forall t, h \in \mathbb{R})$

$$\begin{aligned}
\frac{1}{h}\left(U(t+h)g_F - U(t)g_F\right) &= \frac{1}{h}\int_{\mathbb{R}} F(s)\left(U(t+h)U(s) \;-\; U(t)U(s)\right) g ds \\
&= \frac{1}{h}\int_{\mathbb{R}} F(s)\left(U(t+h+s) \;-\; U(t+s)\right) g ds \\
&= \int_{\mathbb{R}} \frac{1}{h}\left(F(s-t-h) - F(s-t)\right) U(s) g ds \\
\xrightarrow[h\to 0]{} &- \int_{\mathbb{R}} F'(s-t)U(s)gds = g_{-F'(\cdot - t)} = U'(t)g_F.
\end{aligned} \tag{7.8}$$

The limit transition in (7.8) can be easily justified by the following simple estimate (established by limit transition from integral sums):

$$\left\|\int_{\mathbb{R}} G(s)U(s)gds\right\|_H \le \int_{\mathbb{R}} \|U(s)g\|_H \, |G(s)|\, ds \quad (G \in C_0(\mathbb{R}); g \in H). \tag{7.9}$$

As follows from (7.9), the vector function on the right-hand side of (7.8) is strongly continuous. We stress that $U'(t)\, g_F$ has the form (7.7) for all $t \in \mathbb{R}$.

On $\mathcal{D}$, we introduce the operator A in the space H by the formula

$$\mathcal{D} = \mathcal{D}(A) \ni f \mapsto Af = \frac{1}{i}U'(0)f \in \mathcal{D} \subseteq H. \tag{7.10}$$

Note that $\mathcal{D}$ is invariant under the action of A. By applying the operator $U(t)$ to (7.7), using (7.2), and changing the variables in the integral, we establish that $\mathcal{D}$ is also invariant under the action of $U(t)$ for all $t \in \mathbb{R}$.

The operator A introduced above is Hermitian. To prove this, it suffices to show that $(Ag_F, h_G)_H = (g_F, Ah_G)_H$, where $F, G \in C_0^\infty(\mathbb{R})$ and $g, h \in H$. According to (7.8), we have

$$\begin{aligned}
(Ag_F, h_G)_H &= \lim_{h\to 0}\left(\frac{1}{ih}(U(h) - \mathbb{1})g_F, h_G\right)_H \\
&= \lim_{h\to 0}\left(g_F, -\frac{1}{ih}(U(-h) - \mathbb{1})h_G\right)_H = \left(g_F, Ah_G\right)_H.
\end{aligned}$$

Furthermore, we can prove that A is essentially selfadjoint. Let $z \in \mathbb{C} \setminus \mathbb{R}$ and $\varphi \in \mathcal{D}(A^*)$ be such that $A^*\varphi = z\varphi$; it is necessary to prove that $\varphi = 0$. To do this, we first show that

$$U'(t)g_F = iAU(t)g_F \quad (F \in C_0^\infty(\mathbb{R}), g \in H, t \in \mathbb{R}). \tag{7.11}$$

This equality will be proved if we notice that, in this case, $U(t)g_F$ also has the form (7.7) (this vector is equal to $g_{F(\cdot - t)}$) and then compute $AU(t)g_F$ according to (7.10).

In view of (7.11), we obtain, for any $F \in C_0^\infty(\mathbb{R})$ and $g \in H$,

$$\begin{aligned} \frac{d}{dt}(U(t)g_F, \varphi)_H &= (U'(t)g_F, \varphi)_H = i\,(AU(t)g_F, \varphi)_H \\ &= i\,(U(t)g_F, A^*\varphi)_H = i\bar{z}\,(U(t)g_F, \varphi)_H \qquad (t \in \mathbb{R}). \end{aligned}$$

Hence, the complex-valued bounded function $a(t) = \big(U(t)g_F, \varphi\big)_H$ satisfies the equation $a' = i\bar{z}a$. Therefore, $a(t) = e^{i\bar{z}t}a(0)$ $(t \in \mathbb{R})$ and, by virtue of the condition $\operatorname{Im} z \neq 0$, it may be bounded only if $a(0) = 0$. Consequently, $(g_F, \varphi)_H = a(0) = 0$ $\big(F \in C_0^\infty(\mathbb{R}), g \in H\big)$, i.e., $\varphi \perp \mathcal{D} \Rightarrow \varphi = 0$.

Thus, we have proved that the operator $\tilde{A}$ is selfadjoint. In terms of this operator, according to formula (7.1), we can now construct the operator-valued function $\mathbb{R} \ni t \mapsto V(t) = e^{it\tilde{A}} \in \mathcal{L}(H)$. It remains to show that $V(t) = U(t)$ $(t \in \mathbb{R})$.

According to (7.4) and (7.5), $v(t) = V(t)v_0$ $(v_0 \in \mathcal{D}(\tilde{A}))$ is a strong solution of the Cauchy problem

$$v'(t) = i\tilde{A}v(t) \qquad (t \in \mathbb{R}, v(0) = v_0). \tag{7.12}$$

We set $v_0 = g_F$ for some $F \in C_0^\infty(\mathbb{R})$ and $g \in H$. In view of (7.11), the function $u(t) = U(t)g_F$ is also a strong solution of problem (7.12) (in (7.11), A can be replaced by $\tilde{A}$, since $U(t)g_F \in \mathcal{D}$). Later, in Section 8 (Theorem 8.1), it will be shown that strong solutions of equation (7.12) with selfadjoint $\tilde{A}$ that correspond to the same initial data coincide. Therefore, $V(t)g_F = U(t)g_F$ and, by virtue of the denseness of $\mathcal{D}$ in H, we get $V(t) = U(t)$ $(t \in \mathbb{R})$. (It is clear that this coincidence can be proved even without using the general Theorem 8.1)

In order to prove that E is uniquely determined by (7.1), we note that

$$(\forall f, g \in H) : (U(t)f, g)_H = \int_{\mathbb{R}} e^{it\lambda} d\big(E(\lambda)f, g\big)_H \qquad (t \in \mathbb{R}), \tag{7.13}$$

i.e., the function on the left-hand side of (7.13) is the Fourier-Stieltjes transform of the charge $\mathfrak{B}(\mathbb{R}) \mapsto \big(E(\alpha)f, g\big)_H \in \mathbb{C}$, and it is well known (see [Shi1]) that a charge is uniquely determined by its Fourier transform. This proves that E is uniquely determined. □

REMARK 7.1. The strong continuity of a one-parameter unitary group is equivalent to its weak continuity, i.e., to the continuity of the function $\mathbb{R} \ni t \mapsto \big(U(t)f, g\big)_H \in \mathbb{C}$ for all $f, g \in H$.

Indeed, it suffices to prove that weak continuity implies strong continuity. For $f \in H$, we have

$$\begin{aligned}\|U(t)f - U(s)f\|_H^2 &= ((U(t) - U(s))f, (U(t) - U(s))f)_H \\ &= ((U(t) - U(s))^*(U(t) - U(s))f, f)_H \\ &= ((2 \cdot \mathbb{I} - U(t-s) - U(s-t))f, f)_H \xrightarrow[s \to t]{} 0.\end{aligned}$$

□

This remark enables us to introduce appropriate evident changes in the formulation of Theorem 7.1.

REMARK 7.2. A function $\mathbb{R} \ni t \mapsto k(t) \in \mathbb{C}$ is called *positive definite* if it satisfies the inequality

$$\sum_{j,l=1}^{n} k(t_j - t_l)\xi_j \bar{\xi}_l \geq 0 \tag{7.14}$$

for all $t_1, \dots t_n \in \mathbb{R}$, $\xi_1, \dots \xi_n \in \mathbb{C}$, and $n \in \mathbb{N}$.

The well-known Bochner-Khinchin theorem (see [Shi1]) asserts that a continuous positive definite function admits a representation

$$k(t) = \int_{\mathbb{R}} e^{i\lambda t} d\sigma(\lambda) \qquad (t \in \mathbb{R}), \tag{7.15}$$

where $\mathfrak{B}(\mathbb{R}) \ni \alpha \mapsto \sigma(\alpha) \geq 0$ is a finite measure uniquely defined in terms of this function. Conversely, each function of the form (7.15) is positive definite. By comparing (7.13) with (7.15), we conclude that the function

$$\mathbb{R} \ni t \mapsto k(t) = (U(t)f, f)_H \quad (f \in H) \tag{7.16}$$

is positive definite. On this way, one can give another useful proof of Theorem 7.1. Thus, one can easily verify that the function defined by (7.16) is continuous and positive definite. Then one can write for this function a representation of the form (7.15) with a measure $\sigma_{f,f}(\alpha)$ that depends on $f \in H$. Finally, by the reasoning similar to that applied in the proof of Theorem 4.1, one can show that $\sigma_{f,f}(\alpha) = \big(E(\alpha)f, f\big)_H$, and this leads to the required representation (7.1).

7.2 Operator Differential Equations

In studying the operator differential equation (7.4), we have already shown that relations (7.5) and (7.1) give a solution of the corresponding Cauchy problem. Here, we somewhat generalize these arguments.

Consider a densely defined operator B acting on a Hilbert space H and let $I \subseteq \mathbb{R}$ be a finite or infinite closed, open, or half-open interval and $r \in \mathbb{N}$ (in fact, we restrict ourselves to the cases where $r = 1, 2$).

We say that a vector-function $I \ni t \mapsto u(t) \in H$ is a strong solution of the equation

$$\left(\frac{d^r u}{dt^r}\right)(t) + Bu(t) = 0 \quad (t \in I) \tag{7.17}$$

on I if it is r times strongly continuously differentiable (i.e., has r strong derivatives on I, the last of which is continuous), $u(t) \in \mathcal{D}(B)$ for all $t \in I$, and equation (7.17) is satisfied.

An r times strongly continuously differentiable vector function $I \ni t \mapsto u(t) \in H$ is a strong solution of the equation

$$\left(\frac{d^r u}{dt^r}\right)(t) + B^* u(t) = 0 \quad (t \in I) \tag{7.18}$$

if and only if the "weak" equality

$$\left(\left(\frac{d^r u}{dt^r}\right)(t), f\right)_H + \left(u(t), Bf\right)_H = 0 \quad \left(f \in \mathcal{D}(B); t \in I\right) \tag{7.19}$$

holds. This statement immediately follows from the definition of an adjoint operator, since (7.17) implies that $u(t) \in \mathcal{D}(B^*)$ for every $t \in I$ in view of inclusion $u^{(r)}(t) \in H$.

We say that *the Cauchy problem for equation (7.17) on $I = [0, b)$ $(0 < b \leq \infty)$ is uniquely solvable in the strong sense if each strong solution of this equation on $[0, b)$ such that $u(0) = \cdots = u^{(r-1)}(0) = 0$ vanishes for all $t \in (0, b)$.*

REMARK 7.3. If the Cauchy problem is uniquely solvable on $[0, b)$ for some $b > 0$, then it is uniquely solvable on $[0, \infty)$.

Indeed, let $[0, \infty) \ni t \mapsto u(t) \in H$ be a strong solution of equation (7.17) on $[0, \infty)$ such that $u(0) = \cdots = u^{(r-1)}(0) = 0$. In view of the assumed unique solvability on $[0, b)$, we have $u(t) = 0$ for $t \in (0, b)$ and, in particular, $u(t) = 0$ in a neighbourhood of the point $c = b/2$; therefore $u(c) = \cdots = u^{(r-1)}(c) = 0$. The function $[0, \infty) \ni t \mapsto u_1(t) = u(t + c)$ is a strong solution of (7.17) on $[0, \infty)$ such that $u_1(0) = u(c) = 0, \dots, u_1^{r-1}(0) = u^{r-1}(c) = 0$ and, hence, $u_1(t) = 0$ for $t \in (0, b)$. By repeating the same reasoning, we can show that the function $[0, \infty) \ni t \mapsto u_2(t) = u_1(t + c) = u(t + 2c)$ also vanishes for $t \in (0, b)$. Then we construct the function $u_3(t)$, etc. As a final result, we conclude that $u(t) = 0$ $\left(t \in [0, \infty)\right)$.

□

If the operator B in (7.17) has the form $B = \zeta A$, where $\zeta \in \mathbb{C}$ is a fixed number and A is a selfadjoint operator, then, under the corresponding restrictions imposed on the initial conditions, the Cauchy problem for (7.17) is solvable, and one can write a representation of this solution in terms of the resolution of the identity E for the operator A. We consider two examples.

Examples

7.1. The formal solution of the Cauchy problem

$$\left(\frac{du}{dt}\right)(t) + \zeta A u(t) = 0 \quad (t \in [0, \infty); \;\; u(0) = u_0 \in H). \tag{7.20}$$

has the form

$$u(t) = \int_{\mathbb{R}} e^{-\zeta t \lambda} dE(\lambda) u_0 = e^{-\zeta t A} u_0 \quad (t \in [0, \infty)). \tag{7.21}$$

Just as in the case of (7.1), expression (7.21) is a strong solution of the Cauchy problem (7.20) if $u_0 \in \mathcal{D}(Ae^{-\zeta t A})$.

In particular, for $\zeta = -i$ (the "nonstationary Schrödinger equation"), it suffices to require that $u_0 \in \mathcal{D}(A)$ (this case has been considered at the beginning of this section).

For $\zeta = 1$ and $A \geq 0$ (the "heat conduction equation"), it suffices to require that $u_0 \in H$.

7.2. The formal solution of the Cauchy problem

$$\begin{aligned} &\left(\frac{d^2u}{dt^2}\right)(t) + Au(t) = 0 \qquad (t \in [0, \infty); \\ &u(0) = u_0 \in H, \qquad u'(0) = u_1 \in H). \end{aligned} \tag{7.22}$$

has the form

$$\begin{aligned} u(t) &= \int_{\mathbb{R}} \cos\sqrt{\lambda} t dE(\lambda) u_0 + \int_{\mathbb{R}} \frac{\sin\sqrt{\lambda} t}{\sqrt{\lambda}} dE(\lambda) u_1 \\ &= (\cos\sqrt{A} t) u_0 + \left(\frac{\sin\sqrt{A} t}{\sqrt{A}}\right) u_1 \quad (t \in [0, \infty)). \end{aligned} \tag{7.23}$$

In the case of the "hyperbolic" equation (7.22) where the operator A is semibounded below, expression (7.23) is a strong solution of the Cauchy problem (7.22) if $u_0 \in \mathcal{D}(A)$ and $u_1 \in \mathcal{D}(\sqrt{|A|})$. Here, one can easily prove that function (7.23) is twice strongly continuously differentiable.

Exercises

7.1. Let $U(t)$ be a one-parameter group in $L_2(\mathbb{R})$ given by the equality $(U(t)f)(x) = f(x+t)$. Find the infinitesimal generator of the group $U(t)$.

7.2. A one-parameter group of unitary operators in H possesses the property $U(1) = \mathbb{I}$. Prove that the spectrum of its infinitesimal generator lies in $\mathbb{Z}$.

7.3. Is it possible to construct a one-parameter group $U(t)$ such that $U(1) = U$ for an arbitrary unitary operator U?

7.4. Let $V(t), t \geq 0$ be a family of selfadjoint operators in H that satisfies the following conditions:

(i) $(\exists c \in \mathbb{R})(\forall t \geq 0) : \|V(t)\| \leq e^{ct}$;

(ii) $(\forall t, s \geq 0) : V(t)V(s) = V(t+s)$;

(iii) the mapping $[0, \infty) \ni t \mapsto V(t) \in \mathcal{L}(H)$ is strongly continuous;

(iv) $V(0) = \mathbb{I}$.

Following the proof of Stone's theorem, show that $(\exists! A = A^*)(\forall t \geq 0) : V(t) = e^{-tA}$ and $A \geq -c\mathbb{I}$.

8 Evolutionary Criteria of Selfadjointness

In this section, we show that the selfadjointness of operators is closely related to the uniqueness of strong solutions of the Cauchy problems for the corresponding evolutionary equations. Let us first consider the "Schrödinger" criterion of selfadjointness.

8.1 The Schrödinger Criterion of Selfadjointness

Theorem 8.1. *Let A be an Hermitian operator acting on H. For its essential selfadjointness, it is necessary that the Cauchy problems for both equations*

$$\left(\frac{du}{dt}\right)(t) \pm (iA^*)u(t) = 0 \qquad (t \in [0, b)) \tag{8.1}$$

be strongly uniquely solvable on $[0, b)$ for all $b \in (0, \infty]$ and it is sufficient that these problems be uniquely solvable for some b.

Proof. The proof is split into several steps.

I. Let us first establish sufficiency under the assumption that A has equal defect numbers. Assume the contrary: Let $\tilde{A}$ be not selfadjoint. Then A has two different selfadjoint extensions A_1 and A_2 in H. Let E_1 and E_2 be the correspondent resolutions of the identity. For every $g \in \mathcal{D}(A) \subseteq \mathcal{D}(A_1)$, the integral $\int_{\mathbb{R}} \lambda^2 d(E_1(\lambda)g, g)_H$ is convergent. Therefore, the vector function

$$[0, \infty) \ni t \mapsto u_1(t) = \int_{\mathbb{R}} e^{i\lambda t} dE_1(\lambda)g \tag{8.2}$$

is strongly continuously differentiable and $u_1'(t) = i\int_{\mathbb{R}} \lambda e^{i\lambda t} dE_1(\lambda)g$. It is easy to see that it is a strong solution of equation (8.1) with the sign "+" on $[0,\infty)$. Indeed, it is necessary to check the corresponding weak equality (7.19) which now has the form

$$\left(\left(\frac{du_1}{dt}\right)(t), f\right)_H + \big(u_1(t), iAf\big)_H = 0 \quad (f \in \mathcal{D}(A); t \in [0,\infty)).$$

Since

$$\begin{aligned} d\big(E_1(\lambda)g, Af\big)_H &= d\big(E_1(\lambda)g, A_1 f\big)_H \\ &= d\left(\int_{-\infty}^{\lambda} \mu d\big(E_1(\mu)g, f\big)_H\right) = \lambda d\big(E_1(\lambda)g, f\big)_H, \end{aligned}$$

we have

$$\begin{aligned} &\left(\left(\frac{du_1}{dt}\right)(t), f\right)_H + \big(u_1(t), iAf\big)_H \\ &\qquad = i\int_{\mathbb{R}} \lambda e^{i\lambda t} d\big(E_1(\lambda)g, f\big)_H - i\int_{\mathbb{R}} e^{i\lambda t} d\big(E_1(\lambda)g, Af\big)_H = 0 \\ &\qquad\qquad (f \in \mathcal{D}(A); t \in [0,\infty)), \end{aligned}$$

i.e., the required relation is satisfied.

Similarly, the function $u_2(t)$ constructed according to (8.2) for given E_2 is a strong solution of the same equation; $u_1(0) = u_2(0) = g$. Thus, $u(t) = u_1(t) - u_2(t)$ is also a strong solution of equation (8.1) with sign "+" on $[0,\infty)$ such that $u(0) = 0$. By virtue of the condition of the theorem and Remark 7.3, the problem is uniquely solvable on $[0,\infty)$. Therefore, $u(t) = 0$ for $t \in [0,\infty)$, whence

$$\begin{aligned} &\int_{\mathbb{R}} e^{i\lambda t} d\big((E_1(\lambda) - E_2(\lambda))g, h\big)_H = 0 \\ &\big(g \in \mathcal{D}(A), h \in H, t \in [0,\infty)\big). \end{aligned} \tag{8.3}$$

Consider equation (8.1) with sign "−". By repeating the arguments presented above with $e^{i\lambda t}$ replaced by $e^{-i\lambda t}$ in (8.2), we arrive at the relation that differs from (8.3) by the same change. Therefore, if we introduce the charge $\omega(\alpha) = \big((E_1(\alpha) - E_2(\alpha))g, h\big)_H$ $\big(\alpha \in \mathfrak{B}(\mathbb{R})\big)$, then, according to (8.3) and the indicated modification of this relation, $\int_{\mathbb{R}} e^{i\lambda t} d\omega(\lambda) = 0$ for any $t \in \mathbb{R}$. Taking into account the theorem on uniqueness of the Fourier-Stieltjes transform of a charge already applied in Section 7, we conclude that $\omega = 0$, i.e., $((E_1(\alpha) - E_2(\alpha))g, h)_H = 0$ $\big(\alpha \in \mathfrak{B}(\mathbb{R})\big)$. Since $g \in \mathcal{D}(A)$ and $h \in H$ are arbitrary, this implies that $E_1 = E_2$, and we arrive at a contradiction.

II. In the case of an operator A with distinct defect numbers, we use the following lemma:

Lemma 8.1. *Consider the space $H \oplus H$ of vectors $f = \langle f_1, f_2 \rangle$ $(f_1, f_2 \in H)$, an operator C with a dense domain of definition $\mathcal{D}(C) = \mathcal{D}(A) \oplus \mathcal{D}(A)$ acting on this space according to the formula $Cf = \langle Af_1, -Af_2 \rangle$ $(f \in \mathcal{D}(C))$, and the equation $(b \in (0, \infty])$*

$$\left(\frac{du}{dt}\right)(t) + (iC)^* u(t) = 0 \qquad (t \in [0, b)) \tag{8.4}$$

for the vector functions with values in $H \oplus H$.

It is stated that if the Cauchy problem for both equations (8.1) is uniquely solvable in the strong sense on $[0, b)$, then the Cauchy problem for equation (8.4) is also uniquely solvable in the sense of strong solutions, and vice versa.

Proof. Let $[0, b) \ni t \mapsto u(t) = \langle u_1(t), u_2(t) \rangle \in H \oplus H$ be a strong solution of the Cauchy problem for equation (8.4). Since $C^* f = \langle A^* f_1, -A^* f_2 \rangle$ $(f \in \mathcal{D}(C^*) = \mathcal{D}(A^*) \oplus \mathcal{D}(A^*))$, the functions $[0, b) \ni t \mapsto u_1(t) \in H$ and $[0, b) \ni t \mapsto u_2(t) \in H$ are strong solutions of equation (8.1) with signs "+" and "−", respectively. In view of the fact that, by assumption, strong solutions of the Cauchy problem for (8.1) are unique on $[0, b)$, this implies the required uniqueness for (8.4). The converse statement is deduced similarly. □

III. Let us prove sufficiency for an operator A with deficiency index (m, n). As in Lemma 8.1, we construct the operator C. It is easy to verify that the deficiency index of this operator is equal to $(m + n, m + n)$. By virtue of Lemma 8.1, the Cauchy problem for (8.4) is uniquely solvable on $[0, b)$. By applying this lemma to the case where A is replaced by $-A$, we conclude that this uniqueness is preserved for equation (8.4) in which "+" is replaced by "−". In view of the fact that the defect numbers of the operator C coincide and are equal to $m + n$, the reasoning used in step I is applicable in this case, and we conclude that C is essentially selfadjoint. But then $m + n = 0$, whence $m = n = 0$, i.e., A is also essentially selfadjoint.

IV. To prove necessity, we first establish a general lemma that, in our case, reflects the Holmgren principle in the theory of partial differential equations.

Lemma 8.2. *Consider equation (7.18) on $[0, b)$ $(b \in (0, \infty])$. Assume that there exists a set Φ dense in H such that the Cauchy problem*

$$\left(\frac{d^r \varphi}{dt^r}\right)(t) + (-1)^r B \varphi(t) = 0 \qquad (t \in [0, T]);$$

$$\varphi(T) = \varphi_0, \ \dots \ \varphi^{(r-1)}(T) = \varphi_{r-1} \tag{8.5}$$

has a strong solution for all $T \in (0, b)$ and $\varphi_0, \dots \varphi_{r-1} \in \Phi$.

Then the Cauchy problem for (7.8) is uniquely solvable on $[0, b)$ in the sense of strong solutions.

Proof. Let us prove Lemma 8.1, e.g., in the case of $r = 2$. One can easily verify the following formula of integration by parts: Let $[0, T] \ni t \mapsto \alpha(t),\ \beta(t) \in H$ be twice strongly continuously differentiable vector functions. Then

$$\int_0^T \left(\alpha''(t), \beta(t)\right)_H dt = \int_0^T \left(\alpha(t), \beta''(t)\right)_H dt + \left[\left(\alpha'(t), \beta(t)\right)_H - \left(\alpha(t), \beta'(t)\right)_H\right] \Big|_0^T. \quad (8.6)$$

Let $u(t)$ be a strong solution of the Cauchy problem for equation (7.18) with $r = 2$ on $[0, b)$ such that $u((0) = u'(0) = 0$ and let $\varphi(t)$ be a strong solution mentioned in the formulation of the lemma. By using (8.6), we obtain

$$\int_0^T \Big(\left(u''(t), \varphi(t)\right)_H - \left(u(t), \varphi''(t)\right)_H\Big) dt = \left(u'(T), \varphi_0\right)_H - \left(u(T), \varphi_1\right)_H. \quad (8.7)$$

Note that $\varphi(s) \in \mathcal{D}(B)$ for every $s \in [0, T]$. Therefore, according to equality (7.19) with $f = \varphi(s)$, we can write

$$\left(u''(t), \varphi(s)\right)_H + \left(u(t), B\varphi(s)\right)_H = 0 \qquad \left(t \in [0, b)\right).$$

We now set $t = s$ and then replace s by t. This yields

$$\left(u''(t), \varphi(t)\right)_H = -\left(u(t), B\varphi(t)\right)_H \qquad \left(t \in [0, T]\right).$$

By virtue of (8.5) with $r = 2$, we have

$$\left(u(t), \varphi''(t)\right)_H = -\left(u(t), B\varphi(t)\right)_H \qquad \left(t \in [0, T]\right).$$

These two equalities imply that the expression on the left-hand side of (8.7) vanishes. Consequently,

$$\left(u'(T), \varphi_0\right)_H - \left(u(T), \varphi_1\right)_H = 0 \qquad \left(\varphi_0, \varphi_1 \in \Phi\right).$$

Hence, it follows from the denseness of Φ in H that $u(T) = u'(T) = 0$. Since $T \in (0, b)$ is arbitrary, this yields the required assertion.

In the case where $r = 1$, the reasoning is similar, one should only use the following formula for integration by parts:

$$\int_0^T \left(\alpha'(t), \beta(t)\right)_H dt = -\int_0^T \left(\alpha(t), \beta'(t)\right)_H dt + \left[\left(\alpha(t), \beta(t)\right)_H\right] \Big|_0^T, \quad (8.8)$$

which holds for continuously differentiable vector functions $[0, T) \ni t \mapsto \alpha(t),\ \beta(t) \in H$. In the case where r is arbitrary, one must iterate relation (8.8) r times (note that relation (8.7) is, in fact, relation (8.8) iterated twice). □

V. Let us prove necessity. Let $\tilde{A}$ be selfadjoint and let E be its resolution of the identity. We apply Lemma 8.2, setting $r = 1$, $B = (iA)^* = -i\tilde{A}$, and $\Phi = \cup_{n=1}^{\infty} E((-n,n))H$. A strong solution of the Cauchy problem (8.5), which now has the form $\varphi'(t) + i\tilde{A}\varphi(t) = 0$ $(t \in [0,T])$, $\varphi(T) = \varphi_0$, exists and is equal to

$$\varphi(t) = \int_{\mathbb{R}} e^{-i\lambda(t-T)} dE(\lambda)\varphi_0 \qquad (t \in [0,T]) \tag{8.9}$$

(since $\varphi_0 \in \Phi$, the integration in (8.9) is, in fact, carried out over a finite interval and, therefore, the function $[0,T] \ni t \mapsto \varphi(t)$ is continuously differentiable; it is clear that it solves the problem under consideration). Thus, by virtue of this lemma, equation (8.1) with sign "+" is uniquely solvable on $[0,b)$. Equation (8.1) with sign "−" is investigated similarly. Finally, we conclude that $B = -(iA)^* = i\tilde{A}$. □

8.2 The Hyperbolic Criterion of Selfadjointness

The "hyperbolic" criterion of selfadjointness is formulated in the form of two theorems presented below.

Theorem 8.2. *Let A be an Hermitian operator acting on H. For its essential selfadjointness, it is necessary that the Cauchy problem for the equation*

$$\left(\frac{d^2u}{dt^2}\right)(t) + A^*u(t) = 0 \qquad (t \in [0,b)) \tag{8.10}$$

be uniquely solvable on $[0,b)$ for all $b \in (0,\infty]$ (in the sense of strong solutions) and it is sufficient that A be semibounded below and that the indicated Cauchy problem be uniquely solvable in the same sense for some $b > 0$.

Proof. *Sufficiency.* Suppose that $\tilde{A}$ is not selfadjoint. Then A has two different selfadjoint extensions A_1 and A_2 in H bounded below by a number $c > -\infty$ (cf. Section 12.7). Let E_1 and E_2 be the corresponding resolutions of the identity. For every $g \in \mathcal{D}(A) \subseteq \mathcal{D}(A_1)$, the integral $\int_{\mathbb{R}} \lambda^2 d(E_1(\lambda)g, g)_H$ is convergent and, therefore, the vector function

$$[0,\infty) \ni t \mapsto u_1(t) = \int_c^{\infty} \cos(\sqrt{\lambda}t) dE_1(\lambda)g \tag{8.11}$$

is twice strongly continuously differentiable. As in the proof of Theorem 8.1, one can easily show that it is a strong solution of equation (8.10) on $[0,\infty)$. For this purpose, one must check the validity of the corresponding weak equality of the form (7.19). In addition, we have, $u_1(0) = g$ and $u_1'(0) = 0$. Similarly, by changing E_1 by E_2 in (8.11), we construct the function $u_2(t)$. The difference $u(t) = u_1(t) - u_2(t)$ is also a strong solution of equation (8.10) on $[0,\infty)$ such that $u(0) = u'(0) = 0$.

In view of the assumed uniqueness of strong solutions of the Cauchy problem, $u(t) = 0$ for $t \geq 0$. Multiplying this equality scalarly by $h \in H$, we obtain

$$\int_c^\infty \cos(\sqrt{\lambda}t)d\Big((E_1(\lambda) - E_2(\lambda))g, h\Big)_H = 0 \quad (t \in [0,\infty)).$$

Since the charge ω is uniquely determined in terms of its cosine Fourier-Stieltjes transform (see [Shi1]), this enables us to conclude that $E_1 = E_2$, which is absurd.

Necessity. Let $\tilde{A}$ be selfadjoint and let E be its resolution of the identity. Let us apply Lemma 8.2, setting $r = 2$, $B = A^* = \tilde{A}$, and $\Phi = \cup_{n=1}^\infty E((-n,n))H$. A strong solution of the Cauchy problem (8.5), which now has the form $\varphi''(t) + \tilde{A}\varphi(t) = 0$ $(t \in [0,T])$, $\varphi(T) = \varphi_0$, $\varphi'(T) = \varphi_1$, exists and is equal to

$$\varphi(t) = \int_{\mathbb{R}} \cos\left(\sqrt{\lambda}(t-T)\right) dE(\lambda)\varphi_0 + \int_{\mathbb{R}} 1/\sqrt{\lambda} \sin\left(\sqrt{\lambda}(t-T)\right) dE(\lambda)\varphi_1$$

(here, as in (8.9), integration is, in fact, carried out over a finite segment). Therefore, according to Lemma 8.2, we conclude that the Cauchy problem for (8.10) is uniquely solvable on $[0,b)$ for all $b \in (0,\infty]$ in the sense of strong solutions. □

As a rule, it is convenient to use this theorem in a simple combination with Lemma 8.2. Let us formulate the corresponding result.

Theorem 8.3. *Let A be an Hermitian operator acting on H and semibounded below. Assume that there exists a linear set $\Phi \subseteq H$ dense in H and such that the Cauchy problem*

$$\left(\frac{d^2\varphi}{dt^2}\right)(t) + A\varphi(t) = 0 \qquad ((t \in [0,T]);\ \varphi(T) = \varphi_0,\ \ \varphi'(T) = \varphi_1) \tag{8.12}$$

has a strong solution for some $b > 0$ and all $T \in (0,b)$ and $\varphi_0, \varphi_1 \in \Phi$. Then the operator A is essentially selfadjoint.

Proof. By virtue of Lemma 8.2, it follows from the condition of the theorem that the Cauchy problem for equation (8.10) has a unique strong solution on $[0,b)$. But then, according to Theorem 8.2, the operator $\tilde{A}$ is selfadjoint. □

8.3 The Parabolic Criterion of Selfadjointness

Theorem 8.4. *Let A be an Hermitian operator acting on H. For its essential selfadjointness, it is necessary that the Cauchy problem for the equation*

$$\left(\frac{du}{dt}\right)(t) + A^*u(t) = 0 \qquad (t \in [0,\infty)) \tag{8.13}$$

be uniquely solvable in the sense of strong solutions. For an operator semibounded below, this is also a sufficient condition.

Proof. *Necessity.* As in Theorems 8.1 and 8.2, it is proved by using Lemma 8.2 with $r = 1$, $B = \tilde{A}$, and $\Phi = \cup_{n=1}^{\infty} E((-n, n))H$, where E is the resolution of the identity of $\tilde{A}$. A strong solution of the corresponding Cauchy problem which now has the form $\varphi'(t) - \tilde{A}\varphi(t) = 0$ $((t \in [0, T]), \varphi(T) = \varphi_0 \in \Phi)$, exists and is equal to

$$\varphi(t) = \int_{\mathbb{R}} e^{\lambda(t-T)} dE(\lambda)\varphi_0 \qquad (t \in [0, T]).$$

Sufficiency. It is established as in Theorems 8.1 and 8.2. Suppose that $\tilde{A}$ is not selfadjoint. Let A_1 and A_2 be two different selfadjoint extensions of A bounded below by a number $c > -\infty$ and let E_1 and E_2 be the corresponding resolutions of the identity. The vector function

$$[0, \infty) \ni t \mapsto u_1(t) = \int_c^{\infty} e^{-\lambda t} dE_1(\lambda) g \quad (g \in \mathcal{D}(A) \subseteq \mathcal{D}(A_1)) \tag{8.14}$$

is strongly continuously differentiable and $u_1(t) \in \mathcal{D}(A_1) \subseteq \mathcal{D}(A^*)$. The derivative $u_1'(t)$ is expressed by integral (8.14) with the factor $-\lambda$ before $e^{-\lambda t}$. The expression $A^* u_1(t) = A_1 u_1(t)$ also has the same form. Thus, (8.14) is a strong solution of equation (8.13) with $u_1(0) = g$. Further, by the same procedure, we construct $u_2(t)$ in terms of E_2 and consider the difference $u(t) = u_1(t) - u_2(t)$. For this difference, we have $u(0) = 0$ and, therefore, in view of the assumed uniqueness of strong solutions, $u(t) = 0$, whence $(\forall h \in H)$:

$$\int_c^{\infty} e^{-\lambda t} d\Big((E_1(\lambda) - E_2(\lambda)) g, h \Big)_H = 0 \qquad (t \in [0, \infty)). \tag{8.15}$$

This relation means that the Laplace-Stieltjes transform of the charge appearing in (8.15) is equal to zero, but then the charge is also identically equal to zero (see [Shi1]). This leads to the conclusion that $E_1 = E_2$, which is absurd. □

The evolutionary criteria of selfadjointness introduced above will be used in the next section and in Section 16.4.

9 Quasianalytic Criteria of Selfadjointness and Commutability

9.1 The Quasianalytic Criterion of Selfadjointness

First, we recall some facts from the theory of quasianalytic functions (see [Man]).

Let $[a, b] \subset \mathbb{R}$ be a finite segment and let $(m_n)_{n=1}^{\infty}$ be a fixed sequence of positive numbers.

The class $C\{m_n\}$ is defined as the linear set of all functions $f \in C^{\infty}([a, b])$ satisfying the estimates

$$|(D^n f)(t)| \le K_f^n m_n \qquad (t \in [a, b]; n \in \mathbb{N}), \tag{9.1}$$

where K_f is a constant that depends on f.

As is known, the class of analytic functions defined on $[a, b]$ is characterized by estimates (9.1) with $m_n = n!$. It is clear that the class $C\{n!\}$ is characterized by the following property: If $f \in C\{n!\}$ is such that $(D^n f)(t_0) = 0$ for all $n \in \mathbb{N}$ and $f(t_0) = 0$ at a fixed point $t_0 \in [a, b]$, then $f(t) = 0$ for $t \in [a, b]$. In order to generalize this situation, we introduce the following definition:

The class $C\{m_n\}$ is called quasianalytic if the fact that a function $f \in C\{m_n\}$ satisfies the equalities $(D^n f)(t_0) = 0$ $(n \in \mathbb{N})$ and $f(t_0) = 0$ at a fixed point $t_0 \in [a, b]$ implies that $f(t) = 0$ $(t \in [a, b])$.

We have the following Denjoy-Càrleman theorem: *The class $C\{m_n\}$ is quasianalytic if and only if*

$$\sum_{n=1}^{\infty} \left(\inf\left\{m_k^{1/k} | k \geq n\right\}\right)^{-1} = \infty. \tag{9.2}$$

For example, the class $C\{n^{pn}\}$ is quasianalytic $\Leftrightarrow p \leq 1$.

Let H be a Hilbert space and let A be an Hermitian operator in it.

A vector $\varphi \in H$ is called quasianalytic (with respect to A) if $\varphi \in \cap_{n=1}^{\infty} \mathcal{D}(A^n)$ and the class $C\{\|A^n \varphi\|_H\}$ is quasianalytic.

Lemma 9.1. *A vector $\varphi \in \cap_{n=1}^{\infty} \mathcal{D}(A^n)$ is quasianalytic if and only if*

$$\sum_{n=1}^{\infty} \|A^n \varphi\|_H^{-1/n} = \infty. \tag{9.3}$$

Proof. It is clear that $C\{\|A^n \varphi\|_H\} = C\{\|A^n (\lambda\varphi)\|_H\}$, where $\lambda > 0$ is fixed. This implies that it suffices to verify the lemma for a vector φ such that $\|\varphi\|_H = 1$. For a vector of this sort, the sequence

$$\left(\|A^n \varphi\|_H^{1/n}\right)_{n=1}^{\infty} \tag{9.4}$$

is nondecreasing. Indeed,

$$\|A\varphi\|_H^2 = (A\varphi, A\varphi)_H = (A^2\varphi, \varphi)_H \leq \|A^2\varphi\|_H \|\varphi\|_H,$$

i.e., $\|A\varphi\|_H \leq \|A^2\varphi\|_H^{1/2}$.

Assume that the inequality $\|A^n\varphi\|_H^{1/n} \leq \|A^{n+1}\varphi\|_H^{1/(n+1)}$ is already proved and prove that $\|A^{n+1}\varphi\|_H^{1/(n+1)} \leq \|A^{n+2}\varphi\|_H^{1/(n+2)}$ $(n \in \mathbb{N})$. In view of the assumed inequality, we get

$$\begin{aligned} \|A^{n+1}\varphi\|_H^2 &= \left(A^{n+1}\varphi, A^{n+1}\varphi\right)_H = \left(A^{n+2}\varphi, A^n\varphi\right)_H \\ &\leq \|A^{n+2}\varphi\|_H \, \|A^n\varphi\|_H \leq \|A^{n+2}\varphi\|_H \, \|A^{n+1}\varphi\|_H^{n/(n+1)}, \end{aligned}$$

whence $\|A^{n+1}\varphi\|_H^{1+1/(n+1)} \leq \|A^{n+2}\varphi\|_H$. Thus, (9.4) is a nondecreasing sequence.

Let us apply the Denjoy-Càrleman criterion to the class $C\{\|A^n\varphi\|_H\}$ ($\|\varphi\|_H = 1$). Since (9.4) is a nondecreasing sequence, we have

$$\inf\left\{\|A^k\varphi\|_H^{1/k} | k \geq n\right\} = \|A^n\varphi\|_H^{1/n}.$$

Therefore, condition (9.2) for the quasianalyticity of this class, i.e., for the quasianalyticity of the vector φ, can be rewritten in the form (9.3). □

Theorem 9.1. *Let A be a closed Hermitian operator acting on H. It is selfadjoint if and only if H contains a total set that consists of quasianalytic vectors.*

Proof. In one direction, this statement is trivial. Indeed, let A be selfadjoint. Then it suffices to prove the quasianalyticity of each vector φ of the form $\varphi = E((a,b))f$, where E is a resolution of the identity that corresponds to $A, a, b \in \mathbb{R}$ ($a < b$), and $f \in H$. It is obvious that $\varphi \in \bigcap_{n=1}^{\infty}\mathcal{D}(A^n)$. Further, we have

$$\|A^n\varphi\|_H^2 = \int_a^b \lambda^{2n} d(E(\lambda)f,f)_H \leq c^{2n}\|f\|_H^2 \quad (c = \max(|a|,|b|); n \in \mathbb{N}).$$

Therefore, series (9.3) is divergent and, according to Lemma 9.1, the vector φ is quasianalytic.

Suppose that A has a total set M of quasianalytic vectors φ. Since A is closed, it suffices to prove its essential selfadjointness or, according to Theorem 8.1, the uniqueness of strong solutions of the Cauchy problem for equations (8.1) if $b = \infty$. Let $u(t)$ be a strong solution of the problem

$$\left(\frac{du}{dt}\right)(t) - (\zeta A)^* u(t) = 0 \quad (t \in [0,\infty), u(0) = 0), \tag{9.5}$$

where $\zeta = \pm i$. It suffices to establish that $u(t) = 0$ for $t \in [0,T]$ for any $T > 0$.

The "weak" equality (7.19) for (9.5) with a quasianalytic vector $f = \varphi \in \bigcap_{n=1}^{\infty}\mathcal{D}(A^n)$ gives

$$\frac{d}{dt}(u(t),\varphi)_H = \left(\left(\frac{du}{dt}\right)(t),\varphi\right)_H = (u(t),(\zeta A)\varphi)_H \quad (t \in [0,T]).$$

But $(\zeta A)\varphi \in \bigcap_{n=1}^{\infty}\mathcal{D}(A^n)$ and, therefore,

$$\frac{d}{dt}(u(t),(\zeta A)\varphi)_H = (u(t),(\zeta A)^2\varphi)_H \quad (t \in [0,T]), \text{etc.}$$

This implies that $(u(t),\varphi)_H \in C^{\infty}([0,T])$ and

$$D^n(u(t),\varphi)_H = D^{n-1}(u(t),(\zeta A)\varphi)_H = \cdots = (u(t),(\zeta A)^n\varphi)_H \\ (t \in [0,T]; n \in \mathbb{Z}_+). \tag{9.6}$$

Since the values of $u(t)$ on $[0,T]$ are bounded, it follows from (9.6) that

$$\left|D^n \left(u(t),\varphi\right)_H\right| \le c\,\|(\zeta A)^n\varphi\|_H = c\|A^n\varphi\|_H \quad (t\in[0,T];n\in\mathbb{Z}_+),$$

i.e., the scalar function $[0,T]\ni t\mapsto f(t) = \left(u(t),\varphi\right)_H$ belongs to the class $C\{\|A^n\varphi\|_H\}$. Equalities (9.6) and $u(0)=0$ imply that $(D^n f)(0)=0$ $(n\in\mathbb{Z}_+)$. Therefore, by virtue of the quasianalyticity of $C\{\|A^n\varphi\|_H\}$, we get the equality $f(t) = \left(u(t),\varphi\right)_H = 0\big(t\in[0,T]\big)$. Since the set M of vectors φ is total, we have $u(t)=0$ $\big(t\in[0,T]\big)$. □

9.2 Other Criteria of Selfadjointness

The definition given below also proves to be useful. As above, let A be an Hermitian operator acting on a Hilbert space.

A vector $\varphi\in H$ is called analytic (with respect to A) if $\varphi\in\cap_{n=1}^{\infty}\mathcal{D}(A^n)$ and the power series

$$\sum_{n=0}^{\infty}\frac{\|A^n\varphi\|_H}{n!}z^n \tag{9.7}$$

has a nonzero radius of convergence. This vector is called entire if this radius is equal to infinity.

It is clear that each analytic vector is quasianalytic but the converse statements is not true. Note that, in proving the first part of Theorem 9.1, we established the following stronger fact:

If A is selfadjoint, then it possesses a total set of entire vectors.

Indeed, every vector $\varphi = E((a,b))f$ mentioned in this proof satisfies the estimate $\|A^n\varphi\|_H\le c^n\|\varphi\|_H$ $(n\in\mathbb{N})$ and, hence, is entire.

Let us establish a theorem that makes Theorem 9.1 more precise in the case of operators semibounded from below.

Let A be an Hermitian operator acting on H. *A vector $\varphi\in H$ is called a Stieltjes vector (with respect to A) if $\varphi\in\cap_{n=1}^{\infty}\mathcal{D}(A^n)$ and the class $C\left\{\|A^n\varphi\|_H^{1/2}\right\}$ is quasianalytic, or, in other words,*

$$\sum_{n=1}^{\infty}\|A^n\varphi\|_H^{-1/(2n)} = \infty \tag{9.8}$$

(the equivalence of the fact that a vector belongs to the class of Stieltjes vectors and condition (9.8) follows from the fact that the sequence $(\|A^n\varphi\|_H^{1/(2n)})_{n=1}^{\infty}$, where $\|\varphi\|_H=1$, is nondecreasing together with (9.4); to complete the proof, one should use the Denjoy-Càrleman criterion).

It is clear that every quasianalytic vector is a Stieltjes vector but not vice versa. Thus, if we denote the sets of all entire, analytic, quasianalytic, and Stieltjes vectors with respect to the operator A by $\mathcal{E}(A)$, $\mathcal{A}(A)$, $\mathcal{Q}(A)$, and $\mathcal{S}(A)$, respectively, then we get the inclusions

$$\mathcal{E}(A) \subseteq \mathcal{A}(A) \subseteq \mathcal{Q}(A) \subseteq \mathcal{S}(A). \tag{9.9}$$

Theorem 9.2. *Let A be a closed Hermitian operator semibounded below. If H contains a total set that consists of Stieltjes vectors, then A is selfadjoint.*

The converse statement is evident by virtue of the already proved assertions and (9.9).

Proof. According to Theorem 8.2, it suffices to prove the uniqueness of strong solutions of the Cauchy problem for equation (8.10) for $b = \infty$. Let $u(t)$ be a strong solution of this problem such that $u(0) = u'(0) = 0$. Let us show that $u(t) = 0$ $\big(t \in [0,T]\big)$ for all $T > 0$. Assume that M is a total set of Stieltjes vectors φ appearing in the condition of the theorem. We set $[0,T] \ni t \mapsto f(t) = \big(u(t), \varphi\big)_H$, where $\varphi \in M$. It follows from relation (7.19) written for (8.10) that $f \in C^2([0,T])$ and

$$\left(\frac{d^2 f}{dt^2}\right)(t) = -\big(u(t), A\varphi\big)_H \qquad \big(t \in [0,T]\big).$$

Since $A\varphi \in \cap_{n=1}^{\infty} \mathcal{D}(A^n)$, by the same reasoning, we conclude that

$$\big(u(t), A\varphi\big)_H \in C^2\big([0,T]\big), \quad \frac{d^2}{dt^2}\big(u(t), A\varphi\big)_H = -\big(u(t), A^2\varphi\big)_H \quad \big(t \in [0,T]\big), \text{etc.}$$

As a result, we get $f \in C^\infty\big([0,T]\big)$ and

$$\begin{aligned}(D^{2k} f)(t) = D^{2k}\big(u(t), \varphi\big)_H &= -D^{2(k-1)}\big(u(t), A\varphi\big)_H = \dots \\ &= (-1)^k\big(u(t), A^k\varphi\big)_H \quad \big(t \in [0,T]; k \in \mathbb{Z}_+\big).\end{aligned} \tag{9.10}$$

One can also deduce a similar equality for odd derivatives. Indeed, by differentiating (9.10), we obtain

$$(D^{2k+1} f)(t) = (-1)^k\big(u'(t), A^k\varphi\big)_H \quad \big(t \in [0,T]; k \in \mathbb{Z}_+\big). \tag{9.11}$$

The values of $u(t)$ and $u'(t)$ on $[0,T]$ are bounded; therefore, it follows from (9.10), (9.11), the fact that the operator is Hermitian, and the Cauchy-Buniakowski inequality that

$$\big|(D^{2k} f)(t)\big|, \big|(D^{2k+1} f)(t)\big| \le c\,\big\|A^k\varphi\big\|_H \le c\,\|\varphi\|_H^{1/2}\,\big\|A^{2k}\varphi\big\|^{1/2}$$

$$\big(t \in [0,T]; k \in \mathbb{Z}_+\big),$$

i.e., $f \in C\{m_n\}$, where $(m_n)_{n=1}^{\infty}$ is a sequence of numbers $\big(\|\varphi\|_H^{1/2}, \|\varphi\|_H^{1/2}, \|A\varphi\|_H^{1/2}, \|A\varphi\|_H^{1/2}, \|A^2\varphi\|_H^{1/2}, \|A^2\varphi\|_H^{1/2}, \dots\big)$. The class $C\{m_n\}$ will not change if we normalize φ. But then, as already mentioned, the sequence $(\|A^n\varphi\|_H^{1/(2n)})_{n=1}^{\infty}$ is nondecreasing. This means that the Denjoy-Càrleman condition (9.2) for the sequence under consideration can be written in the form (9.8). Hence, the class $C\{m_n\}$ is quasianalytic.

Furthermore, relations (9.10), (9.11), and the condition $u(0) = u'(0) = 0$ imply that $(D^n f)(0) = 0$ $(n \in \mathbb{Z}_+)$. Therefore, $\big(u(t), \varphi\big)_H = f(t) = 0$ $\big(t \in [0,T]\big)$. In view of the fact that M is total, we conclude that $u(t) = 0$ $\big(t \in [0,T]\big)$. □

9.3 Commutability of Operators

Here, we consider another application of the concept of quasianalytic vectors.

Let A_1 and A_2 be two bounded selfadjoint operators. In order that their resolutions of the identity E_1 and E_2 commute, it is necessary and sufficient that these operators be commuting, i.e., $A_1A_2 = A_2A_1$ (see Section 5). In the case of unbounded A_1 and A_2, their formal commutability, i.e., commutability in a certain dense set of vectors from H, does not imply the commutability of E_1 and E_2 (see Example 9.1). The following useful theorem establishes additional conditions that should be imposed on A_1 and A_2 to guarantee the commutability of E_1 and E_2.

Theorem 9.3. *Let A_1 and A_2 be two Hermitian operators acting on H with domains $\mathcal{D}(A_1)$ and $\mathcal{D}(A_2)$, respectively, and let $\mathcal{D}$ be a linear set such that $\mathcal{D} \subseteq \mathcal{D}(A_1) \cap \mathcal{D}(A_2)$. Assume that these operators commute in $\mathcal{D}$, i.e., $A_1\mathcal{D} \subseteq \mathcal{D}(A_2)$, $A_2\mathcal{D} \subseteq \mathcal{D}(A_1)$, and $A_1A_2 = A_2A_1 f$ $(f \in \mathcal{D})$.*

Furthermore, assume that A_1, A_2, and the restriction $A_1 \restriction \big((A_2 - z\mathbb{1})\mathcal{D}\big)$ (for some nonreal z) possess total sets of quasianalytic vectors. Then the operators A_1 and A_2 are essentially selfadjoint and their resolutions of the identity commute.

Proof. By virtue of Theorem 9.1, the operators $\tilde{A}_1$ and $\tilde{A}_2$ are selfadjoint. Let $R_z(\tilde{A}_1)$ and $R_z(\tilde{A}_2)$ be their resolvents. By virtue of Theorem 5.2, to prove the commutability of the resolutions of the identity for the operators $\tilde{A}_1$ and $\tilde{A}_2$, it suffices to verify the commutability of $R_z(\tilde{A}_1)$ and $R_z(\tilde{A}_2)$. For $f \in \mathcal{D}$, in view of the fact that A_1 and A_2 are commuting, we obtain

$$\begin{aligned} R_z(\tilde{A}_1)R_z(\tilde{A}_2)(A_1 - z\mathbb{1})(A_2 - z\mathbb{1})f &= R_z(\tilde{A}_1)R_z(\tilde{A}_2)(A_2 - z\mathbb{1})(A_1 - z\mathbb{1})f \\ &= f = R_z(\tilde{A}_2)R_z(\tilde{A}_1)(A_1 - z\mathbb{1})(A_2 - z\mathbb{1})f. \end{aligned}$$

Therefore, to prove that $R_z(\tilde{A}_1)$ commutes with $R_z(\tilde{A}_2)$, it suffices to show that $(A_1 - z\mathbb{1})(A_2 - z\mathbb{1})\mathcal{D}$ is dense in H. But this set coincides with the range $\mathcal{R}\Big(A_1 \restriction \big((A_2 - z\mathbb{1})\mathcal{D}\big) - z\mathbb{1}\Big)$, and the operator $A_1 \restriction \big((A_2 - z\mathbb{1})\mathcal{D}\big)$ is essentially selfadjoint by the same Theorem 9.1. Therefore, the indicated range is dense in H. □

Example

9.1. Let R be a Riemann surface of the function $\sqrt{z}$, i.e., the planes xOy with cuts along the semiaxis $(-\infty, 0)$ placed one over another and pasted along the cuts crosswise into two sheets R (the upper side of the cut in the upper sheet is pasted to the lower side of the lower sheet, and vice versa). Every point $p \in R$ has two coordinates x and y in the standard local coordinate system. Let $dxdy$ be the Lebesgue measure defined in each sheet and let $L_2 = L_2(R, dxdy)$. For any $t \in \mathbb{R}$, we introduce two operators

$$(U_t f)(x, y) = f(x - t, y) \quad \text{and} \quad (V_t f)(x, y) = f(x, y - t)$$

acting upon the functions f defined in R and belonging to L_2 (this definition is correct because the cuts are sets of measure zero). It is clear that the operator-valued functions $R \ni t \mapsto U_t, V_t$ form two groups of unitary operators in L_2. These groups are not commuting. Indeed, if f is, e.g., the indicator of a small neighbourhood of the point $(-1, 1)$ in the upper sheet, then $V_{-2}U_2 f$ differs from zero in a certain neighbourhood of the point $(1, - - 1)$ in the upper sheet and $U_2 V_{-2} f$ possesses the same property but in the lower sheet.

Denote by $\mathcal{D}$ the linear set of infinitely differentiable functions defined on R and vanishing in certain neighbourhoods of 0 and ∞ (in both sheets). The operators $\mathcal{D} \ni f \mapsto Af = i^{-1}\left(\frac{\partial f}{\partial x}\right)(x, y)$ and $\mathcal{D} \ni f \mapsto Bf = i^{-1}\left(\frac{\partial f}{\partial y}\right)(x, y)$ acting on L_2 are densely defined and Hermitian. Moreover, it is easy to show that they are essentially selfadjoint and $(\forall t \in \mathbb{R})$: $e^{it\tilde{A}} = U_t$, $e^{it\tilde{B}} = V_t$ (in this connection, see Section 16.4).

The operators A and B are formally commuting in the sense that $A\mathcal{D} \subseteq \mathcal{D}$, $B\mathcal{D} \subseteq \mathcal{D}$, and $ABf = BAf$ $(f \in \mathcal{D})$. At the same time, the resolutions of the identity of their closures $\tilde{A}$ and $\tilde{B}$ are not commuting. In fact, if they were commuting, then (clearly) their functions U_{t_1} and V_{t_2} would also commute but we have already seen that, for $t_1 = 2$ and $t_2 = 2$, this is not true.

10 Selfadjointness of Perturbed Operators

The results of the previous sections demonstrate that it is very important to know whether a given Hermitian operator is indeed selfadjoint. In this section, we present some useful facts that enable us to solve the problem of selfadjointness of the operator $A + B$ in the case where A is a selfadjoint operator and B is a Hermitian perturbation. First, we prove the following statement:

REMARK 10.1. Let A be a selfadjoint operator in a Hilbert space H and let B be a bounded selfadjoint operator. Then the operator $A + B$ is selfadjoint.

Indeed, for an arbitrary densely defined operator A and a bounded operator B, we have $(A + B)^* = A^* + B^*$ (Theorem 12.3.2). This equality implies the required assertion. □

However, in the case of an unbounded perturbation B, the situation becomes much more complicated, and only several separate results are available. Here, we prove one of the most widely applied theorems of this sort.

First, we introduce the following definition:

Consider operators A and B acting on H such that $\mathcal{D}(B) \supseteq \mathcal{D}(A)$. We say that B is *subordinated* to A if it satisfies the inequality

$$\|Bf\|_H \leq p\,\|Af\|_H + q\|f\|_H \qquad (f \in \mathcal{D}(A)) \tag{10.1}$$

with constants $p, q \geq 0$ (the constants of subordination). If, for any $p > 0$, there exists $q = q(p)$ such that inequality (10.1) is satisfied, then the operator B is called *infinitely small* as compared to A.

Theorem 10.1 (Rellich-Kato). *Consider a selfadjoint operator A and an Hermitian operator B acting on H such that $\mathcal{D}(B) \supseteq \mathcal{D}(A)$. If B is subordinated to A with a constant of subordination $p \in [0,1)$, then the operator $A+B$ $\big(\mathcal{D}(A+B) = \mathcal{D}(A)\big)$ is selfadjoint.*

Proof. According to Theorem 12.5.1, it suffices to prove that $\mathcal{R}(A+B-iy\mathbb{I}) = H$ and $\mathcal{R}(A+B+iy\mathbb{I}) = H$ for sufficiently large $y > 0$. Let us prove, e.g., the first of these relations. In other words, we shall prove that the equation

$$(A + B - iy\mathbb{I})f = g \qquad (f \in \mathcal{D}(A)) \tag{10.2}$$

is solvable for all $g \in H$ and sufficiently large $y > 0$.

Since A is selfadjoint, $(A - iy\mathbb{I})^{-1} \in \mathcal{L}(H)$ $(y > 0)$ and, therefore, we can transform (10.2) as follows:

$$\big(\mathbb{I} + B(A - iy\mathbb{I})^{-1}\big)(A - iy\mathbb{I})f = g \qquad (f \in \mathcal{D}(A)). \tag{10.3}$$

Denote $C(y) = B(A - iy\mathbb{I})^{-1}$. If we succeed in proving that $\|C(y)\|$ can be made less than one for sufficiently large $y > 0$, then the solvability of (10.3) will be established. Indeed, in this case, the operator $\big(\mathbb{I} + C(y)\big)^{-1}$ exists (see Section 8.3) and (10.3) can be rewritten in the form $(A - iy\mathbb{I})f = \big(\mathbb{I} + C(y)\big)^{-1}g$, but this equation is obviously solvable with respect to $f \in \mathcal{D}(A)$, since the operator A is selfadjoint and $\mathcal{R}(A - iy\mathbb{I}) = H$.

To estimate the norm $\|C(y)\|$, we must first establish several inequalities. Let $f \in \mathcal{D}(A)$. For $y > 0$, we have

$$\begin{aligned}
\|(A - iy\mathbb{I})f\|_H^2 &= \big((A - iy\mathbb{I})f, (A - iy\mathbb{I})f\big)_H \\
&= (Af, Af)_H + iy(Af, f)_H - iy(f, Af)_H + y^2(f,f)_H \\
&= \|Af\|_H^2 + y^2\,\|f\|_H^2 .
\end{aligned}$$

We set $(A - iy\mathbb{I})f = g$. Then the last equality can be rewritten in the form

$$\|g\|_H^2 = \big\|A(A - iy\mathbb{I})^{-1}g\big\|_H^2 + y^2\,\big\|(A - iy\mathbb{I})^{-1}g\big\|_H^2 \quad (g \in H),$$

whence we obtain the following two estimates:

$$\left\|A(A-iy\mathbb{I})^{-1}g\right\|_H \le \|g\|_H, \quad \text{and} \quad \left\|(A-iy\mathbb{I})^{-1}g\right\|_H \le \frac{1}{y}\|g\|_H \quad (g \in H). \tag{10.4}$$

By virtue of (10.1) and (10.4), we get (note that $(A-iy\mathbb{I})^{-1}f \in \mathcal{D}(A)$)

$$\begin{aligned} \|C(y)f\|_H &= \left\|B(A-iy\mathbb{I})^{-1}f\right\|_H \\ &\le p\left\|A(A-iy\mathbb{I})^{-1}\right\|_H + q\left\|(A-iy\mathbb{I})^{-1}f\right\|_H \\ &\le p\|f\|_H + \frac{q}{y}\|f\|_H = \left(p+\frac{q}{y}\right)\|f\|_H \quad (f \in H). \end{aligned} \tag{10.5}$$

Since $p \in [0,1)$, the expression $p+q/y$ can be made less than one for sufficiently large $y>0$. For these y, the required inequality $\|C(y)\| < 1$ is a simple consequence of (10.5). □

REMARK 10.2 (the Wüst theorem). Assume that the conditions of Theorem 10.1 are satisfied with $p=1$. Then the operator $A+B$ is essentially selfadjoint.

The proof is left to the reader. To prove this assertion, one should consider the family of selfadjoint operators $A+tB$, where $t \in [0,1)$, and then pass to the limit as $t \to 1$.

Let us explain that the operator $A+B$ may be not selfadjoint in this case. Indeed, if A is a selfadjoint unbounded operator (therefore, $\mathcal{D}(A) \neq H$), then one can set $B=-A$. All conditions of Remark 10.2 will be satisfied but $A+B=0$ only in $\mathcal{D}(A)$ but not in the whole H. At the same time, the closure of this operator is selfadjoint.

Chapter 14
Rigged Spaces

As shown in Chapter 11, the Sobolev-Schwartz theory of generalized functions deals, in fact, with the following chain of spaces:

$$\mathcal{D}'(\mathbb{R}^N) \supseteq L_2(\mathbb{R}^N) \supseteq \mathcal{D}(\mathbb{R}^N), \tag{0.1}$$

where $L_2(\mathbb{R}^N)$ is constructed according to the Lebesgue measure, $\mathcal{D}(\mathbb{R}^N)$ is the space of test functions that consists of finite infinitely differentiable functions, and $\mathcal{D}'(\mathbb{R}^N)$ is the space of generalized functions, i.e., antilinear continuous functionals defined in $\mathcal{D}(\mathbb{R}^N)$. The role of the space $L_2(\mathbb{R}^N)$ is reduced to the fact that the scalar product $(f, g)_{L_2(\mathbb{R}^N)}$ can be extended by continuity to the bilinear form that determines the action of a generalized function $\alpha \in \mathcal{D}'(\mathbb{R}^N)$ upon a test function $u \in \mathcal{D}'(\mathbb{R}^N)$; this bilinear form is denoted by $(\alpha, u)_{L_2(\mathbb{R}^N)}$. Note that $\mathcal{D}'(\mathbb{R}^N)$ is the closure of $L_2(\mathbb{R}^N)$ in a certain topology. We say that the space $L_2(\mathbb{R}^N)$ in chain (0.1) is rigged by the linear topological spaces $\mathcal{D}(\mathbb{R}^N)$ and $\mathcal{D}'(\mathbb{R}^N)$.

The method of investigation of generalized functions based on the construction of chain (0.1) is quite efficient. In this case, it is also useful to construct chains of the form (0.1) with the spaces $\mathcal{D}(\mathbb{R}^N)$ and $\mathcal{D}'(\mathbb{R}^N)$ replaced by Hilbert spaces. Such "Hilbert riggings" prove to be useful not only in the study of generalized functions but, e.g., in the theory of semibounded bilinear forms. In this chapter, we construct an abstract theory of rigged spaces. Its applications are discussed both in this chapter (Sections 6 and 8) and in Chapters 15 and 16.

1 Hilbert Riggings

1.1 Positive and Negative Norms

Let H_0 be a Hilbert space with scalar product $(\cdot,\cdot)_{H_0}$ and norm $\|\cdot\|_{H_0}$, and let f and g be its elements. Assume that a linear set H_+, which is also a Hilbert space with respect to a new scalar product $(\cdot,\cdot)_{H_+}$, is dense in H_0 and let $\|\cdot\|_{H_+}$ be a norm in H_+ such that

$$\| u \|_{H_0} \leq \| u \|_{H_+} \qquad (u \in H_+). \tag{1.1}$$

The elements of H_+ play the role of test functions and are denoted by $u, v, \ldots$.

Every element $f \in H_0$ generates an antilinear continuous functional l_f in H_+ by the formula

$$l_f(u) = (f, u)_{H_0} \qquad (u \in H_+); \tag{1.2}$$

its continuity follows from the estimate

$$|l_f(u)| = |(f, u)_{H_0}| \leq \|f\|_{H_0}\|u\|_{H_0} \leq \|f\|_{H_0}\|u\|_{H_+}.$$

We now introduce a new norm $\|\cdot\|_{H_-}$ in H_0 by taking the norm of the functional l_f that corresponds to an element f as the norm of this element, i.e.,

$$\|f\|_{H_-} = \|l_f\| = \sup\left\{\frac{|(f,u)_{H_0}|}{\|u\|_{H_+}}\middle| u \in H_+\right\}. \tag{1.3}$$

We must check that if $\|f\|_{H_-} = 0$, then $f = 0$; all the other properties of a norm are evident. If $\|f\|_{H_-} = 0$, then we have $(f,u)_{H_0} = 0$ for all $u \in H_+$. Since H_+ is dense in H_0, this yields $f = 0$.

By completing the space H_0 in the norm (1.3), we arrive at a linear normed space H_-, which is called a *space with negative norm*. Thus, we have constructed the chain of spaces

$$H_- \supseteq H_0 \supseteq H_+ \tag{1.4}$$

with negative, zero, and positive norm, respectively. Their elements are denoted by $\alpha, \beta, \cdots \in H_-$, $f, g, \cdots \in H_0$, and $u, v, \cdots \in H_+$. Sometimes they are called generalized, ordinary, and smooth vectors, respectively. We also say that (1.4) is the (Hilbert) rigging of the space H_0 by the spaces H_+ and H_-.

Since $H_0 \ni f \mapsto l_f \in (H_+)'$ is a one-to-one linear mapping, it is easy to show that H_- can be regarded as a subset of the dual space of antilinear functionals over H_+, namely, $H_- \subseteq (H_+)'$. Therefore, the expression $\alpha(u)$ is meaningful. As in the case of the form $(\alpha, u)_{L_2(\mathbb{R}^N)}$ in the Sobolev-Schwartz theory of generalized functions, we denote this expression by $(\alpha, u)_{H_0} = \overline{(u, \alpha)}_{H_0}$.

The bilinear form

$$H_- \times H_+ \ni \langle \alpha, u \rangle \mapsto (\alpha, u)_{H_0} \in \mathbb{C} \tag{1.5}$$

is an extension of the form $H_+ \times H_+ \ni \langle v, u \rangle \mapsto (v, u)_{H_0} \in \mathbb{C}$ to $H_- \times H_+$ by continuity. Clearly, the Cauchy-Buniakowski inequality admits the following generalization:

$$|(\alpha, u)_{H_0}| \le \|\alpha\|_{H_-} \|u\|_{H_+} \quad (\alpha \in H_-, u \in H_+). \tag{1.6}$$

Let us prove several simple but important facts.

Theorem 1.1. *The negative space H_- is a Hilbert space.*

We stress that the construction of a scalar product in H_- presented below is quite important.

Proof. Let us construct a scalar product in H_-. Consider the bilinear form

$$H_0 \times H_+ \ni \langle f, u \rangle \mapsto b(f, u) = (f, u)_{H_0} \in \mathbb{C}. \tag{1.7}$$

It is continuous. Indeed, $|b(f,u)| \le \|f\|_{H_0} \|u\|_{H_0} \le \|f\|_{H_0} \|u\|_{H_+}$. Therefore, it is representable in the form

$$b(f, u) = (f, Au)_{H_0} = (A^* f, u)_{H_+},$$

where $A\colon H_+ \to H_0$, $A^*\colon H_0 \to H_+$ are mutually adjoint continuous operators (see Section 8.5). According to (1.7), A is equal to the embedding operator $O\colon H_+ \to H_0$. Denote $I = O^*$. Thus,

$$(f,u)_{H_0} = (f,Ou)_{H_0} = (If,u)_{H_+} \quad (f \in H_0, u \in H_+), \quad I\colon H_0 \to H_+. \tag{1.8}$$

In H_0, we introduce a quasiscalar product

$$(f,g)_{H_-} = (If,Ig)_{H_+} = (f,Ig)_{H_0} = (If,g)_{H_0} \qquad (f,g \in H_0). \tag{1.9}$$

According to (1.3), (1.8), and (1.9), we have

$$\begin{aligned}\|f\|_{H_-} &= \sup\left\{\frac{|(f,u)_{H_0}|}{\|u\|_{H_+}} \;\Big|\; u \in H_+\right\} = \sup\left\{\frac{|(If,u)_{H_+}|}{\|u\|_{H_+}} \;\Big|\; u \in H_+\right\}\\ &= \|If\|_{H_+} = \sqrt{(f,f)_{H_-}} \qquad (f \in H_0).\end{aligned}$$

Since $\|\cdot\|_{H_-}$ is a norm in H_0, relation (1.9) actually determines not just a quasiscalar but a scalar product. As a result of completion, this scalar product becomes meaningful not only in H_0 but also in H_- and, therefore, the latter is transformed into a Hilbert space. □

Hence, H_- is now equipped with the scalar product

$$(\alpha,\beta)_{H_-}, \quad \|\alpha\|_{H_-} = \sqrt{(\alpha,\alpha)_{H_-}} \qquad (\alpha,\beta \in H_-).$$

Since $\|I\| = \|O\| = 1$, we have $\|f\|_{H_-} = \|If\|_{H_+} \le \|f\|_{H_0}$ $(f \in H_0)$.

The first equality in (1.9) indicates that I is an isometric operator acting from H_- into H_+ defined on a dense set in the space H_-. Its closure by continuity is an isometric operator $\mathbf{I}\colon H_- \to H_+$ acting from the whole H_- into H_+; $I = \mathbf{I} \restriction H_0$.

It is easy to see that $\mathbf{I}$ is *an isometry between the whole H_- and the whole H_+* (i.e., $\mathcal{R}(\mathbf{I}) = H_+$).

In fact, $\mathcal{R}(\mathbf{I})$ is dense in H_+: If $u \in H_+$ is such that $u \perp \mathcal{R}(\mathbf{I})$ in H_+, then, by virtue of (1.8), for any $f \in H_0$, we have

$$0 = (\mathbf{I}f,u)_{H_+} = (If,u)_{H_+} = (f,u)_{H_0},$$

whence we conclude that $u = 0$. Moreover, $\mathcal{R}(\mathbf{I})$ is closed in H_+.
Therefore, $\mathcal{R}(\mathbf{I}) = H_+$. □

Further, we have

$$(\alpha,u)_{H_0} = (\mathbf{I}\alpha,u)_{H_+} \qquad (\alpha \in H_-, u \in H_+). \tag{1.10}$$

Indeed, assume that $H_0 \ni f_n \to \alpha$ as $n \to \infty$ in H_-. Then, by virtue of (1.6), (1.8), and the continuity of $\mathbf{I}$, we obtain

$$(\alpha,u)_{H_0} = \lim_{n\to\infty}(f_n,u)_{H_0} = \lim_{n\to\infty}(\mathbf{I}f_n,u)_{H_+} = (\mathbf{I}\alpha,u)_{H_+}. \qquad \square$$

Theorem 1.2. *The equality*

$$H_- = (H_+)'$$

holds, i.e., the negative space can be regarded as the space of antilinear functionals dual to the positive space.

Proof. It suffices to show that every functional $l \in (H_+)'$ can be represented in the form $l(u) = (\alpha, u)_{H_0}$ $(u \in H_+)$ with some $\alpha \in H_-$. According to the Riesz theorem (Theorem 7.9.3), there exists $a \in H_+$ such that $l(u) = (a, u)_{H_+}$ $(u \in H_+)$. We set $\alpha = \mathbf{I}^{-1}a \in H_-$. Since $\mathcal{R}(\mathbf{I}) = H_+$, in view of (1.10), we obtain

$$l(u) = (a, u)_{H_+} = (\mathbf{I} \cdot \mathbf{I}^{-1}a, u)_{H_+} = (\mathbf{I}\alpha, u)_{H_+} = (\alpha, u)_{H_0} \quad (u \in H_+). \qquad \square$$

We stress that the identification of the dual space $(H_+)'$ with H_+ (the classical identification) and with H_- depends on the form of a functional $l \in (H_+)'$ $(\forall u \in H_+)$ — it can be written in the form $l\,(u) = (a, u)_{H_+}$ or in the form $l(u) = (\alpha, u)_{H_0}$ $(\alpha \in H_-, a \in H_+)$. In the second case, we say that the spaces H_+ and H_- are coupled by a form $(\cdot\,,\cdot)_{H_0}$ of the type (1.5).

Note that all results discussed above admit a natural generalization to the case of real spaces H_0, H_+, and H_-.

The following theorem gives a useful decomposition of the operator $\mathbf{I}$:

Theorem 1.3. *The isometry* $\mathbf{I}\colon H_- \to H_+$ *admits a decomposition into a product of two isometries, namely*

$$\mathbf{I} = J\mathbf{J}, \quad \textit{where } \mathbf{J}\colon H_- \to H_0, \quad J\colon H_0 \to H_+, \quad \textit{and } OJ = \mathbf{J} \restriction H_0. \tag{1.11}$$

Proof. Let $A = OI\colon H_0 \to H_0$. This operator is bounded and nonnegative. Indeed, according to (1.8), we have

$$(Af, f)_{H_0} = (OIf, f)_{H_0} = (If, If)_{H_+} \geq 0 \qquad (f \in H_0).$$

We set $B = \sqrt{A}\colon H_0 \to H_0$ and consider this operator as acting from a dense set H_0 in the space H_- into H_0. In this case, it is isometric,

$$(Bf, Bg)_{H_0} = (B^2 f, g)_{H_0} = (If, Ig)_{H_+} = (f, g)_{H_-} \quad (f, g \in H_0),$$

and its closure by continuity is an isometric operator $\mathbf{J}\colon H_- \to H_0$.

Let us prove that $\mathcal{R}(\mathbf{J}) = H_0$. It suffices to show that the fact that $f \perp \mathcal{R}(\mathbf{J})$ in H_0 implies that $f = 0$. For an arbitrary $g \in H_0$, we have

$$0 = (\mathbf{J}g, f)_{H_0} = (Bg, f)_{H_0} = (g, Bf)_{H_0},$$

whence $Bf = 0$. Then $OIf = B^2 f = 0$, i.e., $If = 0$ and, hence, $f = 0$.

We now show that $\mathcal{R}(B) \subseteq H_+$. It suffices to prove the equality

$$BJ = OI. \tag{1.12}$$

For $f \in H_0$, we have $B\mathbf{J}f = B^2 f = OIf = O\mathbf{I}f$. Since H_0 is dense in H_- and the operators $\mathbf{J}\colon H_- \to H_0$, $B\colon H_0 \to H_0$, and $O\mathbf{I}\colon H_- \to H_0$ are continuous, this yields (1.12).

Denote by J the operator B regarded as an operator from H_0 into H_+. In this case, (1.12) implies that $J\mathbf{J} = \mathbf{I}$ and it remains to prove that J is an isometric operator and $\mathcal{R}(J) = H_+$. But this follows from the equality $J = \mathbf{I}\mathbf{J}^{-1}$ because $\mathbf{J}^{-1}\colon H_0 \to H_-$ and $\mathbf{I}\colon H_- \to H_+$ are isometries. □

Let us write the most important relations connected with the rigging of the Hilbert space H_0 by the spaces H_+ and H_- constructed above. We have

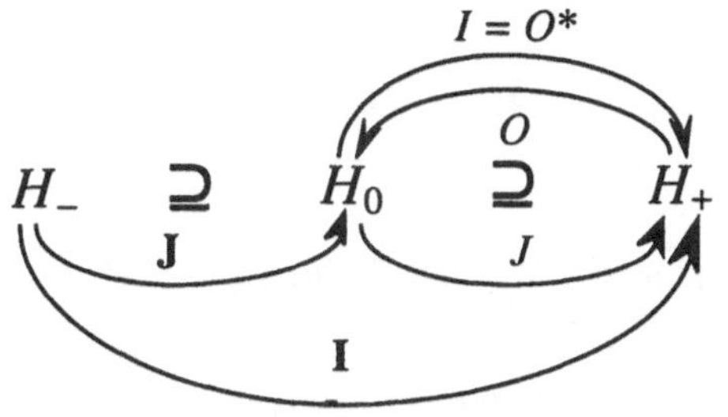

$$
\begin{gathered}
I = O^* \\
H_- \supseteq H_0 \supseteq H_+ \\
\|\cdot\|_{H_-} \le \|\cdot\|_{H_0} \le \|\cdot\|_{H_+}, \\
OJ = \mathbf{J} \restriction H_0, \mathbf{I} = J\mathbf{J}, \\
(\alpha, u)_{H_0} = (\mathbf{I}\alpha, u)_{H_+} = (\alpha, \mathbf{I}^{-1}u)_{H_-}, \\
(\mathbf{I}\alpha, \beta)_{H_0} = (\alpha, \mathbf{I}\beta)_{H_0}, (\mathbf{J}\alpha, f)_{H_0} = (\alpha, Jf)_{H_0} \\
(\alpha, \beta \in H_-; \quad f \in H_0; \quad u \in H_+)
\end{gathered} \tag{1.13}
$$

(the last two equalities in (1.13) are obtained by setting, respectively, $u = \mathbf{I}\beta$ or $u = Jf$ in the last equality but two).

Note that, instead of (1.1), one can require the validity of the inequality $\|u\|_{H_0} \le c\|u\|_{H_+}$ $(u \in H_+)$ with some $c > 0$; by a simple renormalization of H_+, this case is reduced to (1.1).

Let us introduce an important definition.

Rigging (1.4) is called quasinuclear if the embedding operator $O\colon H_+ \to H_0$ is quasinuclear, i.e., a Hilbert-Schmidt operator.

Example

1.1. Let us give a simple but useful example of a Hilbert rigging. Let $H_0 = L_2(R, d\mu(x)) = L_2(R, \mathfrak{R}, \mu)$, where R is a space with measure μ given on the σ-algebra $\mathfrak{R}$ of subsets of R and such that $\mu(R) \leq \infty$. We set $H_+ = L_2(R, p(x)d\mu(x))$, where $p(x) \geq 1$ ($x \in R$) is a measurable weight finite almost everywhere. Clearly, H_+ is dense in H_0 and inequality (1.1) is satisfied.

To construct H_-, we first find the operator I. Equality (1.8) now has the form

$$\int_R f(x)\overline{u(x)}d\mu(x) = \int_R (If)(x)\overline{u(x)}p(x)d\mu(x).$$

This enables us to conclude that

$$(If)(x) = p^{-1}(x)f(x) \quad \big(f \in L_2(R, d\mu(x)), u \in L_2(R, p(x)d\mu(x))\big). \tag{1.14}$$

Relation (1.9) now implies that $H_- = L_2(R, p^{-1}(x)d\mu(x))$. Therefore, chain (1.4) takes the form

$$L_2(R, p^{-1}(x)d\mu(x)) \supseteq L_2(R, d\mu(x)) \supseteq L_2(R, p(x)d\mu(x)). \tag{1.15}$$

The operator $\mathbf{I}$ is given by the same formula (1.14) but on the functions from $L_2(R, p^{-1}(x)d\mu(x))$, the operator $(Jf)(x) = p^{-1/2}(x)f(x)$ $\big(f \in L_2(R, d\mu)\big)$, and $\mathbf{J}$ is determined by the same formula but on $L_2(R, p^{-1}(x)d\mu(x))$.

Let us mention a particular case of chain (1.15) with $R = \mathbb{N}$ and $\mu(\{k\}) = 1$ ($k \in \mathbb{N}$). In this case, the chain takes the form

$$l_2(p^{-1}) \supseteq l_2 \supseteq l_2(p), \tag{1.16}$$

where the space l_2 with a weight $q = (q_k)_{k=1}^{\infty}$ ($q_k > 0$) is denoted by $l_2(q)$, i.e., $l_2(q) = \{f = (f_k)_{k=1}^{\infty}, f_k \in \mathbb{C} \mid \sum_{k=1}^{\infty} |f_k|^2 q_k < \infty\}$. Thus, the "smoothness" of the vector $u \in l_2(p)$ and the fact that $\alpha \in l_2(p^{-1})$ is a "generalized" vector manifest themselves only in the fact that the following series are convergent:

$$\sum_{k=1}^{\infty} |u_k|^2 p_k < \infty \quad \text{and} \quad \sum_{k=1}^{\infty} |\alpha_k|^2 p_k^{-1} < \infty.$$

Since $p_k \geq 1$ ($k \in \mathbb{N}$), smoothness means that a sequence of u_k exhibits a sufficiently rapid decrease as $k \to \infty$; for a generalized vector, the sequence of α_k may even increase as $k \to \infty$.

1.2 Operators in Chains

In this section, we present another procedure for constructing chains (1.4) and generalize the notion of adjointness.

Relations (1.13) imply that

$$(u,v)_{H_+} = (J^{-1}u, J^{-1}v)_{H_0} \qquad (u,v \in H_+),$$
$$(\alpha,\beta)_{H_-} = (\mathbf{J}\alpha, \mathbf{J}\beta)_{H_0} \qquad (\alpha,\beta \in H_-). \tag{1.17}$$

These equalities indicate that it is possible to construct riggings of the space H_0 in terms of a certain operator acting on this space and similar to J or J^{-1} (the latter is more convenient).

Consider a closed operator D acting on H_0 with a dense domain $\mathcal{D}(D)$ such that

$$\|Du\|_{H_0} \geq \|u\|_{H_0} \quad \bigl(u \in \mathcal{D}(D)\bigr). \tag{1.18}$$

By virtue of the closeness of D, the set $\mathcal{D}(D)$ is a Hilbert space with the scalar product

$$(u,v)_{H_+} = (Du, Dv)_{H_0} \quad \bigl(u,v \in \mathcal{D}(D)\bigr). \tag{1.19}$$

This set is regarded as a positive space H_+, and we construct the corresponding Hilbert space H_- dual to H_+ with respect to H_0 according to the procedure described above.

Let J be an isometric operator corresponding to the chain $H_- \supseteq H_0 \supseteq H_+$ just constructed and let $(J^{-1})_{H_0}$ be the operator $J^{-1}\colon H_+ \to H_0$ regarded as an operator in H_0 with the domain H_+. It follows from the proof of Theorem 1.3 that $(J^{-1})_{H_0} = B^{-1}$ and, therefore, $(J^{-1})_{H_0}$ is a positive selfadjoint operator in H_0. By comparing (1.17) with (1.19), we conclude that $\|(J^{-1})_{H_0}f\|_{H_0} = \|Df\|_{H_0}$ ($f \in H_+$), i.e., the operators $(J^{-1})_{H_0}$ and D are metrically equal. This immediately implies that (see Exercises 13.6.5 and 13.6.6)

$$\bigl(J^{-1}\bigr)_{H_0} = \sqrt{D^*D}. \tag{1.20}$$

Relation (1.20) for the operators D and J becomes much simpler if the operator D is selfadjoint. In this case $(J^{-1})_{H_0} = D$. It follows from (1.17) that $(f,g)_{H_-} = (D^{-1}f, D^{-1}g)_{H_0}$ $(f,g \in H_0)$. The negative space H_- is obtained as the completion of H_0 with respect to the last scalar product.

Let us generalize the concept of adjointness for continuous operators acting between the spaces of chain (1.4) ("adjointness with respect to H_0"). For example, let $A \in \mathcal{L}(H_+, H_-)$. Then the operator $A^+ \in \mathcal{L}(H_+, H_-)$ adjoint to A with respect to H_0 is defined by the equality

$$(Au, v)_{H_0} = (u, A^+v)_{H_0} \qquad (u,v \in H_+). \tag{1.21}$$

It is easy to see that the operator A^+ exists and can be expressed in terms of the ordinary adjoint operator $A^* \in \mathcal{L}(H_-, H_+)$. Thus, by using the relation $(\alpha, u)_{H_0} = (\alpha, \mathbf{I}^{-1}u)_{H_-}$ $(\alpha \in H_-, u \in H_+)$, which follows from (1.10), we obtain

$$(\forall u, v \in H_+)\colon (Au, v)_{H_0} = (Au, \mathbf{I}^{-1}v)_{H_-} = (u, A^*\mathbf{I}^{-1}v)_{H_+} = (u, \mathbf{I}^{-1}A^*\mathbf{I}^{-1}v)_{H_0},$$

i.e.,

$$A^+ = \mathbf{I}^{-1}A^*\mathbf{I}^{-1} \quad (A^*\colon H_- \to H_+). \tag{1.22}$$

The concept of adjointness with respect to H_0 can also be introduced for operators acting between the other spaces of chain (1.4). Thus, let $A \in \mathcal{L}(H_+, H_0)$. Then $A^+ \in \mathcal{L}(H_0, H_-)$ is defined by the following equality (similar to (1.21)):

$$(Au, f)_{H_0} = (u, A^+f)_{H_0} \qquad (u \in H_+, f \in H_0). \tag{1.23}$$

It is easy to see that the operator A^+ exists and satisfies the equality $A^+ = \mathbf{I}^{-1}A^*$, where $A^* \in \mathcal{L}(H_0, H_+)$. Indeed, by virtue of (1.10), we have

$$(Au, f)_{H_0} = (u, A^*f)_{H_+} = (u, \mathbf{I}^{-1}A^*f)_{H_0} \quad (u \in H_+, f \in H_0).$$

By using the definitions similar to (1.21) and (1.23), we can write $\mathbf{I}^+ = \mathbf{I}$, $\mathbf{J}^+ = J$, and $J^+ = \mathbf{J}$.

For operators $A \in \mathcal{L}(H_+, H_-)$, the concept of selfadjointness admits the following natural generalization:

We say that A is selfadjoint if $(u, Av)_{H_0} = (Au, v)_{H_0}$ for all $u, v \in H_+$, i.e., $A^+ = A$. Similarly, an operator $A \in \mathcal{L}(H_+, H_-)$ is called nonnegative if $(Au, u)_{H_0} \geq 0$ $(u \in H_+)$.

Nonnegative operators are certainly selfadjoint.

An ordinary bounded selfadjoint operator A in H_0 regarded as an operator acting from H_+ to $H_-$is obviously selfadjoint in the generalized sense. Generally speaking, an operator $A \in \mathcal{L}(H_+, H_-)$ is selfadjoint in the generalized sense if and only if the operator $\mathbf{I}A \in \mathcal{L}(H_+, H_+)$ is selfadjoint in H_+ (or $A\mathbf{I}^{-1}$ is selfadjoint in H_-). This follows from the equality (see (1.13))

$$(\mathbf{I}Au, v)_{H_+} = (Au, v)_{H_0} = (u, Av)_{H_0} = (u, \mathbf{I}Av)_{H_+} \qquad (u, v \in H_+).$$

The operator $A \in \mathcal{L}(H_+, H_-)$ is nonnegative if $\mathbf{I}A \in \mathcal{L}(H_+, H_+)$ is nonnegative in H_+.

Finally, note that the concept of adjointness $A \mapsto A^+$ introduced above is, in fact, a generalization of the construction of a differential expression $\mathcal{L}^+$ formally adjoint to an expression $\mathcal{L}$ (Section 12.2). Both in relation (12.2.7) and in relations (1.21) and (1.23), we "transfer" the action of an operator in a fixed scalar product $(\cdot, \cdot)_{H_0}$ (or $(\cdot, \cdot)_{L_2(G)}$ as in Chapter 12) paying, in fact, no attention to the spaces in which this operator acts.

2 Rigging of Hilbert Spaces by Linear Topological Spaces

We now proceed to the construction of a chain of the type (1.4), where the roles of H_+ and H_- are played by a linear topological space and its dual space, respectively (this construction covers the situation of chain (0.1)). First, we recall some facts from the theory of linear topological spaces used in our presentation (see also Section 6.2).

2.1 Topological Spaces

Let R be an abstract set. It becomes a topological space if we indicate a family (collection) Σ of its subsets $U, V, \cdots \subseteq R$ which are called *neighbourhoods*. Let $x \in R$ and let U be a neighbourhood from Σ which contains x. Then we say that U is a neighbourhood of the point x, and this is denoted by $U = U(x)$. The notion of a neighbourhood plays an important role in the theory if the family Σ satisfies the following two axioms:

(a) Every point $x \in R$ belongs to a certain neighbourhood and, moreover, for all $x, y \in R$ such that $x \neq y$, one can indicate disjoint neighbourhoods $U(x)$ and $U(y)$ of these points (this is the Hausdorff separation axiom; we consider only spaces satisfying this axiom).

(b) For every $x \in R$ and any two neighbourhoods $U(x)$ and $V(x)$, there exists a neighbourhood of this point $W(x)$ such that $W(x) \subseteq U(x) \cap V(x)$.

In topological spaces, one can introduce notions similar to well-known topological notions from analysis on the real axis.

Thus, for any $\alpha \subseteq R$, *its closure* $\tilde{\alpha}$ *is defined as the set that consists of all points* $x \in R$ *such that, for every* $U(x)$, *the intersection* $U(x) \cap \alpha$ *is nonempty.*

It is clear, that $\tilde{\alpha} \supseteq \alpha$.

A set $\alpha \subseteq R$ *is called closed if* $\tilde{\alpha} = \alpha$ *and open if its complement* $R \setminus \alpha$ *is closed.*

(1) *Every neighbourhood* $U \in \Sigma$ *is an open set*, i.e., its complement $R \setminus U$ is closed.

Indeed, it suffices to show that $(R \setminus U)^{\sim} \subseteq R \setminus U$. Assume the contrary, i.e., let $x \in (R \setminus U)^{\sim}$ and $x \notin R \setminus U$. But then $x \in U$, i.e., U is a neighbourhood of x which does not intersect with $R \setminus U$. This means that x does not belong to $(R \setminus U)^{\sim}$. □

(2) *A set* $\alpha \subseteq R$ *is open if and only if each point of this set belongs to* α *together with its neighbourhood.*

Indeed, assume that the conditions of this assertion are satisfied. Then $R \setminus \alpha$ is closed because if $x \in (R \setminus \alpha)^{\sim}$ and $x \notin R \setminus \alpha$, then $x \in \alpha$ and one can find a neighbourhood $U(x) \subseteq \alpha$. This means that $U(x) \cap (R \setminus \alpha) = \emptyset$, i.e., $x \notin (R \setminus \alpha)^{\sim}$. Conversely, assume that α is open, $x \in \alpha$ and, at the same time, $U(x) \cap (R \setminus \alpha)$ is nonempty for any $U(x)$. Consequently, $x \in (R \setminus \alpha)^{\sim}$, which is absurd. □

Other facts well known in the case of the real axis $R = \mathbb{R}$ with the ordinary topology are also established without difficulties. We do not dwell upon these properties and mention only the following fact: By fixing different families Σ of neighbourhoods in R, we arrive, generally speaking, at different topological spaces, and it is convenient to denote these spaces by R_Σ.

We say that the topologies in two spaces R_Σ and $R_{\Sigma'}$ of this sort coincide if, for any $\alpha \subseteq R$, the closures found by using Σ and Σ' coincide. In this case, the families Σ and Σ' are called equivalent and the spaces R_Σ and $R_{\Sigma'}$ are regarded as coinciding, i.e., $R_\Sigma = R_{\Sigma'}$.

It is interesting to clarify under what conditions the families Σ and Σ' are equivalent. The answer is given by the following assertion:

(3) *In order for Σ and Σ' to be equivalent, it is necessary and sufficient that, $(\forall x \in R)$ and $\big(\forall U(x) \in \Sigma\big)$ $\big(\exists U'(x) \in \Sigma'\big)$: $U'(x) \subseteq U(x)$ and the same relation hold with Σ replaced by Σ' and Σ' replaced by Σ.*

In fact, it is clear that if the condition of this assertion is satisfied, then, for any $\alpha \subset R$, its closures in these topologies coincide. We prove necessity. Let $x \in R$, $U(x)$ be a neighbourhood of a point x in the topology Σ. This neighbourhood is an open set in the topology Σ and, hence, in the topology Σ'. But then, according to (2), x must belong to $U(x)$ together with a certain neighbourhood in the topology Σ', i.e., there exists $U'(x) \subseteq U(x)$. □

Let us illustrate this assertion by a simple example: The same (ordinary) topology on the axis $R = \mathbb{R}$ can be defined by the families of neighbourhoods that consist of open intervals $U = (a, b)$ with rational ends, with irrational ends, and with arbitrary ends $a, b \in \mathbb{R}$.

Thus, let $R = R_\Sigma$ be a topological space with a topology generated by a family of neighbourhoods Σ. Denote by Σ' the collection of all open sets of this space; $\Sigma' \supseteq \Sigma$. It is not difficult to show that the family Σ' satisfies axioms (a) and (b) of topological space. Thus, (a) follows from the inclusion $\Sigma' \supseteq \Sigma$ and axiom (a) for Σ. Axiom (b) follows from the fact that the intersection of two open sets is open (this, in turn, follows from (2) and axiom (b) for the family Σ).

This enables us to topologize R by the family Σ', i.e., to regard arbitrary open sets as neighbourhoods. By virtue of (2) and (3), the families of neighbourhoods Σ and Σ' are equivalent. Therefore, we return to the original topological space, i.e., $R_{\Sigma'} = R_\Sigma = R$ (it is worth noting that, in the classical case $R = \mathbb{R}$, we have the well-known fact that any collection of open sets in the axis can be regarded as a system of neighbourhoods without changing the topology). To avoid confusion, in what follows, the neighbourhoods from Σ in the topological space $R = R_\Sigma$ are called *base neighbourhoods* and the family Σ is called the *base of neighbourhoods* or a system of base neighbourhoods.

In topological spaces, one can introduce the notion of convergence. Namely, a sequence $(x_n)_{n=1}^\infty$ of points $x_n \in R$ is called convergent to a point $x \in R$ if, for any $U(x) \in \Sigma$, there exists $N = N(U(x))$ such that $x_n \in U(x)$ whenever

$n > N$. It is well known that, in the classical case of the real axis ($R = \mathbb{R}$), all topological notions can be constructed from the notion of convergence. However, in the general case, this is not true. Indeed, the same set R can be equipped with two different topologies generated by nonequivalent families of neighbourhoods Σ and Σ' and such that the convergence of an arbitrary sequence in one topology implies its convergence in the second topology, and vice versa. In other words, the corresponding supply of convergent sequences does not determine the topology uniquely. This is not true if the families Σ and Σ' are countable (or one can find countable families of base neighbourhoods equivalent to Σ and Σ'). We leave the proof of this assertion to the reader.

In conclusion, we note that if $R_1 = R_{1,\Sigma_1}$ and $R_2 = R_{2,\Sigma_2}$ are topological spaces and $R = R_1 \times R_2$ is their direct product, which consists of the points $x = \langle x_1, x_2 \rangle$, where $x_1 \in R_1$ and $x_2 \in R_2$, then this direct product can always be topologized by the family Σ that consists of all rectangles $U = U_1 \times U_2$ such that $U_1 \in \Sigma_1$ and $U_2 \in \Sigma_2$ (it is easy to show that Σ satisfies axioms (a) and (b) of topological space).

2.2 Projective Limits of Spaces

We now introduce the notion of a linear topological space. Let Φ be a linear space over the field $\mathbb{C}$ of complex numbers (for definiteness), which is, at the same time, a topologic space with a base of neighbourhoods $\Sigma = \{U(\varphi) | \varphi \in \Phi\}$.

If the linear structure of Φ is consistent, in a certain sense, with the topology, then Φ is called a linear topological space.

Consistence is understood in the sense of the validity of the following axiom:

(c) *The mappings $\Phi \times \Phi \ni \langle \varphi, \psi \rangle \mapsto \varphi + \psi \in \Phi$ and $\mathbb{C} \times \Phi \ni \langle \lambda, \varphi \rangle \mapsto \lambda\varphi \in \Phi$ are continuous. In other words, the operations of addition and multiplication by a scalar are continuous.*

It is clear that the continuity of these mapping should be understood as follows: For the first mapping, for any $\varphi, \psi \in \Phi$ and $U(\varphi + \psi)$, one can indicate $U(\varphi)$ and $U(\psi)$ such that $U(\varphi) + U(\psi) \subseteq U(\varphi + \psi)$; for the second mapping, for any $\lambda \in \mathbb{C}$, $\varphi \in \Phi$, and $U(\lambda\varphi)$, one can find $U(\lambda)$ and $U(\varphi)$ such that $U(\lambda)U(\varphi) \subseteq U(\lambda\varphi)$ (here, $U(\lambda)$ denotes a neighbourhood of a point λ in the ordinary topology of $\mathbb{C}$; $(\forall \alpha, \beta \subseteq \Phi)\colon \alpha + \beta \overset{\text{def}}{=} \{\varphi + \psi \mid \varphi \in \alpha, \psi \in \beta\}$; $(\forall \alpha \subseteq \mathbb{C})(\forall \beta \subseteq \Phi)\colon \alpha\beta = \{\lambda\varphi | \lambda \in \alpha, \varphi \in \beta\}$).

Thus, the base neighbourhoods of a linear topological space Φ must satisfy axioms (a)–(c). Due to the fact that Φ is equipped with a linear structure, it is often convenient to introduce first a family Σ_0 of subsets $U, V, \ldots$ of Φ, which are regarded as neighbourhoods of the origin, then construct the sets

$$\{\varphi \in \Phi | \ \varphi - \varphi_1 \in U\} \qquad (U \in \Sigma_0, \varphi_1 \in \Phi) \tag{2.1}$$

shifted by vectors $\varphi_1 \in \Phi$, and, finally, take all possible sets of the form (2.1) as a base of neighbourhoods Σ (as in the case of a linear normed space, where we first consider open spheres centered at the origin and then realize all possible shifts of these spheres). Axioms (a)–(c) can be easily reformulated in terms of the family Σ_0 and we do not dwell upon this case.

Clearly, a linear normed space is an example of a linear topological space with open spheres with arbitrary centers and radii regarded as base neighbourhoods (axioms (a)–(c) were, in fact, verified in Section 6.3). Here, we do not study general properties of linear topological spaces and proceed to the subclass of these spaces most important for our presentation, namely, to projective limits of Banach spaces (and, in particular, of Hilbert spaces).

Consider a family of Banach spaces $(B_\tau)_{\tau\in T}$ parametrized by the elements of an arbitrary indexing set T. Assume that the set $\Phi = \cap_{\tau\in T} B_\tau$ is dense in each B_τ and the family $(B_\tau)_{\tau\in T}$ is directed by embedding, i.e.,

$$(\forall\tau', \tau'' \in T)(\exists\tau''' \in T)\colon B_{\tau'''} \subseteq B_{\tau'}, \quad \text{and} \quad B_{\tau'''} \subseteq B_{\tau''}, \tag{2.2}$$

where the embeddings are dense and continuous. In Φ, we introduce the projective topology with respect to the families of Banach spaces $(B_\tau)_{\tau\in T}$ and natural embeddings $O_\tau\colon \Phi \to B_\tau$. By definition, the family of all possible sets

$$U(\varphi_1;\tau;\varepsilon) = \{\varphi \in \Phi \mid \|\varphi - \varphi_1\|_{B_\tau} < \varepsilon\} \quad (\varphi_1 \in \Phi,\ \tau \in T,\ \varepsilon > 0). \tag{2.3}$$

is a system Σ of base neighbourhoods in this topology.

Thus, the base neighbourhoods (2.3) are intersections of the set Φ with open spheres in the spaces B_τ (with arbitrary index τ, radii, and centers). System (2.3) is obtained according to (2.1) by the shifts of the corresponding spheres centered at the origin by $\varphi_1 \in \Phi$.

System (2.3) satisfies axioms (a)–(c) of a linear topological space.

In fact, axiom (a) is evident. Indeed, let $\varphi, \psi \in \Phi$, $\varphi \neq \psi$; we fix $\tau \in T$ and consider two disjoint spheres in the space B_τ centered at the points φ and ψ; their intersections with Φ are the required separating neighbourhoods from Σ. Thus, axiom (a) is satisfied.

Let us verify axiom (b). Assume that $\varphi \in U(\varphi_1;\tau_1;\varepsilon_1) \cap U(\varphi_2;\tau_2;\varepsilon_2)$ with some $\varphi_j \in \Phi, \tau_j \in T$, and $\varepsilon_j > 0 (j = 1,2)$. According to (2.2), we find $\tau_3 \in T$ for which the embeddings $B_{\tau_3} \subseteq B_{\tau_1}$ and $B_{\tau_3} \subseteq B_{\tau_2}$ are dense and continuous. This implies that there exist $c_1, c_2 > 0$ such that $\|\chi\|_{B_{\tau_1}} \leq c_1\|\chi\|_{B_{\tau_3}}$ and $\|\chi\|_{B_{\tau_2}} \leq c_2\|\chi\|_{B_{\tau_3}}$ $(\chi \in B_\tau)$. Therefore, one can find a sufficiently small $\varepsilon > 0$ such that the following embedding of spheres takes place:

$$\begin{aligned}&\{\chi \in B_{\tau_3} \mid \|\chi - \varphi\|_{B_{\tau_3}} < \varepsilon\} \\ &\qquad \subseteq \{\chi \in B_{\tau_3} \mid \|\chi - \varphi_1\|_{B_{\tau_1}} < \varepsilon_1\} \cap \{\chi \in B_{\tau_3} \mid \|\chi - \varphi_2\|_{B_{\tau_2}} < \varepsilon_2\}\end{aligned} \tag{2.4}$$

Taking the intersection of inclusion (2.4) with Φ, we conclude that $U(\varphi;\tau_3;\varepsilon) \subseteq U(\varphi_1;\tau_1;\varepsilon_1) \cap U(\varphi_2;\tau_2;\varepsilon_2)$. The neighbourhood $U(\varphi,\tau_3,\varepsilon) \in \Sigma$ of the point φ is just the required one because it belongs to the intersection of given neighbourhoods.

Let us check axiom (c) for the first mapping. Let $U(\varphi_1; \tau, \varepsilon)$ be a neighbourhood of the point $\varphi + \psi \in \Phi$. The sphere $\alpha = \{\chi \in B_\tau \mid \|\chi - \varphi_1\|_{B_\tau} < \varepsilon\}$ is a neighbourhood of the vector $\varphi + \psi$ in the Banach space B_τ. By virtue of the continuity of the operator of addition in this space, one can find

$$\beta = \{\chi \in B_\tau \mid \|\chi - \varphi\|_{B_\tau} < \delta\} \quad \text{and} \quad \gamma = \{\chi \in B_\tau \mid \|\chi - \psi\|_{B_\tau} < \delta\}$$

such that $\beta + \gamma \subseteq \alpha$. Taking the intersection of this inclusion with Φ, we obtain the required neighbourhoods from the system Σ in Φ, i.e., $U(\varphi; \tau; \delta) + U(\psi; \tau; \delta) \subseteq U(\varphi_1; \tau; \varepsilon)$. For the second mapping, axiom (c) can be verified similarly. □

The space $\Phi = \cap_{\tau \in T} B_\tau$ *equipped with the topology generated by the base neighbourhoods (2.3) is called the projective limit of the Banach spaces* B_τ *and denoted by*

$$\Phi = \operatorname{pr} \lim_{\tau \in T} B_\tau. \tag{2.5}$$

The projective limit introduced above is sometimes called *reduced* — this stresses the fact that Φ is dense in each B_τ (here, we do not consider the generalizations of limit (2.5) for the case where there are no natural embeddings $O_\tau : \Phi \to B_\tau$ and one should add to the construction a system of operators O_τ ($\tau \in T$)). Note that the convergence of a sequence $(\varphi_n)_{n=1}^\infty$ of vectors $\varphi_n \in \Phi$ to $\varphi \in \Phi$ in terms of space (2.5) means that $\|\varphi_n - \varphi\|_{B_\tau} \to 0$ as $n \to \infty$ for any $\tau \in T$.

If T *is countable, then the projective limit (2.5) is called a countably normed space.*

In the case where $T = \mathbb{Z}_+$, it is also denoted by $\Phi = \operatorname{pr} \lim_{n \to \infty} B_n$. In this case, we often encounter a situation where the norms of the spaces B_n are monotone, i.e.,

$$B_0 \supseteq B_1 \supseteq \ldots, \|\varphi\|_{B_0} \leq \|\varphi\|_{B_1} \leq \ldots \quad \left(\varphi \in \Phi = \cap_{n=0}^\infty B_n\right) \tag{2.6}$$

and the condition of directedness by embedding is clearly satisfied. However, it is possible to show that the general case of countably normed spaces can be reduced to the monotone case (2.6) by a proper renormalization of $B_\tau (\tau \in T)$.

If $B_\tau = H_\tau$ *are Hilbert spaces, then (2.5) is called a projective limit of Hilbert spaces. If* T *is countable,* Φ *is called a countably Hilbert space.*

In the class of projective limits of Hilbert spaces, one can select an important subclass of nuclear spaces (nuclear spaces are extensively used in Chapter 15).

The projective limit of Hilbert spaces $\Phi = \operatorname{pr} \lim_{\tau \in T} H_\tau$ *is called nuclear if, for any* $\tau \in T$, *one can find* $\tau' \in T$ *such that* $H_{\tau'} \subseteq H_\tau$ *and the embedding operator* $H_{\tau'} \to H_\tau$ *is quasinuclear.*

2.3 Riggings Constructed by Using Projective Limits

Let Φ be some linear topological space over the field $\mathbb{C}$.

An antilinear continuous functional l over Φ is defined as a continuous mapping $\Phi \ni \varphi \mapsto l(\varphi) \in \mathbb{C}$ satisfying the condition of antilinearity $l(\lambda\varphi + \mu\psi) = \bar{\lambda}l(\varphi) + \bar{\mu}l(\psi)$ $(\varphi, \psi \in \Phi; \lambda, \mu \in \mathbb{C})$. The collection of all functionals of this sort (for fixed Φ) obviously forms a linear space over $\mathbb{C}$ with the ordinary addition of functions and multiplication by scalars.

As above, this space is called dual to Φ and denoted by Φ'. (Note that, in view of the constructions of Section 1, it is now convenient to consider antilinear functionals instead of linear ones).

There are many methods for introducing a topology in the dual space Φ'. Here, we study only the case of the weak topology.

Weak convergence in Φ' is introduced in a standard way, i.e., a sequence $(l_n)_{n=1}^{\infty}$ of functionals $l_n \in \Phi'$ is called weakly convergent to $l \in \Phi'$ if $l_n(\varphi) \to l(\varphi)$ as $n \to \infty$ for any $\varphi \in \Phi$.

The weak topology in Φ' is generated by the system of base neighbourhoods in Φ' of the form

$$U(l_1; \varphi_1, \ldots, \varphi_n) = \{l \in \Phi' \mid |l(\varphi_k) - l_1(\varphi_k)| < 1; \;\; k = 1, \ldots, n\} \qquad (2.7)$$
$$\big(l_1 \in \Phi'; \varphi_k \in \Phi, k = 1, \ldots, n; n \in \mathbb{N}\big);$$

convergence with respect to this topology is equivalent to weak convergence introduced above.

The fact that Φ' equipped with the system of base neighbourhoods (2.7) satisfies axioms (a)–(c) of a linear topological space can be verified by repeating the reasoning used in the same situation for Banach spaces (Section 7.8). Note that, in this case, the family of neighbourhoods (2.7) can also be obtained as a result of shifts of neighbourhoods of the origin of type (2.1) .

The following simple but important theorem clarifies the structure of Φ' in the case where Φ is a projective limit:

Theorem 2.1 (Schwartz). *Assume that Φ is a projective limit of Banach spaces, i.e., $\Phi = \operatorname{pr\,lim}_{\tau \in T} B_\tau$. Then*

$$\Phi' = \cup_{\tau \in T} B'_\tau, \qquad (2.8)$$

and this equality must be interpreted as follows: For all $l \in \Phi'$, there exists $\tau \in T$ such that l can be extended by continuity from Φ to B_τ to an element of B'_τ and, vice versa, if $l \in B'_\tau$ for some $\tau \in T$, then $l \restriction \Phi \in \Phi'$.

Proof. Let $l \in B'_\tau$. Then $l \restriction \Phi$ is an antilinear functional on Φ. The fact that l is continuous at the origin in the topology of the space B_τ implies that, for any $\varepsilon > 0$, one can find $U(0; \tau; \delta)$ such that $|l(\varphi)| < \varepsilon$ for $\varphi \in U(0; \tau; \delta)$ and, hence, $l \restriction \Phi$ is continuous in the topology of Φ, i.e., $l \restriction \Phi \in \Phi'$.

Conversely, let $l \in \Phi'$. It follows from the continuity of l at the origin that, for $\varepsilon = 1$, one can indicate a base neighbourhood of the origin in Φ of the form $U(0;\tau;\delta)$ $(\tau \in T, \delta > 0)$ such that $|l(\varphi)| < 1$ for $\varphi \in U(0;\tau;\delta)$. Consider the space B_τ. The antilinear functional l is defined on the linear set Φ dense in B_τ and is bounded by one in the intersection of Φ with a sphere of the form $\{\chi \in B_\tau \mid \|\chi\|_{B_\tau} < \delta\}$ in the space B_τ. Thus, this functional is continuous in Φ equipped with the norm of the space B_τ. Extending this functional by continuity to the whole B_τ, we obtain the required functional from B'_τ. □

Decomposition (2.8) enables one to introduce in Φ', parallel with the weak topology, so called *topology of the inductive limit of spaces*. Thus, relation (2.2) yields the following property of directedness of the family $(B'_\tau)_{\tau\in T}$:

$$(\forall \tau', \tau'' \in T)\ (\exists \tau''' \in T)\colon B'_{\tau'''} \supseteq B'_{\tau'}, \quad \text{and} \quad B'_{\tau'''} \supseteq B'_{\tau''}, \tag{2.9}$$

where the embeddings are dense and continuous. By using (2.9), one can easily prove that the collection of all possible sets of the form

$$U\big(l_1;\varepsilon(\cdot)\big) = \cup_{\tau\in T}\left\{l \in \Phi' \mid l - l_1 \in B'_\tau, \quad \|l - l_1\|_{B'_\tau} < \varepsilon(\tau)\right\} \tag{2.10}$$
$$\big(l_1 \in \Phi'; T \ni \tau \mapsto \varepsilon(\tau) > 0\big)$$

forms a system of base neighbourhoods in Φ'. In other words, for $l_1 = 0$, all possible unions of open spheres with positive radii centered at the origin form systems of base neighbourhoods at the origin in the spaces B'_τ $(\tau \in T)$. Then these neighbourhoods are shifted by arbitrary $l_1 \in \Phi'$.

Note that, in fact, we have given the general definition of an *inductive limit of Banach spaces*. Thus, let $(B_\tau)_{\tau\in T}$ be a given family of Banach spaces with the following property of directedness of the type (2.9): For any $\tau', \tau'' \in T$, there exists $\tau''' \in T$ such that $B_{\tau'''} \supseteq B_{\tau'}$ and $B_{\tau'''} \supseteq B_{\tau''}$.

In this case, the union $\cup_{\tau\in T} B_\tau = \Psi$ can be equipped with a natural linear structure. Indeed, let

$$\varphi, \psi \in \cup_{\tau\in T} B_\tau \Rightarrow (\exists \tau', \tau'' \in T)\colon \varphi \in B_{\tau'}, \psi \in B_{\tau'} \Rightarrow (\exists \tau''' \in T)\colon \varphi, \psi \in B_{\tau'''}.$$

Therefore, the expression $\lambda\varphi + \mu\psi$ is meaningful as a vector in $B_{\tau'''}$ and, hence, in Ψ.

A linear space Ψ is called the inductive limit of the spaces B_τ (notation: $\Psi = \operatorname{ind}\lim_{\tau\in T} B_\tau$) if it is equipped with a system of base neighbourhoods of the form (2.10), where Φ' and B'_τ are replaced by Ψ and B_τ, respectively.

After a brief survey of simple properties of linear topological spaces, we now proceed to the construction of the rigging of a Hilbert space by linear topological spaces.

Let H_0 be a Hilbert space with elements $f, g, \ldots$ and let Φ be a linear topological space with elements $\varphi, \psi, \ldots$ densely and continuously embedded in H_0. As in Section 1, every element $f \in H_0$ generates an antilinear continuous functional in Φ by the formula $l_f(\varphi) = (f, \varphi)_{H_0} (\varphi \in \Phi)$. By identifying f with l_f (this is possible in view of the fact that Φ is dense in H_0), we arrive at the embedding of H_0 in the space Φ' of antilinear continuous functionals on Φ. It is clear that if Φ' is equipped with the weak topology, then the embedding $H_0 \to \Phi'$ is continuous. Thus, we have constructed a chain that generalizes (1.4), namely,

$$\Phi' \supseteq H_0 \supseteq \Phi. \tag{2.11}$$

We also say that (2.11) is the rigging of the space H_0 by the spaces Φ and Φ' or, in other words, that we have defined the pairing of the spaces Φ and Φ' by the space H_0. If Φ is a nuclear space, then rigging (2.11) is called nuclear.

In what follows, we consider only riggings of the form (2.11) with Φ defined as a projective limit of Hilbert spaces, i.e.,

$$\Phi = \operatorname{pr} \lim_{\tau \in T} H_\tau. \tag{2.12}$$

In addition, we suppose that each H_τ is densely and continuously embedded in H_0 and, moreover, $\|\varphi\|_{H_0} \leq \|\varphi\|_{H_\tau}$ $(\varphi \in H_\tau, \tau \in T)$, where T is assumed to contain the index 0 (in fact, this is not a restriction because any general situation can always be reduced to this case).

Thus, for any $\tau \in T$, one can construct a chain of the form (1.4)

$$H_{-\tau} \supseteq H_0 \supseteq H_\tau, \tag{2.13}$$

where H_τ is a positive space and $H_{-\tau}$ is the corresponding negative space. Since $H_{-\tau} = (H_\tau)'$ (see Theorem 1.2), equality (2.8) implies that

$$\Phi' = \cup_{\tau \in T} H_{-\tau}. \tag{2.14}$$

This construction takes a fairly simple form in the case where Φ is a countably Hilbert space and the norms of the spaces are monotone (see (2.6)). In this case, we arrive at the following chain:

$$\begin{aligned} \Phi' = \cup_{n=1}^{\infty} H_{-n} \supseteq \cdots \supseteq H_{-2} \supseteq H_{-1} \supseteq H_0 \\ \supseteq H_1 \supseteq H_2 \supseteq \cdots \supseteq \cup_{n=1}^{\infty} H_n = \Phi, \end{aligned} \tag{2.15}$$

where each space H_{m+1} is densely embedded in H_m and

$$\|\varphi\|_{H_m} \leq \|\varphi\|_{H_{m+1}} \qquad (\varphi \in H_{m+1}; m \in \mathbb{Z}).$$

Note that all results of this section remain true for real spaces.

Let us give a simple but important example of a projective limit of Hilbert spaces and construct the corresponding rigging.

Example

2.1. Let $H_0 = l_2$ be the space of square summable sequences $f = (f_k)_{k=1}^{\infty}$, $f_k \in \mathbb{C}$. Denote by T the set of all possible weights $\tau = (\tau_k)_{k=1}^{\infty}$, $\tau_k \geq 1$. Every weights τ is associated with the Hilbert space $l_2(\tau)$ (see Example 1.1) as follows:

$$H_\tau = l_2(\tau) = \left\{ f \in l_2 \,\middle|\, \sum_{k=1}^{\infty} |f_k|^2 \tau_k = \|f\|_{H_\tau}^2 < \infty \right\}, \tag{2.16}$$

$$(f, g)_{H_\tau} = \sum_{k=1}^{\infty} f_k \bar{g}_k \tau_k; \quad H_1 = H_0.$$

Clearly, H_τ is dense in H_0 and $\|\varphi\|_{H_\tau} \geq \|\varphi\|_{H_0}$ $(\tau \in T)$. The intersection $\cap_{\tau \in T} H_\tau$ contains the collection $\mathbb{C}_0^\infty$ of all finite sequences $\varphi = (\varphi_k)_{k=1}^{\infty}$, $\varphi_k \in \mathbb{C}$ and, therefore, is dense in H_τ for all $\tau \in T$. The family of Hilbert spaces $(H_\tau)_{\tau \in T}$ is directed by embedding. Indeed, for given $\tau' = (\tau_k')_{k=1}^{\infty} \in T$ and $\tau'' = (\tau_k'')_{k=1}^{\infty} \in T$, one can choose, for example, $\tau''' = (\tau_k''' = \tau_k' + \tau_k'')_{k=1}^{\infty} \in T$. Then $H_{\tau'''} \subseteq H_{\tau'}$ and $H_{\tau'''} \subseteq H_{\tau''}$ densely and continuously. Thus, we can construct the projective limit

$$\Phi = \operatorname{pr} \lim_{\tau \in T} H_\tau. \tag{2.17}$$

Let us show that $\Phi = \mathbb{C}_0^\infty$, i.e., that the space just constructed is the collection of all finite complex-valued sequences topologized in a certain way.

In fact, let $\varphi = (\varphi_k)_{k=1}^{\infty} \in \Phi$ be infinite, i.e., there exists an infinite sequence of indices k such that $\varphi_k \neq 0$. Consider a weight $\tau \in T$ such that $\tau_k = |\varphi_k|^{-2} + 1$ for $\varphi_k \neq 0$ and $\tau_k = 1$, otherwise. Then $\varphi \in H_\tau$, but this is impossible because the series $\sum_{k=1}^{\infty} |\varphi_k|^2 \tau_k$ is clearly divergent. □

Space (2.17) is nuclear.

Indeed, let $\tau \in T$. We take a weight $\tau' = (2^k \tau_k)_{k=1}^{\infty} \in T$ and consider the embedding operator $O_{\tau',\tau}\colon H_{\tau'} \to H_\tau$. This operator is quasinuclear. Indeed, let $(e_k)_{k=1}^{\infty}$ be a natural basis in l_2 $(e_k = (\delta_{jk})_{j=1}^{\infty})$. Then the vectors $\left(\tau_k^{-1/2} e_k\right)_{k=1}^{\infty}$ form a basis in H_τ and, therefore, for the Hilbert norm of the embedding operator $O_{\tau',\tau}$, we have

$$O_{\tau',\tau}{}^2 = \sum_{k=1}^{\infty} \left\| (\tau_k')^{-1/2} e_k \right\|_{H_\tau}^2 = \sum_{k=1}^{\infty} (\tau_k')^{-1} \tau_k = \sum_{k=1}^{\infty} 2^{-k} < \infty. \qquad \square$$

Let us clarify the exact meaning of the convergence of a sequence of vectors in Φ.

We say that a sequence of vectors $\left(\varphi^{(n)}\right)_{n=1}^{\infty}$, $\varphi^{(n)} \in \Phi$, *converges in* Φ *to* $\varphi \in \Phi$ *if and only if the vectors* $\varphi^{(n)}$ *are uniformly finite and, for any* $k \in \mathbb{N}$, *their coordinates* $\varphi_k^{(n)} \to \varphi_k$ *as* $n \to \infty$ *(recall that we say that a sequence is uniformly finite if one can indicate* $l \in \mathbb{N}$ *such that* $\varphi_k^{(n)} = 0$ *for all* $n \in \mathbb{N}$ *and* $k \geq l$*).*

In fact, the neighbourhoods (2.3) for (2.17) are obtained by the shifts of the neighbourhoods of the origin. Therefore, it suffices to consider the case where $\varphi = 0$. Clearly, if a sequence converges in the indicated sense, then, by virtue of (2.16), $\|\varphi^{(n)}\|_{H_\tau}$ becomes arbitrarily small for any $\tau \in T$, i.e., $\varphi^n \to 0$ in Φ.

Let us prove the converse statement. Assume that $\varphi^{(n)} \to 0$ in Φ as $n \to \infty$. First, we prove that the vectors $\varphi^{(n)}$ are uniformly finite. Assume the contrary. Then one can find sequences of indices $(k_m)_{m=1}^\infty$ and $(n_m)_{m=1}^\infty$, $k_1 < k_2 < \ldots$, $n_1 < n_2 < \ldots$ such that $\varphi_{k_m}^{(n_m)} \neq 0$ $(m \in \mathbb{N})$. Consider a weight $\tau = \left(\tau_k\right)_{k=1}^\infty \in T$ such that $\tau_{k_m} = \left|\varphi_{k_m}^{(n_m)}\right|^{-2} + 1$ for all $m \in \mathbb{N}$ and $\tau_k = 1$ for all other $k \in \mathbb{N}$. According to (2.16), we have

$$\left\|\varphi^{(n_m)}\right\|_{H_\tau}^2 = \sum_{k=1}^\infty \left|\varphi_k^{(n_m)}\right|^2 \tau_k \geq \left|\varphi_{k_m}^{(n_m)}\right|^2 \tau_{k_m} > 1 \quad (m \in \mathbb{N}),$$

and, therefore, $\|\varphi^{(n)}\|_{H_\tau}$ does not approach 0 as $n \to \infty$, i.e., $\varphi^{(n)}$ is not convergent to zero. Hence, the vectors $\varphi^{(n)}$ are uniformly finite. Assume that their coordinates are equal to zero for $k > l$. But then

$$\sum_{k=1}^{l} \left|\varphi_k^{(n)}\right|^2 = \left\|\varphi^{(n)}\right\|_{H_0}^2 \to 0, \qquad n \to \infty,$$

whence we conclude that $\varphi_k^{(n)} \to 0$ as $n \to \infty$ $(k \in \mathbb{N})$. □

For every $\tau \in T$, the Hilbert space $H_{-\tau} = l_2(\tau^{-1})$, $\tau^{-1} = (\tau_k^{-1})_{k=1}^\infty$, is dual to $H_\tau = l_2(\tau)$ with respect to $H_0 = l_2$ (see Example 1.1). It follows from the arguments presented above that the space Φ' coincides with $\cup_{\tau \in T} H_{-\tau}$ and, therefore, $\Phi' = \mathbb{C}^\infty$ ($\mathbb{C}^\infty$ is the set of all complex-valued sequences). Indeed, for any vector $\alpha = \left(\alpha_k\right)_{k=1}^\infty \in \mathbb{C}^\infty$, we take $\tau \in T$, where $\tau_k = \left(|\alpha_k| + 1\right)^2 2^k \quad (k \in \mathbb{N})$. Then

$$\|\alpha\|_{H_{-\tau}}^2 = \sum_{k=1}^\infty |\alpha_k|^2 \left(1 + |\alpha_k|\right)^{-2} 2^{-k} < \infty,$$

i.e., $\alpha \in H_{-\tau}$. The scalar product in $H_0 = l_2$ defines a natural pairing of the elements of $\mathbb{C}_0^\infty$ and $\mathbb{C}^\infty$, namely, $(\alpha, \varphi)_{H_0} = \sum_{k=1}^\infty \alpha_k \bar{\varphi}_k \quad (\alpha \in \mathbb{C}_0^\infty,\ \varphi \in \mathbb{C}^\infty)$. Thus, we have constructed the nuclear rigging

$$\mathbb{C}^\infty \supseteq l_2(\tau^{-1}) \supseteq l_2 \supseteq l_2(\tau) \supseteq \mathbb{C}_0^\infty \qquad (\tau \in T). \tag{2.18}$$

3 Sobolev Spaces in Bounded Domains

3.1 The δ-Function

We consider an important class of rigged spaces connected with the notion of Sobolev spaces. For this purpose, we recall the corresponding definitions and facts (see Section 6.8).

Let G be a bounded domain in the space $\mathbb{R}^N$ with sufficiently smooth boundary ∂G. The (positive) Sobolev space $W_2^l(G)$, where $l \in \mathbb{Z}_+$, is defined as the completion of $C^\infty(\tilde{G})$ ($\tilde{G} = G \cup \partial G$) with respect to the scalar product

$$(u, v)_{W_2^l(G)} = \sum_{|\mu| \le l} \Big(D^\mu u, D^\mu v\Big)_{L_2(G)}$$

$$\Big(u, v \in C^\infty(\tilde{G}), \quad \mu = (\mu_1, \dots \mu_N), \quad \mu_k \in \mathbb{Z}_+, \quad |\mu| = \mu_1 + \cdots + \mu_N;$$

$$D^\mu = D_1^{\mu_1} \dots D_N^{\mu_N}, \quad D_k = \partial/\partial x_k\Big) \tag{3.1}$$

(the space $L_2(G)$ is constructed according to the Lebesgue measure on G; $W_2^0(G) = L_2(G)$). The elements of the Sobolev space $W_2^l(G)$ are functions defined on G, which may be even smooth if l is sufficiently large. Thus, the following embedding theorem holds:

Let $l, k \in \mathbb{Z}_+$ be such that

$$l - k > N/2. \tag{3.2}$$

Then the Sobolev space $W_2^l(G)$ is embedded in the Banach space $C^k(\tilde{G})$ of k times continuously differentiable functions defined on $\tilde{G}$, i.e., $W_2^l(G) \subset C^k(\tilde{G})$, and, moreover, the embedding operator is compact.
(In particular, for $l > N/2$, $W_2^l(G) \subset C(\tilde{G})$ and the embedding operator is compact).

The space $W_2^l(G)$ is clearly dense in $C_k(\tilde{G})$. It is not difficult to show that, for $l'' \ge l'$ ($l', l'' \in \mathbb{Z}_+$), $W_2^{l''}(G) \subseteq W_2^{l'}(G)$ densely and continuously (see Example 7.4).

Clearly, $\|u\|_{L_2(G)} \le \|u\|_{W_2^l(G)}$ ($u \in W_2^l(G)$) and $W_2^l(G)$ is dense in $L_2(G)$. Therefore, one can construct chain (1.4) by setting $H_0 = L_2(G)$ and $H_+ = W_2^l(G)$. If we denote the corresponding negative space H_- by $W_2^{-l}(G)$, then we get

$$W_2^{-l}(G) \supseteq L_2(G) \supseteq W_2^l(G) \qquad (l \in \mathbb{Z}_+). \tag{3.3}$$

The space of generalized functions $W_2^{-l}(G)$ is called negative Sobolev space.

We study this space by using a rather indirect approach (the direct way is more complicated because it involves the investigation of the properties of the operator I associated with (3.3)).

We now give a general definition of the δ-function (cf. Section 11.1). Let R be an abstract space and let Φ be a linear topological space which consists of functions $R \ni x \mapsto \varphi(x) \in \mathbb{C}$ and is such that the convergence $\Phi \ni \varphi_n \xrightarrow[n\to\infty]{} \varphi \in \Phi$ implies that $\varphi_n(x) \xrightarrow[n\to\infty]{} \varphi(x)$ for any $x \in R$. We fix $x \in R$ and consider an antilinear functional l on Φ such that $l(\varphi) = \overline{\varphi(x)}$ ($\varphi \in \Phi$). According to assumption just made, it is continuous, i.e., $l \in \Phi'$. This functional is called a δ-function concentrated at the point x and denoted by δ_x. In particular, the same situation is observed if R is compact and $\Phi \subseteq C(R)$ densely and continuously. Note that $\|\delta_x\|_{C'(R)} = 1$ ($x \in R$), which follows, e.g., from the expression for the norm of a functional in $C(R)$ (see Section 7.5).

Theorem 3.1. *If $l > N/2$, then the space $W_2^{-l}(G)$ contains the δ-function δ_x concentrated at the point $x \in \tilde{G}$. Furthermore, the vector function $\tilde{G} \ni x \mapsto \delta_x \in W_2^{-l}(G)$ is strongly continuous.*

Proof. By virtue of the embedding theorem mentioned above, $W_2^l \subset C(\tilde{G})$ is dense and continuous (and even compact). Therefore, δ_x ($x \in \tilde{G}$) is defined as an element of $W_2^{-l}(G)$.

Let us prove that δ_x is a continuous function of x. First, note that the following evident relations are true:

$$W_2^{-l}(G) \supset C'(\tilde{G}) \supset L_2(G) \supset C(\tilde{G}) \supset W_2^l(G), \tag{3.4}$$

where all embeddings are dense and continuous. Here, we have used the simple fact that if E_1 and E_2 are two Banach spaces and $E_1 \subseteq E_2$ densely and continuously then $E_2' \subseteq E_1'$ densely and continuously (cf. Exercise 7.2.13).

Let S be the compact operator that realizes the embedding $W_2^l(G) \to C(\tilde{G})$. The operator $S^*\colon C'(\tilde{G}) \to W_2^{-l}(G)$ adjoint to S in ordinary sense is also a compact embedding.

Assume the contrary, i.e., let the vector function $\tilde{G} \ni x \mapsto \delta_x \in W_2^{-l}(G)$ be not strongly continuous at a point $x_0 \in \tilde{G}$. Then there exists a sequence of points $x_n \in \tilde{G}$ convergent to x_0 and $\varepsilon_0 > 0$ such that $\|\delta_{x_n} - \delta_{x_0}\|_{W_2^{-l}(G)} \geq \varepsilon_0$ ($n \in \mathbb{N}$). Since $\|\delta_{x_n}\|_{C'(\tilde{G})} = 1$ ($n \in \mathbb{N}$) and the embedding $C'(\tilde{G}) \to W_2^{-l}(G)$ is compact, the sequence $(\delta_{x_n})_{n=1}^{\infty}$ is precompact in $W_2^{-l}(G)$ and, therefore, contains a subsequence $(\delta_{x_{n(k)}})_{k=1}^{\infty}$ such that $\delta_{x_{n(k)}}$ strongly converges in $W_2^{-l}(G)$ to a certain element $\alpha \in W_2^{-l}(G)$ as $k \to \infty$. It is easy to see that $\alpha = \delta_{x_0}$. Indeed, for $u \in W_2^l(G)$, we obtain

$$(\alpha, u)_{L_2(G)} = \lim_{k\to\infty} \left(\delta_{x_{n(k)}}, u\right)_{L_2(G)} = \lim_{k\to\infty} \overline{u(x_{n(k)})} = \overline{u(x_0)} = \left(\delta_{x_0}, u\right)_{L_2(G)}.$$

Thus, $\delta_{x_{n(k)}} \xrightarrow[k\to\infty]{} \delta_{x_0}$ in the sense of strong convergence, which contradicts the choice of the points x_n. □

3.2 Embeddings of Sobolev Spaces

The following theorem is quite important:

Theorem 3.2. *Let $W_2^{l'}(G)$ and $W_2^{l''}(G)$ be two Sobolev spaces such that $l'' - l' > \frac{N}{2}$ ($l', l'' \in \mathbb{Z}_+$). Then the embedding $W_2^{l''}(G) \to W_2^{l'}(G)$ is quasinuclear.*

Proof. First, we consider the principal case where $l' = 0$. We must prove that the embedding $O\colon W_2^l(G) \to L_2(G)$ in chain (3.3) is quasinuclear for $l > N/2$. Let $J\colon L_2(G) \to W_2^l(G)$ be an isometry associated with chain (3.3). The quasinuclearity of O is equivalent to the quasinuclearity of $OJ\colon L_2(G) \to L_2(G)$. In fact, let $(e_j)_{j=1}^\infty$ be an orthonormal basis in $L_2(G)$. Then $(Je_j)_{j=1}^\infty$ is an orthonormal basis in $W_2^l(G)$ and

$$O^2 = \sum_{j=1}^{\infty} \|OJe_j\|^2_{L_2(G)} = OJ^2.$$

Let us establish the quasinuclearity of OJ. For $f \in L_2(G)$, by using (1.13), we obtain

$$\begin{aligned}(OJf)(x) &= (Jf)(x) = \left(Jf, \delta_x\right)_{L_2(G)} \\ &= \left(f, J^+\delta_x\right)_{L_2(G)} = \left(f, \mathbf{J}\delta_x\right)_{L_2(G)} = \int_G f(y)\overline{(\mathbf{J}\delta_x)(y)}dy\end{aligned} \tag{3.5}$$

(note that $\mathbf{J}\delta_x \in L_2(G)$). We set $K(x,y) = \overline{(\mathbf{J}\delta_x)(y)}$. Then

$$\begin{aligned}\int_G \int_G |K(x,y)|^2\, dxdy &= \int_G \|\mathbf{J}\delta_x\|^2_{L_2(G)}\, dx = \int_G \|\delta_x\|^2_{W_2^{-l}(G)}\, dx \\ &\le \max\left\{\|\delta_x\|^2_{W_2^{-l}(G)} \,\middle|\, x \in \tilde{G}\right\} m(G) < \infty.\end{aligned} \tag{3.6}$$

Here, we have used the continuity of the scalar function $\tilde{G} \ni x \mapsto \|\delta_x\|_{W_2^{-l}(G)} \in \mathbb{R}$, which follows from Theorem 3.1. Relations (3.5) and (3.6) show that OJ is a Hilbert-Schmidt operator.

In the general case, the proof of Theorem 3.2 can be reduced to the already proved part and the following general lemmas:

Lemma 3.1. *Let E be a linear set with two scalar products $(f,g)_{H_1}$ and $(f,g)_{G_1}$ ($f,g \in E$) and let $E \ni f \mapsto Tf \in E$ be a linear operator. Consider the new scalar products*

$$(f,g)_{H_2} = (f,g)_{H_1} + (Tf,Tg)_{H_1} \quad \text{and} \quad (f,g)_{G_2} = (f,g)_{G_1} + (Tf,Tg)_{G_1} \quad (f,g \in E).$$

Denote by H_1, G_1, H_2, and G_2 the completions of E with respect to the corresponding scalar products. Assume that $H_1 \subseteq G_1$ densely and continuously and, *moreover,* $H_2 \subseteq H_1$. *It is stated that if the embedding* $O_1\colon H_1 \to G_1$ *is quasinuclear, then* $H_2 \subseteq G_2$ *and the embedding* $O\colon H_2 \to G_2$ *is also quasinuclear.*

Proof. Since $\|f\|_{H_1} \le \|f\|_{H_2}$ $(f \in E)$, the space H_2 is continuously embedded in H_1. In its turn, by assumption, the embedding $H_1 \to G_1$ is quasinuclear. Therefore, the embedding $H_2 \to G_1$ is quasinuclear and, consequently, if $\left(e_j\right)_{j=1}^{\infty}$ is an orthonormal basis in H_2 (which can be composed of vectors $e_j \in E$), then

$$\sum_{j=1}^{\infty} \|e_j\|_{G_1}^2 < \infty. \tag{3.7}$$

Let $(f,g)_{H_3} = (Tf,Tg)_{H_1}$ and $(f,g)_{G_1}$ $(f,g \in E)$ be, generally speaking, quasiscalar products in E. The identification of E with respect to each of these products gives the same linear set $\hat{E}$ of complete pre-images $\hat{f} = T^{-1}f' = \{f \in E \mid Tf = f'\}$ of the vectors $f' \in E$. As a result of subsequent completion, we arrive at Hilbert spaces H_3 and G_3 such that $H_3 \subseteq G_3$ densely and continuously. The embedding $H_3 \to G_3$ is quasinuclear. Indeed, let $\left(\hat{l}_j\right)_{j=1}^{\infty}$ be an orthonormal basis in H_3 composed of vectors $\hat{l}_j \in \hat{E}$ and let $l_j \in E$ be a representative of the class $\hat{l}_j$. Then

$$(Tl_j, Tl_k)_{H_1} = (l_j, l_k)_{H_3} = (\hat{l}_j, \hat{l}_k)_{H_3} = \delta_{jk} \quad (j,k \in \mathbb{N}),$$

i.e., $(Tl_j)_{j=1}^{\infty}$ is an orthonormal system in H_1 and $\sum_{j=1}^{\infty} \|Tl_j\|_{G_1}^2 < \infty$ by virtue of the quasinuclearity of the embedding $H_1 \to G_1$. But $\| Tl_j \|_{G_1} = \| \hat{l}_j \|_{G_3}$ and, hence, the last condition means that the embedding $A\colon H_3 \to G_3$ is quasinuclear.

By virtue of the inequality $\|\hat{f}\|_{H_3} = \|Tf\|_{H_1} \le \|f\|_{H_2}$, the mapping $E \ni f \mapsto Bf = \hat{f} \in \hat{E}$ can be extended by continuity to the continuous mapping $B\colon H_2 \to H_3$. The mapping $AB\colon H_2 \to G_3$ is quasinuclear. Therefore, if $\left(e_j\right)_{j=1}^{\infty}$ is an orthonormal basis in H_2 composed of vectors from E, then

$$\infty > \sum_{j=1}^{\infty} \|ABe_j\|_{G_3}^2 = \sum_{j=1}^{\infty} \|\hat{e}_j\|_{G_3}^2 = \sum_{j=1}^{\infty} \|Te_j\|_{G_1}^2.$$

This and (3.7) enable us to conclude that

$$\sum_{j=1}^{\infty} \|e_j\|_{G_2}^2 = \sum_{j=1}^{\infty} \left(\|e_j\|_{G_1}^2 + \|Te_j\|_{G_1}^2\right) < \infty.$$

It is easy to see that this inequality yields the embedding $H_2 \subseteq G_2$ (for more details, see Section 7). It is also clear that O_2 is quasinuclear. □

Remark 3.1. Assume that the operator T in the formulation of Lemma 3.1 is invertible, i.e., $\operatorname{Ker} T = \{0\}$. Then the assertion of the lemma remains true with scalar products $(\cdot,\cdot)_{H_2}$ and $(\cdot,\cdot)_{G_2}$ in its formulation replaced by $(f,g)_{H_2} = (Tf,Tg)_{H_1}$ and $(f,g)_{G_2} = (Tf,Tg)_{G_1}$ $(f,g \in E)$.

This fact follows from the arguments presented above.

Lemma 3.2. *Let E be a linear set with scalar products $(\cdot,\cdot)_{H_k}$ and $(\cdot,\cdot)_{G_k}$ $(k = 1, \dots, n)$ and let*

$$(\cdot,\cdot)_H = \sum_{k=1}^{n} (\cdot,\cdot)_{H_k} \quad \textit{and} \quad (\cdot,\cdot)_G = \sum_{k=1}^{n} (\cdot,\cdot)_{G_k}.$$

Assume that H_k, G_k, H, and G are the corresponding completions of E. If, in addition, $H \subseteq H_k \subseteq G_k$ and the embeddings $H_k \to G_k$ are quasinuclear $(k = 1, \dots, n)$, then $H \subseteq G$ and this embedding is also quasinuclear.

Proof. We fix $k = 1, \dots, n$. Since $H \subseteq H_k$ and $\|f\|_{H_k} \le \|f\|_H$ $(f \in E)$, this embedding is continuous. But then the embedding $H \to G_k$ is quasinuclear and, therefore, if $(e_j)_{j=1}^{\infty}$ is an orthonormal basis in H, then $\sum_{j=1}^{\infty} \|e_j\|_{G_k}^2 < \infty$. Taking the sum of these inequalities over all k, we find that $\sum_{j=1}^{\infty} \|e_j\|_G^2 < \infty$. Thus, as above, one can easily conclude that $H \subseteq G$; it is also clear that this embedding is quasinuclear. □

Let us complete the proof of Theorem 3.2. We must show that if the integer $l > N/2$, then the embedding $W_2^{m+l}(G) \to W_2^m(G)$ is quasinuclear for any $m \in \mathbb{Z}_+$. To apply Lemma 3.1, we assume that $E = C^\infty(\tilde{G})$ and $(Tf)(x) = (D^\nu f)(x)$, where D^ν is a fixed derivative of the order $|\nu| \le m$,

$$(f,g)_{H_1} = (f,g)_{W_2^l(G)}, \quad \text{and} \quad (f,g)_{G_1} = (f,g)_{L_2(G)} \ (f,g \in E).$$

According to the already proved part of the theorem, the embedding $H_1 = W_2^l(G) \to L_2(G) = G_1$ is quasinuclear. Therefore, in view of Lemma 3.1, the embedding $H_2 \to G_2$ is also quasinuclear; here, H_2 and G_2 are the completions of E with respect to the scalar products

$$\begin{aligned} (f,g)_{H_2} &= (f,g)_{W_2^l(G)} + (D^\nu f, D^\nu g)_{W_2^l(G)}, \\ (f,g)_{G_2} &= (f,g)_{L_2(G)} + (D^\nu f, D^\nu g)_{L_2(G)} \quad (f,g \in E) \end{aligned} \tag{3.8}$$

(the embedding $H_2 \subseteq H_1$ can be established without difficulties; see Example 7.4).

Denote the scalar products $(\cdot,\cdot)_{H_2}$ and $(\cdot,\cdot)_{G_2}$ in (3.8) by $(\cdot,\cdot)_{H_\nu}$ and $(\cdot,\cdot)_{G_\nu}$, respectively. We now apply Lemma 3.2 under the assumption that $E = C^\infty(\tilde{G})$ and n is equal to the number of vector indices $\nu = (\nu_1, \dots, \nu_N)$ such that $|\nu| \le m$.

According to what has been proved, $H_\nu \to G_\nu$ is quasinuclear for all ν satisfying this inequality. Therefore, the embedding $H \to G$, where H and G are the completions of E with respect to the scalar products

$$(f,g)_H = n(f,g)_{W_2^l(G)} + \sum_{|\nu|\le m} (D^\nu f, D^\nu g)_{W_2^l(G)},$$
$$(f,g)_G = n(f,g)_{L_2(G)} + \sum_{|\nu|\le m} (D^\nu f, D^\nu g)_{L_2(G)} \quad (f,g \in E),$$

is also quasinuclear (it is not difficult to establish the embedding $H \subseteq H_k$, see Example 7.4). The first of these products is equivalent to $(\cdot,\cdot)_{W_2^{m+l}(G)}$, while the second one is equivalent to $(\cdot,\cdot)_{W_2^m(G)}$. Therefore, $H = W_2^{m+l}(G)$ and $G = W_2^m(G)$. □

Note that, for $l > N/2$, the operator OI associated with chain (3.3) is an integral operator with kernel $K(x,y) \in C(\tilde{G} \times \tilde{G})$ (it is clear that its trace is finite). This can be proved just as the principal case of Theorem 3.2 but with the operator I instead of the operator J. Then $K(x,y) = \overline{(\mathbf{I}\delta_x)(y)}$. Since $\mathbf{I}$ is an isometry between $W_2^{-l}(G)$ and $W_2^l(G) \subset C(\tilde{G})$, and $G \ni x \mapsto \delta_x \in W_2^{-l}(G)$ is a continuous vector function, this gives the required continuity of K.

Example

3.1. Let us present a simple but important example of a countably Hilbert nuclear function space. It serves as a model of the classical spaces $\mathcal{S}(\mathbb{R}^N)$ and $\mathcal{D}(\mathbb{R}^N)$ considered in Section 4.

Let G be a bounded domain in the space $\mathbb{R}^N$ with sufficiently smooth boundary. For $\tau \in \mathbb{Z}_+$, we set $B_\tau = C^\tau(\tilde{G})$, where $C^\tau(\tilde{G})$ is a space of τ times continuously differentiable functions defined on $\tilde{G}$ with the standard norm

$$\|u\|_{C^\tau(\tilde{G})} = \max\left\{ \sum_{|\mu|\le\tau} |(D^\mu u)(x)| \;\middle|\; x \in \tilde{G} \right\} \tag{3.9}$$
$$(u \in C^\tau(\tilde{G}), \quad C^0(\tilde{G}) = C(\tilde{G})).$$

It is clear that norms (3.9) are monotone, $\cap_{\tau=0}^\infty C^\tau(\tilde{G}) = C^\infty(\tilde{G})$, and the last linear set is dense in each $C^\tau(\tilde{G})$. Thus, we can consider the projective limit of the spaces $C^\tau(\tilde{G})$, i.e., a countably normed space $C^\infty(\tilde{G}) = \operatorname{pr\,lim}_{\tau\to\infty} C^\tau(\tilde{G})$. The base of neighbourhoods of zero in $C^\infty(\tilde{G})$ is formed by the sets $U(0;\tau;\varepsilon) = \left\{\varphi \in C^\infty(\tilde{G}) \mid \|\varphi\|_{C^\tau(\tilde{G})} < \varepsilon\right\}$ with arbitrary $\tau \in \mathbb{Z}_+$ and $\varepsilon > 0$. The convergence of a sequence $(\varphi_n)_{n=1}^\infty$, $\varphi_n \in C^\infty(\tilde{G})$, to $\varphi \in C^\infty(\tilde{G})$ means that $\|\varphi_n - \varphi\|_{C^\tau(G)} \to 0$ as $n \to \infty$ $(\forall \tau \in \mathbb{Z}_+)$.

We now consider the Sobolev spaces $W_2^\tau(G)$ with $\tau \in \mathbb{Z}_+$. It follows from (3.1) that the norms in these spaces are monotone. Further, $\cap_{\tau=0}^{\infty} W_2^\tau(G) = C^\infty(\tilde{G})$ because $W_2^\tau(G) \subset C^k(\tilde{G})$ for all $\tau > N/2 + k$ by virtue of the embedding theorem. Therefore,

$$\cap_{\tau=0}^{\infty} W_2^\tau(G) = \cap_{k=0}^{\infty} C^k(\tilde{G}) = C^\infty(\tilde{G}).$$

The last linear set is dense in $W_2^\tau(G)$ for any $\tau \in \mathbb{Z}$ by the definition of this space. Thus, we can construct the projective limit $\operatorname{pr}\lim_{\tau \in \infty} W_2^\tau(G) = C^\infty(\tilde{G})$. In this space, neighbourhoods are defined just as above but the norms $\|\cdot\|_{C^\tau(\tilde{G})}$ must be replaced by $\|\cdot\|_{W_2^\tau(G)}$.

It is not difficult to show that these projective limits coincide not only as sets but also as topological spaces. For this purpose, it is necessary to check the validity of conditions (3) in Section 2, which guarantee coincidence of topologies, for the systems of neighbourhoods in both projective limits under consideration. Taking the form of the neighbourhoods into account, we can reduce this problem to the following inequalities:

$$\begin{gathered}(\forall \tau \in \mathbb{Z}_+) \quad (\exists \tau' \in \mathbb{Z}_+) \quad (\exists c_{\tau\tau'} > 0): \ \|\varphi\|_{W_2^\tau(G)} \le c_{\tau\tau'} \|\varphi\|_{C^{\tau'}(\tilde{G})}; \\ (\forall \tau \in \mathbb{Z}_+) \quad (\exists \tau' \in \mathbb{Z}_+) \quad (\exists d_{\tau\tau'} > 0): \ \|\varphi\|_{C^\tau(\tilde{G})} \le d_{\tau\tau'} \|\varphi\|_{W_2^{\tau'}(G)} \\ (\varphi \in C^\infty(\tilde{G})). \end{gathered} \tag{3.10}$$

The first inequality in (3.10) is elementary and can be established if we estimate the integrals in (3.1) by taking the maxima of their integrands; here, $\tau' = \tau$. The second inequality in (3.10) follows from the embedding theorem because its validity is equivalent to the continuity of the embedding operator. We may set $\tau' = \tau + [N/2] + 1$.

Thus, the space $C^\infty(\tilde{G})$, which was constructed as countably normed, is, in fact, countably Hilbert. This space is nuclear. Indeed, according to Theorem 3.2, for any $\tau \in \mathbb{Z}_+$, there exists $\tau' \in \mathbb{Z}_+$ such that the embedding $W_2^{\tau'}(G) \to W_2^\tau(G)$ is quasinuclear (one can set $\tau' = \tau + [N/2] + 1$).

4 Sobolev Spaces in Unbounded Domains. Classical Spaces of Test Functions

4.1 The δ-Function

We now proceed to the investigation of Sobolev spaces in unbounded domains. Let $G \subseteq \mathbb{R}^N$ be, generally speaking, an unbounded domain in the space $\mathbb{R}^N (N \in \mathbb{N})$ with sufficiently smooth boundary. For the functions $u, v \in \left(C_0^\infty(\mathbb{R}^N)\right) \restriction \tilde{G}$, where $C_0^\infty(\mathbb{R}^N)$ is the collection of infinitely differentiable finite functions in $\mathbb{R}^N$, we introduce, by using expression (3.1), the scalar product $(u, v)_{W_2^l}(G)$.

The completion of $\left(C_0^\infty(\mathbb{R}^N)\right) \restriction \tilde{G}$ with respect to this scalar product is called the Sobolev space $W_2^l(G)$ $(l \in \mathbb{Z}_+)$ (as in the case of bounded G).

It is clear that, in this case, $W_2^l(G)$ and $L_2(G)$ can also be regarded as positive and zero spaces, respectively, and, hence, we can construct the chain. For this chain, we preserve the same notation as in (3.3). For an unbounded domain G, all assertions made in Subsection 3.1 concerning the embeddings of Sobolev spaces with an increase in the index l remain true.

Let G' be a bounded subdomain of G. Then $\|u \restriction G'\|_{W_2^l(G')} \leq \|u\|_{W_2^l(G)}$ ($u \in (C_0^\infty(\mathbb{R}^N)) \restriction G$) and, therefore, as a result of the completion, we find that, for $u \in W_2^l(G)$, the restriction $(u \restriction G') \in W_2^l(G')$. Thus, the functions from the space $W_2^l(G)$ possess the same local properties as the functions from the analogous space in the case of a bounded domain. The global properties of $W_2^l(G)$ may differ from the case of a bounded domain. Thus, in particular, Theorem 3.2 on the quasinuclearity of embeddings is, as a rule, violated.

In introducing the space $W_2^l(\mathbb{R}^N)$, it is convenient to pass to the Fourier transforms considered in Section 11.3, namely,

$$C_0^\infty(\mathbb{R}^N) \ni u(x) \mapsto \hat{u}(s) = (2\pi)^{-N/2} \int_{\mathbb{R}^N} u(x) e^{-i(s,x)} dx \tag{4.1}$$
$$\big(s \in \mathbb{R}^N, (s,x) = s_1 x_1 + \cdots + s_N x_N\big).$$

For $u \in C_0^\infty(\mathbb{R}^N)$, clearly, $(D^\mu \hat{u})(s) = i^{|\mu|} s^\mu \hat{u}(s)$, where $\mu = (\mu_1, \ldots, \mu_N)$, $s^\mu = s_1^{\mu_1} \ldots s_N^{\mu_N}$ $(s \in \mathbb{R}^N)$. Therefore, by virtue of the Parseval equality,

$$(u,v)_{W_2^l(\mathbb{R}^N)} = \int_{\mathbb{R}^N} \hat{u}(s)\overline{\hat{v}(s)} p_l(s) ds,$$
$$p_l(s) = \sum_{|\mu| \leq l} s^{2\mu} \geq 1 \quad (l \in \mathbb{Z}_+; u, v \in C_0^\infty(\mathbb{R}^N)). \tag{4.2}$$

Since $\big(C_0^\infty(\mathbb{R}^N)\big)^\wedge$ is dense in $L_2(\mathbb{R}^N)$, it follows from (4.2) that $W_2^l(\mathbb{R}^N)$ is isometric to the space $L_2\big(\mathbb{R}^N, p_l(s)ds\big)$, whence we conclude that the closure of (4.1) by continuity establishes the isometry between the spaces of the chain $W_2^{-l}(\mathbb{R}^N) \supseteq L_2(\mathbb{R}^N) \supseteq W_2^l(\mathbb{R}^N)$ and the corresponding spaces of a chain of the type (1.15)

$$L_2\left(\mathbb{R}^N, p_l^{-1}(s)ds\right) \supseteq L_2(\mathbb{R}^N) \supseteq L_2\left(\mathbb{R}^N, p_l(s)ds\right) \quad (l \in \mathbb{Z}_+). \tag{4.3}$$

It is easy to prove that the embedding of the positive space into the zero space in (4.3) cannot be quasinuclear for any choice of the weight. This proves the remark made above that Theorem 3.2 is not true in this case. Let us show what should be changed in the definition of the Sobolev space to guarantee the quasinuclearity of the embedding.

Let $q(x) \in C^l(\tilde{G}), q(x) > 0 (x \in \tilde{G})$ be fixed. On the functions $u, v \in \big(C_0^\infty(\mathbb{R}^N)\big) \restriction \tilde{G}$, we introduce a scalar product

$$(u,v)_{W_2^{(l,q)}(G)} = (u(x)q(x), v(x)q(x))_{W_2^l(G)}\,. \tag{4.4}$$

The completion of $(C_0^\infty(\mathbb{R}^N)) \upharpoonright \tilde{G}$ with respect to this scalar product is denoted by $W_2^{(l,q)}(G)$. It is clear that $u(x) \in W_2^{(l,q)}(G)$ if and only if $u(x)q(x) \in W_2^l(G)$. This means that functions from $W_2^{(l)}(G)$ have the same local properties as functions from $W_2^{(l,q)}(G)$, i.e., as the corresponding functions defined in a bounded region.

Assume in addition, that $q(x) \geq 1$ $(x \in \tilde{G})$. Then

$$\|u\|_{W_2^{(l,q)}(G)} = \|uq\|_{W_2^l(G)} \geq \|u\|_{L_2(G)} \quad \left(u \in W_2^{(l,q)}(G)\right).$$

Therefore, $W_2^{(l,q)}(G)$ and $L_2(G)$ can be taken as the positive and zero spaces, respectively. By constructing the relevant negative space, we arrive at the chain

$$W_2^{-(l,q)}(G) \supseteq L_2(G) \supseteq W_2^{(l,q)}(G). \tag{4.5}$$

Theorem 4.1. *If $l > N/2$, then the δ-function δ_x concentrated in $x \in \mathbb{R}^N$ is defined in the space $W_2^{-(l,q)}(\mathbb{R}^N)$. Moreover, the vector function $\mathbb{R}^N \ni x \mapsto \delta_x \in W_2^{-(l,q)}(\mathbb{R}^N)$ is continuous and*

$$\|\delta_x\|_{W_2^{-(l,q)}(\mathbb{R}^N)} \leq cq^{-1}(x) \quad (x \in \mathbb{R}^N; c > 0). \tag{4.6}$$

Proof. Let $x \in \mathbb{R}^N$. Denote by B some ball with radius one such that $x \in B$. According to the embedding theorems, $W_2^l(B) \subset C(\tilde{B})$ and

$$|v(y)| \leq c\|v\|_{W_2^l(B)} \quad (y \in \tilde{B}, v \in W_2^l(B)). \tag{4.7}$$

Therefore, $W_2^{(l,q)}(\mathbb{R}^N) \subset C(\mathbb{R}^N)$ and, by virtue of (4.7), the inequality

$$\begin{aligned} |u(x)| &= q^{-1}(x)|u(x)q(x)| \leq cq^{-1}(x)\, \|(uq) \upharpoonright B\|_{W_2^l(B)} \\ &\leq cq^{-1}(x)\, \|u\|_{W_2^{(l,q)}(\mathbb{R}^N)} \quad \left(x \in \mathbb{R}^N; u \in W_2^{(l,q)}(\mathbb{R}^N)\right). \end{aligned} \tag{4.8}$$

holds for $v = uq$ and $y = x$. This inequality demonstrates that δ_x is defined as an element of $W_2^{-(l,q)}(\mathbb{R}^N)$ and satisfies (4.6).

To prove that δ_x is continuous, we note that the inequality

$$\|\delta_x - \delta_y\|_{W_2^{-(l,q)}(\mathbb{R}^N)} \leq \left\|q^{-1}(x)\delta_x - q^{-1}(y)\delta_y\right\|_{W_2^{-l}(B)} \quad (x, y \in B) \tag{4.9}$$

is true for an arbitrary open ball B with radius one. Indeed, for $u \in W_2^{(l,q)}(\mathbb{R}^N)$, by analogy with (4.8), we obtain

$$\begin{aligned} \left|(\delta_x - \delta_y, u)_{L_2(\mathbb{R}^N)}\right| &= |(u(x) - u(y)| = \left|(q^{-1}(x)\delta_x - q^{-1}(y)\delta_y,\ (uq) \upharpoonright B)_{L_2(B)}\right| \\ &\leq \left\|q^{-1}(x)\delta_x - q^{-1}(y)\delta_y\right\|_{W_2^{-l}(B)} \|(uq) \upharpoonright B\|_{W_2^l(B)} \\ &\leq \left\|q^{-1}(x)\delta_x - q^{-1}(y)\delta_y\right\|_{W_2^{-l}(B)} \|uq\|_{W_2^l(\mathbb{R}^N)} \\ &= \left\|q^{-1}(x)\delta_x - q^{-1}(y)\delta_y\right\|_{W_2^{-l}(B)} \|u\|_{W_2^{(l,q)}(\mathbb{R}^N)}, \end{aligned}$$

whence we arrive at (4.9).

The required continuity of δ_x follows from (4.9) and from the continuity of $q^{-1}(x)$ and the vector function $\tilde{B} \ni x \mapsto \delta_x \in W_2^{-l}(B)$ (see Theorem 3.1). □

The assertion of Theorem 4.1 is true not only for $G = \mathbb{R}^N$ but also for a broader class of domains. A domain $G \subset \mathbb{R}^N$ with piecewise-continuously differentiable boundary is called *regular* if there exist a bounded domain $K \subset \mathbb{R}^N$ with boundary of the same type and a number $R > 0$ such that, for any point $x \in \tilde{G}, |x| \geq R$, one can indicate a domain K_x obtained from K by means of an orthogonal rotation combined with translation and satisfying the condition $x \in \tilde{K}_x, \tilde{K}_x \subseteq \tilde{G}$ (this, in particular, excludes the domains with infinitely stretched "cusps").

If $G \subseteq \mathbb{R}^N$ is a regular domain, then the assertion of Theorem 4.1 holds for the space $W_2^{-(l,q)}(G)$ with $x \in \tilde{G}$ in its formulation.

Indeed, inequality (4.7) now holds with c which does not depend on $x \in \tilde{G}$, $|x| \geq R$, whence we get an inequality of the form (4.8)

$$|u(x)| \leq cq^{-1}(x)\|u\|_{W_2^{(l,q)}(G)} \quad (x \in \tilde{G}, |x| \geq R).$$

This enables us to deduce an estimate similar to (4.6). The continuity of δ_x is established similarly. □

Theorem 4.2. *Assume that $G = \mathbb{R}^N$ or, more generally, that G is a regular domain and $q(x) \geq 1$ $(x \in \tilde{G})$. The embedding $W_2^{(l,q)}(G) \to L_2(G)$ is quasinuclear provided that*

$$l > N/2, \quad \text{and} \quad \int_G q^{-2}(x)dx < \infty. \tag{4.10}$$

Proof. The theorem is proved by analogy with the principal case of Theorem 3.2 with insignificant modifications. Indeed, consider chain (4.5) and the corresponding operators. It is necessary to prove that the operator $OJ\colon L_2(G) \to L_2(G)$ is quasinuclear. It is clear that this operator is representable in the form (3.5). We set $K(x,y) = \overline{(\mathbf{J}\delta_x)(y)}$. By virtue of (4.6) for G and (4.10), this enables us to conclude that

$$\begin{aligned}\int_G\int_G |K(x,y)|^2 dxdy &= \int_G \|\mathbf{J}\delta_x\|^2_{L_2(G)}\, dx \\ &= \int_G \|\delta_x\|^2_{W_2^{-(l,q)}(G)}\, dx \leq c^2 \int_G q^{-2}(x)dx < \infty.\end{aligned}$$

□

As in the case of a bounded domain, for a regular domain G and $l > N/2$, OJ is an integral operator whose kernel belongs to $C(\tilde{G} \times \tilde{G})$. Moreover, this fact can be proved by using only the continuity of the vector function $\tilde{G} \ni x \mapsto \delta_x \in W_2^{-(l,q)}(G)$ and the validity of (4.6) is not required. Therefore, this assertion remains true for nonregular domains, as follows from the proof of Theorem 4.1.

4.2 Embeddings of Weighted Sobolev Spaces

As a rule, the spaces $W_2^{(l,q)}(G)$ are regarded as auxiliary. Much more often one encounters weighted Sobolev spaces $W_2^l(G, p(x)dx)$, which can be defined as follows:

Let $G \subseteq \mathbb{R}^N$ be, generally speaking, an unbounded domain in the space $\mathbb{R}^N (N \in \mathbb{N})$ with a sufficiently smooth boundary and let $p \in C(\tilde{G}), p(x) > 0$ $(x \in \tilde{G})$ be a fixed weight. A scalar product of functions $u, v \in (C_0^\infty(\mathbb{R}^N)) \upharpoonright \tilde{G}$ is introduced by the formula

$$(u, v)_{W_2^l(G,p(x)dx)} = \sum_{|\mu| \leq l} \int_G (D^\mu u)(x)\overline{(D^\mu v)(x)} p(x) dx \qquad (l \in \mathbb{Z}_+). \tag{4.11}$$

The space $W_2^l(G, p(x)dx)$ is defined as the complement of $(C_0^\infty(\mathbb{R}^N)) \upharpoonright \tilde{G}$ with respect to (4.11). It is clear that the local properties of the functions of $W_2^l(G, p(x)dx)$ coincide with the corresponding properties of the functions from $W_2^l(G')$ with bounded G'. By comparing (4.4) with (4.11), we easily arrive at the following estimate for the functions $u \in (C_0^\infty(\mathbb{R}^N)) \upharpoonright \tilde{G}$:

$$\|u\|_{W_2^{(l,q)}(G)} \leq c_l \|u\|_{W_2^l(G, q_{(l)}^2(x)dx)} \qquad (c_l > 0),$$

where the following notation is used:

$$q_{(l)}(x) = \max \{|(D^\mu q)(x)| \mid |\mu| \leq l\} \quad (x \in \tilde{G}). \tag{4.12}$$

This implies the continuity of the embedding

$$W_2^l(G, q_{(l)}^2(x)dx) \subseteq W_2^{(l,q)}(G) \qquad (l \in \mathbb{Z}_+); \tag{4.13}$$

moreover, the first space is dense in the second one.

Assume that $p(x) \geq 1$ $(x \in \tilde{G})$, then $W_2^l(G, p(x)dx)$ and $L_2(G)$ can be regarded as a positive space and a zero space, respectively, and we can construct the corresponding negative space $W_2^{-l}(G, p(x)dx)$. As a result, we arrive at the chain, which is frequently used in what follows, namely,

$$W_2^{-l}(G, p(x)dx) \supseteq L_2(G) \supseteq W_2^l(G, p(x)dx) \qquad (l \in \mathbb{Z}_+). \tag{4.14}$$

The role of Theorem 3.2 is played by the following theorem:

Theorem 4.3. *Assume that $G = \mathbb{R}^N$ or, more generally, that G is a regular domain, $m \in \mathbb{Z}_+$, and integer $l > N/2$. If weights $q_1, q_2 \in C^l(\tilde{G})$ are such that $0 < q_1(x) \leq q_2(x)$ $(x \in \tilde{G})$ and*

$$\int_G q_1^2(x)/q_2^2(x) dx < \infty, \tag{4.15}$$

then the embedding $W_2^{m+l}(G, q_{2,(l)}^2(x)dx) \to W_2^m(G, q_1^2(x)dx)$ is quasinuclear.

Proof. By comparing the conditions of this theorem and Theorem 4.2, we conclude that the embedding $H_1 = W_2^{\left(l, \frac{q_2}{q_1}\right)}(G) \to L_2(G) = G_1$ is quasinuclear. Let us now apply Remark 3.1 to these spaces, assuming that $E = (C_0^\infty(\mathbb{R}^N)) \restriction \tilde{G}$ and $(Tf)(x) = q_1(x)f(x)$ $(x \in \tilde{G}; f \in E)$. It is easy to see that all necessary conditions are satisfied. As a result, we conclude that the embedding $H_2 = W_2^{(l,q_2)}(G) \to L_2(G, q_1^2(x)dx) = G_2$ is quasinuclear. By virtue of (4.13), the embedding $W_2^l(G, q_{2,(l)}^2(x)dx) \to W_2^{(l,q_2)}(G)$ is continuous; therefore, the embedding $W_2^l(G, q_{2,(l)}^2(x)dx) \to L_2(G, q_1^2(x)dx)$ is quasinuclear.

We now act just as in the proof of Theorem 3.2. Thus, we use Lemma 3.1, setting $E = \left(C_0^\infty(\mathbb{R}^N)\right) \restriction \tilde{G}$, $(Tf)(x) = (D^\nu f)(x)$, where D^ν is a fixed derivative of the νth order, $|\nu| \le m$, $(f,g)_{H_1} = (f,g)_{W_2^l(G,q_{2,(l)}^2(x)dx)}$, and $(f,g)_{G_1} = (f,g)_{L_2(G,q_1^2(x)dx)}$ $(f,g \in E)$. Taking into account the quasinuclearity of the embedding $H_1 \to G_1$ established above, we conclude that the embedding $H_2 \to G_2$, where H_2 and G_2 are, respectively, the completions of E with respect to the scalar products

$$(f,g)_{H_2} = (f,g)_{W_2^l(G,q_{2,(l)}^2(x)dx)} + (D^\nu f, D^\nu g)_{W_2^l(G,q_{2,(l)}^2(x)dx)}$$

$$\text{and} \quad (f,g)_{G_2} = (f,g)_{L_2(G,q_1^2(x)dx)} + (D^\nu f, D^\nu g)_{L_2(G,q_1^2(x)dx)} \quad (f,g \in E),$$

is also quasinuclear. Then, by applying Lemma 3.2, we establish the quasinuclearity of the embedding $H \to G$, where H and G are, respectively, the completions of E with respect to the scalar products

$$(f,g)_H = n\,(f,g)_{W_2^l(G,q_{2,(l)}^2(x)dx)} + \sum_{|\nu|\le m} (D^\nu f, D^\nu g)_{W_2^l(G,q_{2,(l)}^2(x)dx)}$$

$$\text{and} \quad (f,g)_G = n\,(f,g)_{L_2(G,q_1^2(x)dx)} + \sum_{|\nu|\le m} (D^\nu f, D^\nu g)_{L_2(G,q_1^2(x)dx)} \quad (f,g \in E).$$

It is clear that $H = W_2^{m+l}(G, q_{2,(l)}^2(x)dx)$ and $G = W_2^m(G, q_1^2(x)dx)$, whence we get the assertion of the theorem. As in the proof of Theorem 3.2, the application of Lemmas 3.1 and 3.2 can be easily justified by using the results presented in Section 7. $\square$

To investigate the Schwartz space $\mathcal{S}(\mathbb{R}^N)$ of test functions, it is necessary to consider Sobolev spaces with special weights. Assume that

$$S_l(\mathbb{R}^N) = W_2^l(\mathbb{R}^N, (1+|x|^2)^l dx).$$

Then

$$(u,v)_{S_l(\mathbb{R}^N)} = \sum_{|\mu|\le l} \int_{\mathbb{R}^N} (D^\mu u)(x)\overline{(D^\mu v)(x)}(1+|x|^2)^l dx \tag{4.16}$$

$$\left(l \in \mathbb{Z}_+; u, v \in S_l(\mathbb{R}^N)\right),$$

and the sequence of norms $\|\cdot\|_{S_l(\mathbb{R}^N)}$ is monotone, i.e., $\|\cdot\|_{S_0(\mathbb{R}^N)} \leq \|\cdot\|_{S_1(\mathbb{R}^N)} \leq \dots$. The formulation of the following theorem is similar to Theorem 3.2 for a bounded domain:

Theorem 4.4. *If $l'' - l' > N/2$ (l', $l'' \in \mathbb{Z}_+$), then the embedding $S_{l''}(\mathbb{R}^N) \to S_{l'}(\mathbb{R}^N)$ is quasinuclear.*

Proof. Let $m \in \mathbb{Z}_+$ and let $l > N/2$ be integer. It is necessary to show that the embedding $S_{m+l}(\mathbb{R}^N) \to S_m(\mathbb{R}^N)$ is quasinuclear. Let us apply Theorem 4.3 with $q_1(x) = \left(1 + |x|^2\right)^{m/2}$ and $q_2(x) = \left(1 + |x|^2\right)^{\frac{1}{2}(m+l)}$; condition (4.15) is obviously satisfied. This implies that the embedding $W_2^{m+l}(\mathbb{R}^n, q_{2,(l)}^2(x)\, dx) \to S_m(\mathbb{R}^N)$ is quasinuclear.

The weight $q_2(x)$ satisfies the estimate $q_{2,(l)}(x) \leq c_{m,l} q_2(x)$ ($x \in \mathbb{R}^N, c_{m,l} > 0$). Therefore, $S_{m+l}(\mathbb{R}^N) \subseteq W_2^{m+l}(\mathbb{R}^N, q_{2,(l)}^2(x) dx)$ topologically. Taking the superposition of the last two embeddings, we conclude that the embedding $S_{m+l}(\mathbb{R}^N) \to S_m(\mathbb{R}^N)$ is quasinuclear. □

4.3 The Classical Spaces of Test Functions

Let us show that the two classical spaces $\mathcal{S}(\mathbb{R}^N)$ and $\mathcal{D}(\mathbb{R}^N)$ considered in Sections 11.1 and 11.3 are projective limits of Sobolev spaces. First, we consider the Schwartz space $\mathcal{S}(\mathbb{R}^N)$. It is traditionally defined as a countably normed space as follows: Let us construct a monotone sequence of norms in $C_0^\infty(\mathbb{R}^N)$ ($N \in \mathbb{N}$) by setting

$$\|u\|_{\mathcal{S}_\tau(\mathbb{R}^N)} = \max\left\{ \left(1 + |x|^2\right)^{\tau/2} \sum_{|\mu| \leq \tau} |(D^\mu u)(x)| \;\Big|\; x \in \mathbb{R}^N \right\} \quad (\tau \in \mathbb{Z}_+). \quad (4.17)$$

Let $\mathcal{S}_\tau(\mathbb{R}^N)$ be the completion of $C_0^\infty(\mathbb{R}^N)$ with respect to (4.17). Then $\mathcal{S}(\mathbb{R}^N) = \operatorname{pr\,lim}_{\tau \in T} \mathcal{S}_\tau(\mathbb{R}^N)$, where $T = \mathbb{Z}_+$, i.e., $\mathcal{S}(\mathbb{R}^N) = \cap_{\tau=0}^\infty \mathcal{S}_\tau(\mathbb{R}^N)$, and the base of neighbourhoods of zero in $\mathcal{S}(\mathbb{R}^N)$ is formed by the sets

$$U(0; \tau; \varepsilon) = \left\{\varphi \in \mathcal{S}(\mathbb{R}^N) \mid \|\varphi\|_{\mathcal{S}_\tau(\mathbb{R}^N)} < \varepsilon\right\}$$

for arbitrary $\tau \in T$ and $\varepsilon > 0$. Thus, $\mathcal{S}(\mathbb{R}^N)$ consists of infinitely differentiable functions defined on $\mathbb{R}^N$ and decreasing as $|x| \to \infty$ together with all their derivatives faster than any power $|x|^{-l}$. Convergence in $\mathcal{S}(\mathbb{R}^N)$ is introduced by the indicated base of neighbourhoods. Let us show that this space admits a similar construction in terms of the Sobolev spaces $S_l(\mathbb{R}^N)$.

Theorem 4.5. *The space $\mathcal{S}(\mathbb{R}^N)$ coincides with the projective limit of the spaces $\mathcal{S}_\tau(\mathbb{R}^N)$, i.e., $\mathcal{S}(\mathbb{R}^N) = \operatorname{pr\,lim}_{\tau \in T} S_\tau(\mathbb{R}^N)$. This space is nuclear.*

Proof. To deduce the relation $\mathcal{S}(\mathbb{R}^N) = \operatorname{pr\,lim}_{\tau \in T} S_\tau(\mathbb{R}^N)$, it suffices to establish the following two inequalities:

For each $\tau \in T$, one can indicate $\tau' \in T$ such that $\|\varphi\|_{S_\tau(\mathbb{R}^N)} \le c_{\tau\tau'} \|\varphi\|_{\mathcal{S}_{\tau'}(\mathbb{R}^N)}$ ($\varphi \in C_0^\infty(\mathbb{R}^N)$) for some $c_{\tau\tau'} > 0$; the second inequality is obtained from this by the transposition of $S_\tau(\mathbb{R}^N)$ and $\mathcal{S}_{\tau'}(\mathbb{R}^N)$ (with some other τ' and $c_{\tau\tau'}$).

The first inequality is trivial. Indeed, let $l > N/2$ be integer. Then, according to (4.16), we have

$$\|\varphi\|^2_{S_\tau(\mathbb{R}^N)} \le \max\{(1+|x|^2)^{\tau+l} \sum_{|\mu|\le\tau} |(D^\mu\varphi)(x)|^2 \,|\, x \in \mathbb{R}^N\} \int_{\mathbb{R}^N} (1+|x|^2)^{-l} dx$$
$$\le c^2_{\tau\tau'} \|\varphi\|^2_{\mathcal{S}_{\tau'}(\mathbb{R}^N)} \quad (\tau \in T; \tau' = \tau + l)$$

for $\varphi \in C_0^\infty(\mathbb{R}^N)$.

Let us show that, for any $\tau \in T$, one can find $\tau' \in T$ such that $\|\varphi\|_{\mathcal{S}_\tau(\mathbb{R}^N)} \le c_{\tau\tau'}\|\varphi\|_{S_{\tau'}(\mathbb{R}^N)}$ ($\varphi \in C_0^\infty(\mathbb{R}^N)$) for some $c_{\tau\tau'} > 0$. We fix an integer $l > N/2$ and denote by B_x an open sphere with unit radius centered at a point $x \in \mathbb{R}^N$. According to the embedding theorems, $\|\varphi\|_{C(\tilde{B}_x)} \le c_1 \|\varphi\|_{W_2^l(B_x)}$ ($\varphi \in W_2^l(B_x)$) with a constant c_1 independent of x. This implies that

$$|\varphi(x)| \le c_1 \|\varphi \restriction B_x\|_{W_2^l(B_x)} \le c_1 \|\varphi\|_{W_2^l(\mathbb{R}^N)} \quad (x \in \mathbb{R}^N) \tag{4.18}$$

for $\varphi \in C_0^\infty(\mathbb{R}^N)$. By substituting the function $(1+|x|^2)^{\frac{\tau}{2}}(D^\mu\varphi)(x)$, where $\varphi \in C_0^\infty(\mathbb{R}^N)$ and $|\mu| \le \tau$, for $\varphi(x)$ in (4.18), we obtain

$$(1+|x|^2)^{\tau/2}|(D^\mu\varphi)(x)| \le c_1 \left\|(1+|x|^2)^{\tau/2}(D^\mu\varphi)(x)\right\|_{W_2^l(\mathbb{R}^N)}$$
$$= c_1 \Big(\sum_{|\nu|\le l} \int_{\mathbb{R}^N} \left|D^\nu(1+|x|^2)^{\tau/2}(D^\mu\varphi)(x)\right|^2 dx\Big)^{1/2}$$
$$= c_1 \Big(\sum_{|\nu|\le l} \int_{\mathbb{R}^N} \Big| \sum_{|\kappa|\le l, |\lambda|\le\tau+l} c_{\mu\kappa\nu\lambda}(D^\kappa(1+|x|^2)^{\tau/2})(D^\lambda\varphi)(x)\Big|^2 dx\Big)^{1/2}, \tag{4.19}$$

where $c_{\mu\kappa\nu\lambda}$ are coefficients obtained from the Leibniz formula. In view of the fact that

$$|D^\kappa(1+|x|^2)^{\tau/2}| \le c_2(1+|x|^2)^{\tau/2} \quad (x \in \mathbb{R}^N; |\kappa| \le m),$$

by using elementary inequalities, we obtain

$$(1+|x|^2)^{\tau/2}|(D^\mu\varphi)(x)| \le c_3 \sum_{|\lambda|\le\tau+l} \left(\int_{\mathbb{R}^N} |(D^\lambda\varphi)(x)|^2 (1+|x|^2)^\tau dx\right)^{1/2}$$
$$\le c_4 \|\varphi\|_{S_{\tau+l}(\mathbb{R}^N)}.$$

This inequality and (4.17) imply that $\|\varphi\|_{\mathcal{S}_\tau(\mathbb{R}^N)} \le c_{\tau\tau'}\|\varphi\|_{S_{\tau'}(\mathbb{R}^N)}$, where $\varphi \in C_0^\infty(\mathbb{R}^N)$, $\tau' = \tau + l$.

Thus, $\mathcal{S}(\mathbb{R}^N) = \operatorname{pr\,lim}_{\tau\in T} S_\tau(\mathbb{R}^N)$. The nuclearity of $\mathcal{S}(\mathbb{R}^N)$ easily follows from Theorem 4.4. $\square$

Note that Theorem 2.1 yields the following important equality:

$$\mathcal{S}'(\mathbb{R}^N) = \cup_{\tau=0}^{\infty} S_{-\tau}(\mathbb{R}^N), S_{-\tau}(\mathbb{R}^N) = W_2^{-\tau}(\mathbb{R}^N, (1+|x|^2)^\tau dx). \tag{4.20}$$

This enables us to conclude that every generalized function from $\mathcal{S}'(\mathbb{R}^N)$ has "finite order", or, more precisely, is an element of a certain negative Sobolev space dual to the space of finitely differentiable functions.

Consider the space $\mathcal{D}(\mathbb{R}^N)$ ($N \in \mathbb{N}$). It is usually defined as the linear set of functions $C_0^\infty(\mathbb{R}^N)$ with the following classical convergence: $C_0^\infty(\mathbb{R}^N) \ni \varphi_n \xrightarrow[n\to\infty]{} \varphi \in C_0^\infty(\mathbb{R}^N)$ if the functions φ_n are uniformly finite (i.e., there exists $r > 0$ that depends on $(\varphi_n)_{n=1}^\infty$ such that $\varphi_n(x) = 0$ for $|x| > r$ and all $n \in \mathbb{N}$) and $(D^\mu \varphi_n)(x) \xrightarrow[n\to\infty]{} (D^\mu \varphi)(x)$ uniformly for each derivative. This convergence appears if $C_0^\infty(\mathbb{R}^N)$ is equipped with a proper inductive or projective topology (these topologies are not equivalent). Here, we study only the case of the projective topology which is more important for our presentation. First, we construct $\mathcal{D}(\mathbb{R}^N)$ as the projective (not countably normed) limit of Banach spaces.

Denote by T the collection of all pairs $\tau = \langle \tau_1, \tau_2(x)\rangle$ such that $\tau_1 \in \mathbb{Z}_+, \tau_2 \in C^\infty(\mathbb{R}^N)$, and $\tau_2(x) \geq 1$ ($x \in \mathbb{R}^N$). For each $\tau \in T$, we define the Banach space $\mathcal{D}_\tau(\mathbb{R}^N)$ as the completion of $C_0^\infty(\mathbb{R}^N)$ with respect to the norm

$$\|u\|_{\mathcal{D}_\tau(\mathbb{R}^N)} = \max\left\{\tau_2(x) \sum_{|\mu|\leq\tau_1} |(D^\mu u)(x)| \mid x \in \mathbb{R}^N\right\} \quad (u \in C_0^\infty(\mathbb{R}^N)). \tag{4.21}$$

If $\tau' = \langle \tau_1', \tau_2'(x)\rangle \in T$ is such that $\tau_1' \geq \tau_1$, $\tau_2'(x) \geq \tau_2(x)$ ($x \in \mathbb{R}^N$), then, obviously, $\|\cdot\|_{\mathcal{D}_\tau(\mathbb{R}^N)} \leq \|\cdot\|_{\mathcal{D}_{\tau'}(\mathbb{R}^N)}$. It is not difficult to show that $\mathcal{D}_{\tau'}(\mathbb{R}^N) \subseteq \mathcal{D}_\tau(\mathbb{R}^N)$ and the embedding operator is continuous (see Example 7.4). This means that the family $\left(\mathcal{D}_\tau(\mathbb{R}^N)\right)_{\tau\in T}$ is directed by embedding, i.e., for all $\tau', \tau'' \in T$, there exists $\tau''' \in T$ such that $\mathcal{D}_{\tau'''}(\mathbb{R}^N) \subseteq \mathcal{D}_{\tau'}(\mathbb{R}^N)$ and $D_{\tau'''}(\mathbb{R}^N) \subseteq \mathcal{D}_{\tau''}(\mathbb{R}^N)$ and, moreover, the embeddings are dense and continuous. Thus, it suffices, e.g., to set $\tau_1''' = \tau_1' + \tau_1''$ and $\tau_2'''(x) = \tau_2'(x) + \tau_2''(x)$ ($x \in \mathbb{R}^N$).

The intersection

$$\cap_{\tau\in T} \mathcal{D}_\tau(\mathbb{R}^N) \supseteq C_0^\infty(\mathbb{R}^N) \tag{4.22}$$

is clearly dense for all $\tau \in T$ in $\mathcal{D}_\tau(\mathbb{R}^N)$. It is easy to understand that (4.22) *is, in fact, an equality.* Indeed, assume that φ belongs to the left-hand side of (4.22). Then $\varphi \in C_0^\infty(\mathbb{R}^N)$. The function φ is finite. Indeed, assuming the contrary, we can find a sequence of points $(x_n)_{n=1}^\infty \subset \mathbb{R}^N$ such that $|x_n| \xrightarrow[n\to\infty]{} \infty$ and $\varphi(x_n) \neq 0$. Let B_n be an open sphere in $\mathbb{R}^N$ centered at the point x_n and such that $|\varphi(x)| > \varepsilon_n > 0$ ($x \in B_n, n \in \mathbb{N}$). Let us construct a function $\tau_2 \in C^\infty(\mathbb{R}^N), \tau_2(x) \geq 1$ ($x \in \mathbb{R}^N$) such that $\tau_2(x) \geq \frac{n}{\varepsilon_n}$ for $x \in B_n$ ($n \in \mathbb{N}$). Then $\varphi \notin \mathcal{D}_\tau(\mathbb{R}^N)$, where $\tau = \langle 0, \tau_2(x)\rangle$, and we arrive at a contradiction. □

Thus, we can consider $\operatorname{pr}\lim_{\tau\in T}\mathcal{D}_\tau(\mathbb{R}^N)$; as a set, it coincides with $C_0^\infty(\mathbb{R}^N)$. We regard this projective limit as the space $\mathcal{D}(\mathbb{R}^N)$ (the fact that convergence in this space coincides with the required classical convergence is proved below). The base of neighbourhoods of zero in this space is formed by the sets $U(0;\tau;\varepsilon) = \{\varphi\in\mathcal{D}(\mathbb{R}^N) \mid \|\varphi\|_{\mathcal{D}_\tau(\mathbb{R}^N)} < \varepsilon\}$ with all possible $\tau\in T$ and $\varepsilon > 0$.

Let us show that the space $\mathcal{D}(\mathbb{R}^N)$ can also be constructed as the projective limit of Sobolev spaces. For the same T as above, we set

$$D_\tau(\mathbb{R}^N) = W_2^{\tau_1}(\mathbb{R}^N, \tau_2(x)dx) \quad (\tau = \langle\tau_1,\tau_2(x)\rangle\in T). \tag{4.23}$$

The collection of norms (4.23) is also directed. Indeed, as above, for all $\tau',\tau''\in T$, we can set $\tau''' = \langle\tau_1'+\tau_1'',\tau_2'(x)+\tau_2''(x)\rangle\in T$. Then $D_{\tau'''}(\mathbb{R}^N)\subseteq D_{\tau'}(\mathbb{R}^N)$ and $D_{\tau'''}(\mathbb{R}^N)\subseteq D_{\tau''}(\mathbb{R}^N)$ densely and continuously.

Theorem 4.6. *The space $\mathcal{D}(\mathbb{R}^N)$ coincides with the projective limit of the spaces $D_\tau(\mathbb{R}^N)$, i.e., $\mathcal{D}(\mathbb{R}^N) = \operatorname{pr}\lim_{\tau\in T} D_\tau(\mathbb{R}^N)$. This space is nuclear.*

Proof. The proof is similar to the proof of Theorem 4.5. Let $p(x)\in C^\infty(\mathbb{R}^N)$ be such that $p(x)\geq 1$ $(x\in\mathbb{R}^N)$ and $\int_{\mathbb{R}^N} p^{-1}(x)dx < \infty$. Then, according to (4.11), for any $\tau\in T$ and $\varphi\in C_0^\infty(\mathbb{R}^N)$, we obtain

$$\begin{aligned}\|\varphi\|^2_{D_\tau(\mathbb{R}^N)} &\leq \max\Big\{\tau_2(x)p(x)\sum_{|\mu|\leq\tau_1}|(D^\mu\varphi)(x)|^2 \mid x\in\mathbb{R}^N\Big\}\int_{\mathbb{R}^N}p^{-1}(x)dx\\ &\leq c^2_{\tau\tau'}\,\|\varphi\|^2_{\mathcal{D}_{\tau'}(\mathbb{R}^N)} \quad \Big(\tau' = \langle\tau_1,\big(\tau_2(x)p(x)\big)^{1/2}\rangle\Big).\end{aligned}$$

Let us prove the inverse inequality. We fix $\tau = \langle\tau_1,\tau_2(x)\rangle\in T$ and an integer $l > N/2$ and substitute the function $\tau_2(x)(D^\mu\varphi)(x)$, where $\varphi\in C_0^\infty(\mathbb{R}^N)$ and $|\mu|\leq\tau_1$, for $\varphi(x)$ in (4.18). As a result, we arrive at an inequality similar to (4.19), i.e.,

$$\tau_2(x)|(D^\mu\varphi)(x)| \leq c_1\Big(\sum_{|\nu|\leq l}\int_{\mathbb{R}^N}\Big|\sum_{|\kappa|\leq l,|\lambda|\leq\tau_1+l} c_{\mu\nu\kappa\lambda}(D^\kappa\tau_2)(x)(D^\lambda\varphi)(x)\Big|^2dx\Big)^{1/2}. \tag{4.24}$$

Denote by $\tau_2(x)$ a function from $C^\infty(\mathbb{R}^N)$ such that $|(D^\kappa\tau_2)(x)|^2\leq\tau_2'(x)$ $(x\in\mathbb{R}^N)$ for all $|\kappa|\leq l$. Estimating the right-hand side of (4.24) from above, we get

$$\begin{aligned}\tau_2(x)|(D^\mu\varphi)(x)| &\leq c_2\sum_{|\lambda|\leq\tau_1+l}\Big(\int_{\mathbb{R}^N}|(D^\lambda\varphi)(x)|^2\tau_2'(x)dx\Big)^{1/2}\\ &\leq c_3\,\|\varphi\|_{W_2^{\tau_1+l}(\mathbb{R}^N,\tau_2'(x)dx)}.\end{aligned}$$

This inequality and (4.21) imply that $\|\varphi\|_{\mathcal{D}_\tau(\mathbb{R}^N)}\leq c'_{\tau\tau'}\|\varphi\|_{D_{\tau'}(\mathbb{R}^N)}$ $(\varphi\in C_0^\infty(\mathbb{R}^N))$, where $\tau' = \langle\tau_1+l,\tau_2'(x)\rangle\in T$.

Thus, $\mathcal{D}(\mathbb{R}^N) = \operatorname{pr}\lim_{\tau\in T} D_\tau\mathbb{R}^N)$. The nuclearity of $\mathcal{D}(\mathbb{R}^N)$ follows from Theorem 4.3. $\square$

Theorem 2.1 now yields the following equality similar to (4.20):

$$\mathcal{D}'(\mathbb{R}^N) = \cup_{\tau\in T} D_{-\tau}(\mathbb{R}^N), \quad D_{-\tau}(\mathbb{R}^N) = W_2^{-\tau_1}(\mathbb{R}^N, \tau_2(x)dx). \tag{4.25}$$

It has the same meaning as before: The order of every generalized function from $\mathcal{D}'(\mathbb{R}^N)$ is finite.

Thus, the space $\mathcal{D}(\mathbb{R}^N)$ is defined as the projective limit of Banach spaces $\mathcal{D}_\tau(\mathbb{R}^N)$ or Hilbert spaces $D_\tau(\mathbb{R}^N)$.

Let us show that *convergence in this space coincides with classical convergence introduced in* $\mathcal{D}(\mathbb{R}^N)$.

Indeed, assume that $C_0^\infty(\mathbb{R}^N) \ni \varphi_n \xrightarrow[n\to\infty]{} \varphi \in C_0^\infty(\mathbb{R}^N)$ in the classical sense. Then, by virtue of (4.21), $\|\varphi_n - \varphi\|_{\mathcal{D}_\tau(\mathbb{R}^N)} \to 0$ as $n \to \infty$ for all $\tau \in T$. Let us prove the converse assertion: Let $C_0^\infty(\mathbb{R}^N) \ni \varphi_n \xrightarrow[n\to\infty]{} \varphi \in C_0^\infty(\mathbb{R}^N)$ in $\mathcal{D}(\mathbb{R}^N)$; then this sequence is also convergent in the classical sense. As above, we use the collections of neighbourhoods defined by norms (4.21). The fact that the φ_n converge to φ in each of these norms implies that $(D^\mu\varphi_n)(x)$ converges to $(D^\mu\varphi)(x)$ for any derivative D^μ uniformly in each bounded subset of $\mathbb{R}^N$. Therefore, the required assertion will be proved if we show that the functions φ_n are uniformly finite. Clearly, it suffices to consider the case $\varphi = 0$.

Assume the contrary. Then one can find a sequence of indices $(n_m)_{m=1}^\infty$ and a sequence of points $(x_m)_{m=1}^\infty$ from $\mathbb{R}^N$ such that $n_1 < n_2 < \dots$, $\lim_{m\to\infty} |x_m| = \infty$, and $\varphi_{n_m}(x_m) \neq 0 \ (m \in \mathbb{N})$. Denote by B_m an open sphere in $\mathbb{R}^N$ centered at a point x_m and such that $|\varphi_{n_m}(x)| > \varepsilon_m > 0 \ (x \in B_m, m \in \mathbb{N})$. Let us construct a function $\tau_2 \in C^\infty(\mathbb{R}^N), \tau_2(x) \geq 1 \ (x \in \mathbb{R}^N)$ for which $\tau_2(x) \geq \varepsilon_m^{-1}$ whenever $x \in B_m (m \in \mathbb{N})$ and consider the corresponding space $\mathcal{D}_\tau(\mathbb{R}^N)$, where $\tau = \langle 0, \tau_2(x)\rangle$. Then, according to (4.21), $\|\varphi_{n_m}\|_{\mathcal{D}_\tau(\mathbb{R}^N)} \geq 1$ for all $m \in \mathbb{N}$ and, therefore, the sequence $(\varphi_n)_{n=1}^\infty$ does not approach $\varphi = 0$ in the topology of the space $\mathcal{D}(\mathbb{R}^N)$. We arrive at a contradiction. □

Thus, we have explained in what way the classical spaces of the theory of generalized functions are connected with positive and negative Sobolev spaces.

Here, we do not discuss the problem of introducing the inductive topology in the space $\mathcal{D}(\mathbb{R}^N)$ in detail and only mention the following fact: Let B_r be an open sphere in $\mathbb{R}^N$ with radius $r > 0$ centered at the origin. Denote by $C_0^\infty(B_r)$ the collection of all functions from $C_0^\infty(\mathbb{R}^N)$ that annul outside the sphere B_r and some neighbourhood of its boundary. Clearly, $C_0^\infty(\mathbb{R}^N) = \cup_{n=1}^\infty C_0^\infty(B_n)$. Each $C_0^\infty(B_n)$ is equipped with the relative topology induced by the space $C^\infty(\tilde{B}_n) \supset C_0^\infty(B_n)$ (i.e., neighbourhoods in $C_0^\infty(B_n)$ are obtained as the intersections of the neighbourhoods from $C^\infty(\tilde{B}_n)$ constructed in Example 3.1 with $C_0^\infty(B_n)$). The "inductive" neighbourhoods in $\mathcal{D}(\mathbb{R}^N)$ are similar to (2.10) but, instead of the spheres on the right-hand side of (2.10), one should consider the indicated neighbourhoods from the space $C_0^\infty(B_n)$. Hence, from this point of view, $\mathcal{D}(\mathbb{R}^N)$ is the inductive (countable) limit of projective limits.

5 Tensor Products of Spaces

In functional analysis, an important role is played by the concept of the tensor product of spaces because it gives an abstract description of the procedure of the construction of a space of functions of several variables in terms of a given space of functions of a single variable. To make our presentation as simple and clear as possible, we study this concept and introduce all relevant structures only for the case of separable Hilbert spaces, their riggings, and projective limits. Here, we do not dwell upon the case of infinitely many variables.

5.1 Tensor Products of Spaces

Let $(H_k)_{k=1}^n$ be a finite sequence of separable Hilbert spaces and let $(e_j^{(k)})_{j=0}^\infty$ be an orthonormal basis in H_k. Consider the formal product

$$e_\alpha = e_{\alpha_1}^{(1)} \otimes \cdots \otimes e_{\alpha_n}^{(n)}, \tag{5.1}$$

where $\alpha = (\alpha_1, \ldots, \alpha_n) \in \mathbb{Z}_+^n = \mathbb{Z}_+ \times \cdots \times \mathbb{Z}_+$ (n times), i.e., we consider the ordered sequence $(e_{\alpha_1}^{(1)}, \ldots, e_{\alpha_n}^{(n)})$ and construct a Hilbert space spanned by the formal vectors (5.1) which are assumed to be an orthonormal basis of this space. The separable Hilbert space thus constructed is called the tensor product of the spaces $H_1, \ldots, H_n$ and is denoted by $H_1 \otimes \cdots \otimes H_n = \otimes_{k=1}^n H_k$. Its vectors have the form

$$f = \sum_{\alpha \in \mathbb{Z}_+^n} f_\alpha e_\alpha \quad (f_\alpha \in \mathbb{C}), \quad \|f\|^2_{\otimes_{k=1}^n H_k} = \sum_{\alpha \in \mathbb{Z}_+^n} |f_\alpha|^2 < \infty, \tag{5.2}$$

$$(f, g)_{\otimes_{k=1}^n H_k} = \sum_{\alpha \in \mathbb{Z}_+^n} f_\alpha \bar{g}_\alpha, \quad g = \sum_{\alpha \in \mathbb{Z}_n^+} g_\alpha e_\alpha \in \bigotimes_{k=1}^n H_k.$$

Let $f^{(k)} = \sum_{j=0}^\infty f_j^{(k)} e_j^{(k)} \in H_k$ ($k = 1, \ldots, n$) be some vectors. By definition,

$$f = f^{(1)} \otimes \cdots \otimes f^{(n)} = \sum_{\alpha \in \mathbb{Z}_+^n} f_{\alpha_1}^{(1)} \ldots f_{\alpha_n}^{(n)} e_\alpha. \tag{5.3}$$

The coefficients $f_\alpha = f_{\alpha_1}^{(1)} \ldots f_{\alpha_n}^{(n)}$ of decomposition (5.3) satisfy condition (5.2). Therefore, vector (5.3) belongs to $\otimes_{k=1}^n H_k$ and, in addition,

$$\|f\|_{\otimes_{k=1}^n H_k} = \prod_{k=1}^n \|f_k\|_{H_k}. \tag{5.4}$$

Clearly, the function

$$H_1 \oplus \cdots \oplus H_n \ni \langle f^{(1)}, \ldots, f^{(n)} \rangle \mapsto f^{(1)} \otimes \cdots \otimes f^{(n)} \in \bigotimes_{k=1}^n H_k$$

is linear in each argument and the linear span L of vectors (5.3) is dense in $\otimes_{k=1}^n H_k$. This linear span is called an algebraic (noncompleted) tensor product of the spaces $H_1, \dots, H_n$ and is denoted by $\text{a.}\otimes_{k=1}^n H_k$. If L_k is a linear set in H_k $(k = 1, \dots, n)$ then, by analogy,

$$\text{a.} \bigotimes_{k=1}^{n} L_k = \text{l.s.}\{f^{(1)} \otimes \cdots \otimes f^{(n)} \mid f^{(k)} \in L_k, \quad k = 1, \dots, n\},$$
$$\bigotimes_{k=1}^{n} L_k = \text{c.l.s.}\{f^{(1)} \otimes \cdots \otimes f^{(n)} \mid f^{(k)} \in L_k, \quad k = 1, \dots, n\}.$$

This definition of the tensor product depends, clearly, on the choice of an orthonormal basis $(e_j^{(k)})_{j=0}^\infty$ in each H_k. However, it is easy to show that, by changing the basis, one always arrives at a tensor product which is isomorphic to the original one with preservation of the structure.

In fact, for the case of two Hilbert spaces H_1 and H_2, the concept of tensor product introduced above has the following meaning: We consider the linear span L of the formal products $f^{(1)} \otimes f^{(2)}$ and suppose that

$$\begin{gathered}
(f^{(1)} + g^{(1)}) \otimes f^{(2)} = f^{(1)} \otimes f^{(2)} + g^{(1)} \otimes f^{(2)}, \\
f^{(1)} \otimes (f^{(2)} + g^{(2)}) = f^{(1)} \otimes f^{(2)} + f^{(1)} \otimes g^{(2)}, \\
(\lambda f^{(1)}) \otimes f^{(2)} = \lambda(f^{(1)} \otimes f^{(2)}), \\
f^{(1)} \otimes (\lambda f^{(2)}) = \lambda(f^{(1)} \otimes f^{(2)}) \\
(f^{(1)}, g^1 \in H_1; \quad f^{(2)}, g^{(2)} \in H_2; \quad \lambda \in \mathbb{C}).
\end{gathered} \tag{5.5}$$

In other words, the linear space L is factorized by its linear subspace spanned by all possible vectors representable as differences between right-hand and left-hand sides of equalities (5.5).

Then L is equipped with a scalar product. For vectors of the form $f^{(1)} \otimes f^{(2)}$, it is defined by the formula

$$\begin{gathered}
(f^{(1)} \otimes f^{(2)}, g^{(1)} \otimes g^{(2)})_{H_1 \otimes H_2} = (f^{(1)}, g^{(1)})_{H_1} (f^{(2)}, g^{(2)})_{H_2} \\
(f^{(1)}, g^{(1)} \in H_1; \quad f^{(2)}, g^{(2)} \in H_2)
\end{gathered}$$

and then bilinearly extended to the other elements of the factorized space L.

Example

5.1. Let $H_k = L_2(R_k, \mathfrak{R}_k, d\mu_k(x_k))$, where R_k is a measurable space with a measure μ_k given on a σ-algebra $\mathfrak{R}_k$; $\mu_k(R_k) \le +\infty$ $(k = 1, \dots, n)$. Then

$$\begin{gathered}
\bigotimes_{k=1}^{n} H_k = L_2\big(\mathop{\times}_{k=1}^{n} R_k, \quad \mathop{\times}_{k=1}^{n} \mathfrak{R}_k, \quad d(\mathop{\times}_{k=1}^{n} \mu_k)(x)\big) = L_2 \\
(x = (x_1, \dots x_n) \in \mathop{\times}_{k=1}^{n} R_k).
\end{gathered} \tag{5.6}$$

Indeed, to prove (5.6), one must associate a vector of the form (5.1) $e_\alpha = e^{(1)}_{\alpha_1}\otimes\cdots\otimes e^{(n)}_{\alpha_n} \in \otimes^n_{k=1} H_k$ with a function $e_\alpha(x) = e^{(1)}_{\alpha_1}(x_1)\ldots e^{(n)}_{\alpha_n}(x_n) \in L_2$. These functions form an orthonormal basis of the space L_2 (see Lemma 8.7.1). Therefore, the indicated correspondence generates the required isomorphism between $\otimes^n_{k=1} H_k$ and L_2.

5.2 Tensor Products of Operators

Here, we give a definition of the tensor product of bounded operators.

Theorem 5.1. *Let $(H_k)^n_{k=1}$ and $(G_k)^n_{k=1}$ be two sequences of Hilbert spaces and let $(A_k)^n_{k=1}$ be a sequence of operators $A_k \in \mathcal{L}(H_k, G_k)$. The tensor product $A_1 \otimes \cdots \otimes A_n = \otimes^n_{k=1} A_k$ is defined by the formula*

$$\Big(\overset{n}{\underset{k=1}{\otimes}} A_k\Big) f = \Big(\overset{n}{\underset{k=1}{\otimes}} A_k\Big)\Big(\sum_{\alpha\in\mathbb{Z}^n_+} f_\alpha e_\alpha\Big) = \sum_{\alpha\in\mathbb{Z}^n_+} f_\alpha \Big(A_1 e^{(1)}_{\alpha_1}\Big)\otimes\cdots\otimes\Big(A_n e^{(n)}_{\alpha_n}\Big)$$

$$\Big(f \in \overset{n}{\underset{k=1}{\otimes}} H_k\Big). \tag{5.7}$$

It is stated that the series on the right-hand side of (5.7) is weakly convergent in $\otimes^n_{k=1} G_k$ and defines the operator $\otimes^n_{k=1} A_k \in \mathcal{L}\Big(\otimes^n_{k=1} H_k, \otimes^n_{k=1} G_k\Big)$. Furthermore,

$$\Big\|\overset{n}{\underset{k=1}{\otimes}} A_k\Big\| = \prod_{k=1}^{n} \|A_k\|. \tag{5.8}$$

Proof. It suffices to consider the case of $n = 2$ because, in view of the equality $H_1 \otimes \cdots \otimes H_n = (H_1 \otimes \cdots \otimes H_{n-1}) \otimes H_n$ (associativity of the tensor product), the general case can be then obtained by induction.

Thus, let $n = 2$. Denote by $(l^{(k)}_j)^\infty_{j=0}$ an orthonormal basis in $G_k (k = 1, 2)$ and assume that $g = \sum_{\beta\in\mathbb{Z}^2_+} g_\beta l^{(1)}_{\beta_1}\otimes l^{(2)}_{\beta_2} \in G_1\otimes G_2$. Let f be a vector from $H_1\otimes H_2$ with finitely many nonzero coordinates f_α. We fix $\alpha_2, \beta_1 \in \mathbb{Z}_+$ and denote the vectors $f(\alpha_2) = \sum^\infty_{\alpha_1=0} f_\alpha e^{(1)}_{\alpha_1}$ and $g(\beta_1) = \sum^\infty_{\beta_2=0} g_\beta l^{(2)}_{\beta_2}$ by $f(\alpha_2) \in H_1$ and $g(\beta_1) \in G_2$, respectively. As a result, we obtain

$$\Big|\Big(\sum_{\alpha\in\mathbb{Z}^2_+} f_\alpha A_1 e^{(1)}_{\alpha_1} \otimes A_2 e^{(2)}_{\alpha_2}, g\Big)_{G_1\otimes G_2}\Big|^2$$

$$= \Big|\sum_{\alpha,\beta\in\mathbb{Z}^2_+} f_\alpha \bar{g}_\beta \Big(A_1 e^{(1)}_{\alpha_1}, l^{(1)}_{\beta_1}\Big)_{G_1} \Big(A_2 e^{(2)}_{\alpha_2}, l^{(2)}_{\beta_2}\Big)_{G_2}\Big|^2$$

$$= \Big|\sum_{\alpha_2=0}^{\infty}\sum_{\beta_1=0}^{\infty} \Big(A_1 f(\alpha_2), l^{(1)}_{\beta_1}\Big)_{G_1} \overline{\Big(A_2^* g(\beta_1), e^{(2)}_{\alpha_2}\Big)_{H_2}}\Big|^2$$

$$
\begin{aligned}
&\le \sum_{\alpha_2=0}^{\infty}\sum_{\beta_1=0}^{\infty}\left|\left(A_1 f(\alpha_2), l_{\beta_1}^{(1)}\right)_{G_1}\right|^2 \sum_{\alpha_2=0}^{\infty}\sum_{\beta_1=0}^{\infty}\left|\left(A_2^* g(\beta_1), e_{\alpha_2}^{(2)}\right)_{H_2}\right|^2 \\
&\le \sum_{\alpha_2=0}^{\infty}\left\|A_1 f(\alpha_2)\right\|_{G_1}^2 \sum_{\beta_1=0}^{\infty}\left\|A_2^* g(\beta_1)\right\|_{H_2}^2 \\
&\le \|A_1\|^2\|A_2^*\|^2 \sum_{\alpha_2=0}^{\infty}\left\|f(\alpha_2)\right\|_{H_1}^2 \sum_{\beta_1=0}^{\infty}\left\|g(\beta_1)\right\|_{G_2}^2 \\
&= \|A_1\|^2\|A_2\|^2 \sum_{\alpha\in\mathbb{Z}_+^2}|f_\alpha|^2 \sum_{\beta\in\mathbb{Z}_+^2}|g_\beta|^2.
\end{aligned}
$$

This relation implies that the series $\sum_{\alpha\in\mathbb{Z}_+^2} f_\alpha A_1 e_{\alpha_1}^{(1)} \otimes e_{\alpha_2}^{(2)}$ is weakly convergent in $G_1 \otimes G_2$ for arbitrary $f \in H_1 \otimes H_2$ and its norm in $G_1 \otimes G_2$ is majorized by the number $\|A_1\|\,\|A_2\|\,\|f\|_{H_1\otimes H_2}$. Thus, the operator $A_1 \otimes A_2 \colon H_1 \otimes H_2 \to G_1 \otimes G_2$ is well-defined by (5.7); moreover, it is bounded and its norm does not exceed $\|A_1\|\,\|A_2\|$.

On the other hand, according to (5.4) and (5.7), we have

$$\left\|(A_1 \otimes A_2)\,(f_1 \otimes f_2)\right\|_{G_1\otimes G_2} = \|A_1 f_1\|_{G_1}\,\|A_2 f_2\|_{G_2} \quad (f_k \in H_k;\ k = 1, 2).$$

By choosing proper unit vectors f_1 and f_2, one can make the last product as close to $\|A_1\|\,\|A_2\|$ as desired. Therefore, the inequality $\|A_1 \otimes A_2\| < \|A_1\|\,\|A_2\|$ is impossible, i.e., for $n = 2$, relation (5.8) is established. □

Remark 5.1. Definition (5.7) yields the equality

$$\Big(\bigotimes_{k=1}^{n} A_k\Big)\ \left(f^{(1)} \otimes \cdots \otimes f^{(n)}\right) = A_1 f^{(1)} \otimes \cdots \otimes A_n f^{(n)} \qquad (5.9)$$
$$\left(f^{(k)} \in H_k;\quad k = 1, \ldots, n\right),$$

which uniquely determines the operator $\otimes_{k=1}^{n} A_k$. The mapping

$$\mathop{\times}_{k=1}^{n} \mathcal{L}(H_k, G_k) \ni \langle A_1, \ldots, A_n\rangle \mapsto \bigotimes_{k=1}^{n} A_k \in \mathcal{L}\Big(\bigotimes_{k=1}^{n} H_k, \otimes_{k=1}^{n} G_k\Big)$$

is linear in each variable. Note that, by using (5.7), one can get the following relations:

$$\Big(\bigotimes_{k=1}^{n} B_k\Big)\Big(\bigotimes_{k=1}^{n} A_k\Big) = \bigotimes_{k=1}^{n} (B_k A_k), \quad \Big(\bigotimes_{k=1}^{n} A_k\Big)^* = \bigotimes_{k=1}^{n} A_k^* \qquad (5.10)$$

for $A_k \in \mathcal{L}(H_k, G_k)$ and $B_k \in \mathcal{L}(G_k, F_k)$ $(k = 1, \ldots, n)$.

Remark 5.2. Suppose that each A_k in Theorem 5.1 is a Hilbert-Schmidt operator. Then $\otimes_{k=1}^{n} A_k$ is also a Hilbert-Schmidt operator and

$$\bigotimes_{k=1}^{n} A_k = \prod_{k=1}^{n} A_k. \qquad (5.11)$$

Indeed, according to (5.7) and (5.4),

$$\mathop{\otimes}_{k=1}^{n} A_k{}^2 = \sum_{\alpha\in\mathbb{Z}_+^n} \|(\mathop{\otimes}_{k=1}^{n} A_k)e_\alpha\|^2_{\otimes_{k=1}^n G_k}$$

$$= \sum_{\alpha\in\mathbb{Z}_+^n} \|A_1 e^{(1)}_{\alpha_1} \otimes \cdots \otimes A_n e^{(n)}_{\alpha_n}\|^2_{\otimes_{k=1}^n G_k} \qquad \square$$

$$= \prod_{k=1}^{n} \Big(\sum_{\alpha_k=0}^{\infty} \|A_k e^{(k)}_{\alpha_k}\|^2_{G_k}\Big) = \prod_{k=1}^{n} A_k{}^2.$$

Corollary 5.1. *Let $H_k \subseteq G_k$ be Hilbert spaces such that the embedding operator $O_k \colon H_k \to G_k$ is continuous ($k = 1, \ldots, n$). Then $\otimes_{k=1}^n H_k \subseteq \otimes_{k=0}^n G_k$ and, for the corresponding embedding operator, we have $O = \otimes_{k=0}^n O_k$. If the operators O_k are quasinuclear for all k, the operator O is also quasinuclear.*

This statement immediately follows from Theorem 5.1 and Remark 5.2 if we consider the mapping

$$\mathop{\otimes}_{k=1}^{n} H_k \ni \sum_{\alpha\in\mathbb{Z}_+^n} f_\alpha e_\alpha \mapsto \sum_{\alpha\in\mathbb{Z}_+^n} f_\alpha O_1 e^{(1)}_{\alpha_1} \otimes \cdots \otimes O_n e^{(n)}_{\alpha_n} \in \mathop{\otimes}_{k=1}^{n} G_k. \qquad \square$$

5.3 Tensor Products of Chains

Let us now study tensor products of the chains of Hilbert spaces introduced in Section 1. Consider a collection of chains of the form (1.4)

$$H_{-,k} \supseteq H_{0,k} \supseteq H_{+,k} \quad (k = 1, \ldots, n). \tag{5.12}$$

According to Corollary 5.1 of Theorem 5.1, we have

$$\mathop{\otimes}_{k=1}^{n} H_{-,k} \supseteq \mathop{\otimes}_{k=1}^{n} H_{0,k} \supseteq \mathop{\otimes}_{k=1}^{n} H_{+,k}. \tag{5.13}$$

Theorem 5.2. *The Hilbert space $\otimes_{k=1}^n H_{-,k}$ can be regarded as the negative space with respect to the zero space $\otimes_{k=1}^n H_{0,k}$ and positive space $\otimes_{k=1}^n H_{+,k}$, i.e., (5.13) is a chain.*

Proof. Denote by G_- the negative space with respect to the zero space $\otimes_{k=1}^n H_{0,k}$ and the positive space $\otimes_{k=1}^n H_{+,k}$. Let O and I be the operators connected with the chain $G_- \supseteq \otimes_{k=1}^n H_{0,k} \supseteq \otimes_{k=1}^n H_{+,k}$ and let O_k and I_k be the operators of the same sort for (5.12). Then $O = \otimes_{k=1}^n O_k$ and, according to (5.10), $I = O^* = \otimes_{k=1}^n I_k$. But G_- is the completion of $\otimes_{k=1}^n H_{0,k}$ with respect to the scalar product $(f, g)_{G_-} = (If, g)_{\otimes_{k=1}^n H_{0,k}}$ which, by virtue of the equality $I = \otimes_{k=1}^n I_k$, coincides with a scalar product in $\otimes_{k=1}^n H_{-,k}$ on the dense set a. $\otimes_{k=1}^n H_{0,k}$. $\square$

In proving Theorem 5.2, we in fact established the following equalities for the operators connected with chains (5.12) and (5.13):

$$I = \mathop{\otimes}_{k=1}^{n} I_k, \quad \mathbf{I} = \mathop{\otimes}_{k=1}^{n} \mathbf{I}_k, \quad J = \mathop{\otimes}_{k=1}^{n} J_k, \quad \mathbf{J} = \mathop{\otimes}_{k=1}^{n} \mathbf{J}_k. \tag{5.14}$$

5.4 Projective Limits

By using the properties of the tensor products of Hilbert spaces established above, we can investigate the same circle of problems for riggings by linear topological spaces studied in Section 2. Consider a collection of riggings of the form (2.11)

$$\Phi_k' \supseteq H_{0,k} \supseteq \Phi_k \quad (k = 1, \ldots, n), \tag{5.15}$$

where $\Phi_k = \operatorname{pr} \lim_{\tau_k \in T_k} H_{+,\tau_k}$ is the projective limit of a family of Hilbert spaces $(H_{+,\tau_k})_{\tau_k \in T_k}$ $(k = 1, \ldots, n)$ directed by embedding and satisfying the required conditions indicated in Section 2. For every multiindex $\tau = (\tau_1, \ldots, \tau_n) \in T = \times_{k=1}^n T_k$, we consider the collection of Hilbert riggings

$$H_{-,\tau_k} \supseteq H_{0,\tau_k} \supseteq H_{+,\tau_k} \quad (k = 1, \ldots, n), \tag{5.16}$$

where H_{-,τ_k} is the Hilbert space dual to H_{+,τ_k} with respect to $H_{0,k}$. According to Lemma 5.2, for fixed $\tau \in T$, the tensor product of chain (5.16) is also a chain

$$\bigotimes_{k=1}^n H_{-,\tau_k} \supseteq \bigotimes_{k=1}^n H_{0,\tau_k} \supseteq \bigotimes_{k=1}^n H_{+,\tau_k}. \tag{5.17}$$

Since each family $\left(H_{+,\tau_k}\right)_{\tau_k \in T_k}$ $(k = 1, \ldots, n)$ is directed, the family of Hilbert spaces $\left(\otimes_{k=1}^n H_{+,\tau_k}\right)_{\tau_k \in T_k}$ is also directed by embedding as follows from (5.8). Furthermore, in addition, the set $\bigcap_{\tau \in T} \otimes_{k=1}^n H_{+,\tau_k}$ is dense in each space of this family. According to our assumption, for any $\tau_k \in T_k$, we have $\|\cdot\|_{H_{0,k}} \le \|\cdot\|_{H_{+,\tau_k}}$. Then, by virtue of Corollary 5.1, we conclude that

$$(\forall \tau \in T)\colon \|\cdot\|_{\otimes_{k=1}^n H_{0,k}} \le \|\cdot\|_{\otimes_{k=1}^n H_{+,\tau_k}}.$$

The tensor product $\otimes_{k=1}^n \Phi_k$ of the spaces Φ_k $(k = 1, \ldots, n)$ is defined as the following projective limit:

$$\bigotimes_{k=1}^n \Phi_k = \operatorname{pr} \lim_{\tau = (\tau_1, \ldots, \tau_n) \in T} \bigotimes_{k=1}^n H_{+,\tau_k} \quad (T = \mathop{\times}_{k=1}^n T_k). \tag{5.18}$$

Hence, as the multiindex $\tau = (\tau_1, \ldots, \tau_n)$ in (5.17) runs over the indexing set T, we obtain the family of chains of the form (2.13) with $H_0 = \otimes_{k=1}^n H_{0,k}$. This enables us to apply the scheme described in Section 2 to construct the chain

$$\Phi' \supseteq H_0 = \bigotimes_{k=1}^n H_{0,k} \supseteq \operatorname{pr} \lim_{\tau \in T} \bigotimes_{k=1}^n H_{+,\tau_k} = \bigotimes_{k=1}^n \Phi_k = \Phi.$$

The space Φ' can be topologized by the topology of the inductive limit of negative spaces $\otimes_{k=1}^n H_{-,\tau_k}$ of chain (5.17). By definition,

$$\bigotimes_{k=1}^n \Phi_k' = \operatorname{ind} \lim_{\tau = (\tau_1, \ldots, \tau_n) \in T} \bigotimes_{k=1}^n H_{-,\tau_k}. \tag{5.19}$$

Finally, we arrive at the chain

$$\bigotimes_{k=1}^{n} \Phi'_k \supseteq \bigotimes_{k=1}^{n} H_{0,k} \supseteq \bigotimes_{k=1}^{n} \Phi_k. \tag{5.20}$$

Corollary 5.1 implies that if each Φ_k $(k = 1, \ldots, n)$ is a nuclear space, then the space $\otimes_{k=1}^{n} \Phi_k$ is also nuclear.

In fact, we have proved the following theorem:

Theorem 5.3. *The tensor product of the chains*

$$\Phi'_k \supseteq H_{0,k} \supseteq \Phi_k \quad (k = 1, \ldots, n),$$

where $\Phi_k = \operatorname{pr}\lim_{\tau_k \in T_k} H_{+,\tau_k}$*, is also a chain*

$$\bigotimes_{k=1}^{n} \Phi'_k \supseteq \bigotimes_{k=1}^{n} H_{0,k} \supseteq \bigotimes_{k=1}^{n} \Phi_k,$$

where the spaces $\otimes_{k=1}^{n} \Phi_k$ *and* $\otimes_{k=1}^{n} \Phi'_k$ *are defined by equalities (5.18) and (5.19).*

Furthermore, assume that each rigging (5.16) is nuclear. Then rigging (5.20) constructed as indicated above is also nuclear.

6 The Kernel Theorem

The question as to whether it is possible to represent the bilinear form of a bounded operator (or, more generally, an arbitrary continuous polylinear form) as an "integral operator with kernel" is closely related to the concept of the tensor product of spaces. We shall demonstrate that this representation is always possible but the kernel is, generaly speaking, a generalized function.

The first version of this result belongs to Schwartz and relates to the space of test functions $\mathcal{S}(\mathbb{R}^N)$. It states that every continuous bilinear form $a(\varphi, \psi)\big(\varphi, \psi \in \mathcal{S}(\mathbb{R}^N)\big)$ can be represented as

$$a(\varphi, \psi) = \langle \alpha, \varphi \otimes \bar{\psi} \rangle, \tag{6.1}$$

where $\alpha \in \mathcal{S}'(\mathbb{R}^{2N})$ is a "generalized kernel" and

$$(\varphi \otimes \bar{\psi})(x_1, \ldots, x_{2N}) = \varphi(x_1, \ldots, x_N)\, \overline{\psi(x_{N+1}, \ldots, x_{2N})}.$$

It is convenient to present results of this type in an abstract form, using tensor products of spaces introduced in Section 5. In this section, we immediately consider the case of polylinear forms; bilinear forms are studied at the end of the section. For the classical spaces $\mathcal{S}(\mathbb{R}^N)$ and $\mathcal{D}(\mathbb{R}^N)$ of test functions, the indicated results turn into well-known facts in the theory of generalized functions.

6.1 Hilbert Riggings

First, we introduce the notion of a generalized kernel. Consider a collection of n chains

$$H_{-,k} \supseteq H_{0,k} \supseteq H_{+,k} \quad (k = 1, \dots, n). \tag{6.2}$$

and their tensor product

$$\bigotimes_{k=1}^{n} H_{-,k} \supseteq \bigotimes_{k=1}^{n} H_{0,k} \supseteq \otimes H_{+,k}. \tag{6.3}$$

Elements $F, G, \dots \in \otimes_{k=1}^{n} H_{0,k}$, $U, V, \dots \in \otimes_{k=1}^{n} H_{+,k}$, and $A, B, \dots \in \otimes_{k=1}^{n} H_{-,k}$ are called ordinary, smooth, and generalized kernels, respectively.

Also consider a continuous n-linear form $a(f^{(1)}, \dots, f^{(n)})$ regarded as a continuous function

$$H_{0,1} \oplus \dots \oplus H_{0,n} \ni \langle f^{(1)}, \dots, f^{(n)} \rangle \mapsto a(f^{(1)}, \dots, f^{(n)}) \in \mathbb{C} \tag{6.4}$$

linear in each $f^{(k)}$ provided that all other variables are fixed. The continuity of (6.4) is equivalent to the existence of the estimate

$$|a(f^{(1)}, \dots, f^{(n)})| \le c \prod_{k=1}^{n} \|f^{(k)}\|_{H_{0,k}} \quad (f^{(k)} \in H_{0,k}; \quad k = 1, \dots, n) \tag{6.5}$$

with some $c > 0$ (this can be proved as in the case where $n = 1$, see Section 7.2). In every $H_{0,k}$, we fix an orthonormal basis $(e_j^{(k)})_{j=0}^{\infty}$. Let $f^{(k)} = \sum_{\alpha_k=0}^{\infty} f_{\alpha_k} e_{\alpha_k}^{(k)}$ be the decomposition of a vector $f^{(k)} \in H_{0,k}$ in this basis. We set $\alpha = (\alpha_1, \dots, \alpha_n) \in \mathbb{Z}_+^n$. In view of the continuity and polylinearity of a, it can be represented in the form of a convergent series in its coordinates a_α and the coordinates of the vectors $f^{(k)}$

$$a(f^{(1)}, \dots, f^{(n)}) = \sum_{\alpha \in \mathbb{Z}_+^n} a_\alpha f_{\alpha_1}^{(1)} \dots f_{\alpha_n}^{(n)}, a_\alpha = a(e_{\alpha_1}^{(1)}, \dots, e_{\alpha_n}^{(n)}). \tag{6.6}$$

The proof of the kernel theorem is based on the following two lemmas:

Lemma 6.1. *Let a be the continuous n-linear form (6.4) and let $A_k \in \mathcal{L}(H_{0,k})$ $(k = 2, \dots, n)$ be Hilbert-Schmidt operators. Consider the continuous n-linear form*

$$H_{0,1} \oplus \dots \oplus H_{0,n} \ni \langle f^{(1)}, \dots, f^{(n)} \rangle \mapsto b(f^{(1)}, \dots, f^{(n)}) = a(f^{(1)}, A_2 f^{(2)}, \dots, A_n f^{(n)}).$$

It is stated that the coordinates $(b_\alpha)_{\alpha \in \mathbb{Z}_+^n}$ of the form b are such that

$$\sum_{\alpha \in \mathbb{Z}_+^n} |b_\alpha|^2 < \infty. \tag{6.7}$$

Conversely, if $A_k \in \mathcal{L}(H_{0,k})$, $A_k \neq 0$ $(k = 2, \dots, n)$ and, for any continuous form a, the coordinates of the form b satisfy condition (6.7), then all A_k are Hilbert-Schmidt operators.

Proof. We fix $f^{(k)} \in H_{0,k}$ $(k = 2, \dots, n)$. Then $H_{0,1} \ni f^{(1)} \mapsto l(f^{(1)}) = a(f^{(1)}, \dots, f^{(n)})$ is a linear continuous functional on $H_{0,1}$, whose norm does not exceed $c \prod_{k=2}^{n} \|f^{(k)}\|_{H_{0,k}}$ (by (6.5)). Since $l(f^{(1)}) = (f^{(1)}, h)_{H_{0,1}}$, where the coordinates $(h_{\alpha_1})_{\alpha_1=0}^{\infty}$ of the vector $h \in H_{0,1}$ in the basis $(e_{\alpha_1}^{(1)})_{\alpha_1=0}^{\infty}$ have the form $a(e_{\alpha_1}^{(1)}, f^{(2)}, \dots, f^{(n)})$, we obtain

$$\sum_{\alpha_1=0}^{\infty} |a(e_{\alpha_1}^{(1)}, f^{(2)}, \dots, f^{(n)})|^2 = \|l\|^2 \leq c^2 \prod_{k=2}^{n} \|f^{(k)}\|_{H_{0,k}}^2$$
$$\left(f^{(k)} \in H_{0,k}, k = 2, \dots, n\right).$$

This estimate implies that

$$\begin{aligned}\sum_{\alpha \in \mathbb{Z}_+^n} |b_\alpha|^2 &= \sum_{\alpha \in \mathbb{Z}_+^n} |b\left(e_{\alpha_1}^{(1)}, \dots, e_{\alpha_n}^{(n)}\right)|^2 \\ &= \sum_{\alpha \in \mathbb{Z}_+^n} |a\left(e_{\alpha_1}^{(1)}, A_2 e_{\alpha_2}^{(2)}, \dots, A_n e_{\alpha_n}^{(n)}\right)|^2 \\ &\leq c^2 \sum_{\alpha_1, \dots, \alpha_n = 0}^{\infty} \prod_{k=2}^{n} \|A_k e_{\alpha_k}^{(k)}\|_{H_{0,k}}^2 = c^2 \prod_{k=2}^{n} A_k{}^2 < \infty.\end{aligned}$$

The converse statement can be obtained if, for given $k = 2, \dots, n$, we construct a continuous form a such that $\sum_{\alpha \in \mathbb{Z}_+^n} |b_\alpha|^2 = c_k\, A_k{}^2$ with some $c_k > 0$ (b is constructed in terms of the form a). Then the fact that the last series is convergent enables us to conclude that $A_k < \infty$. The form a is determined by its coordinates a_α, which have the form $\delta_{\alpha_1 \alpha_k} \delta_{\alpha_2 \beta_2} \cdots \delta_{\alpha_{k-1} \beta_{k-1}} \delta_{\alpha_{k+1} \beta_{k+1}} \cdots \delta_{\alpha_n \beta_n}$ $(\alpha \in \mathbb{Z}_+^n)$, where δ_{jk} is the Kronecker symbol and $\beta_2, \dots, \beta_{k-1}, \beta_{k+1}, \dots, \beta_n = 0, 1, \dots$ are fixed. By computing the value of $\sum_{\alpha \in \mathbb{Z}_+^n} |b_\alpha|^2$, we now easily arrive at the required equality with the constant

$$c_k = \prod_{j \neq k, j=2}^{n} \|A_j^* e_{\beta_j}^{(j)}\|_{H_{0,j}}.$$

Since $A_j \neq 0 (j = 2, \dots, n)$, we can always find β_j such that $c_k > 0$. □

Lemma 6.2. *Let $H_{0,1} \oplus \cdots \oplus H_{0,n} \ni \langle f^{(1)}, \dots, f^{(n)} \rangle \mapsto b(f^{(1)}, \dots, f^{(n)})$ be a continuous n-linear form. It can be represented in the form $b(f^{(1)}, \dots, f^{(n)}) = (f^{(1)} \otimes \cdots \otimes f^{(n)}, K)_{\otimes_{k=1}^n H_{0,k}}$, where $K \in \otimes_{k=1}^n H_{0,k}$, if and only if its coordinates $(b_\alpha)_{\alpha \in \mathbb{Z}_+^n}$ satisfy condition (6.7).*

Proof. Assume that (6.7) is true. We set

$$K = \sum_{\alpha \in \mathbb{Z}_+^n} \bar{b}_\alpha e_\alpha \in \bigotimes_{k=1}^{n} H_{0,k},$$

where $(e_\alpha)_{\alpha \in \mathbb{Z}_+^n}$ is a basis in $\otimes_{k=1}^n H_{0,k}$ of the form (5.1). Then, clearly,

$$\big(f^{(1)} \otimes \cdots \otimes f^{(n)}, K\big)_{\otimes_{k=1}^n H_{0,k}} = \sum_{\alpha \in \mathbb{Z}_+^n} f_{\alpha_1}^{(1)} \cdots f_{\alpha_n}^{(n)} b_\alpha = b\big(f^{(1)}, \dots, f^{(n)}\big)$$

$$\big(f^{(k)} \in H_{0,k}, k = 1, \dots, n\big).$$

Conversely, if the required representation of the form b takes place, then

$$b\Big(e_{\alpha_1}^{(1)}, \dots, e_{\alpha_n}^{(n)}\Big) = \Big(e_\alpha, K\Big)_{\otimes_{k=1}^n H_{0,k}} = \overline{K}_\alpha (\alpha \in \mathbb{Z}_+^n).$$

Thus, by virtue of the inclusion $K \in \otimes_{k=1}^n H_{0,k}$, condition (6.7) is satisfied. □

Theorem 6.1. *Assume that chains (6.2) are such that the embeddings $O_k \colon H_{+,k} \mapsto H_{0,k}$ $(k = 2, \dots, n)$ are quasinuclear. Then every continuous n-linear form*

$$H_{0,1} \oplus \cdots \oplus H_{0,n} \ni \langle f^{(1)}, \dots, f^{(n)} \rangle \mapsto a\big(f^{(1)}, \dots, f^{(n)}\big) \in \mathbb{C}$$

can be associated with a unique generalized kernel $\mathrm{A} \in H_{0,k} \otimes H_{-,2} \otimes \cdots \otimes H_{-,n}$ such that

$$a\big(f^{(1)}, u^{(2)}, \dots, u^{(n)}\big) = \big(f^{(1)} \otimes u^{(2)} \otimes \cdots \otimes u^{(n)}, \mathrm{A}\big)_{\otimes_{k=1}^n H_{0,k}}$$

$$\big(f^{(1)} \in H_{0,1}; \quad u^{(k)} \in H_{+,k}; \quad k = 2, \dots, n\big). \tag{6.8}$$

Conversely, if every form a of the indicated type admits representation (6.8) with $\mathrm{A} \in H_{0,1} \otimes H_{-,2} \otimes \cdots \otimes H_{-,n}$, then the embeddings O_k $(k = 2, \dots, n)$ are quasinuclear.

Proof. Let O_k and $J_k (k = 2, \dots, n)$ be the operators connected with chain (6.2). For $k = 1$, we assume that $H_{+,1} = H_{0,1} = H_{-,1}$. For $f^{(1)} \in H_{0,1}, u^{(k)} \in H_{+,k}$ $(k = 2, \dots, n)$, we have

$$\begin{aligned} a\big(f^{(1)}, u^{(2)}, \dots, u^{(n)}\big) &= a\big(f^{(1)}, O_2 J_2 J_2^{-1} u^{(2)}, \dots, O_n J_n J_n^{-1} u^{(n)}\big) \\ &= b\big(f^{(1)}, J_2^{-1} u^{(2)}, \dots, J_n^{-1} u^{(n)}\big), \end{aligned} \tag{6.9}$$

where

$$b\big(f^{(1)}, \dots, f^{(n)}\big) = a\big(f^{(1)}, O_2 J_2 f^{(2)}, \dots, O_n J_n f^{(n)}\big)$$

$$\big(f^{(k)} \in H_{0,k}, k = 1, \dots, n\big). \tag{6.10}$$

For $k = 2, \dots, n$, the operators O_k are Hilbert-Schmidt operators. The same is true for the operators $A_k = O_k J_k \colon H_{0,k} \to H_{0,k}$. According to Lemma 6.1, in this case, the coordinates b_α of the form b satisfy condition (6.7) and, hence, by virtue of Lemma 6.2, we can write the following representation:

$$b\big(f^{(1)}, \dots, f^{(n)}\big) = \big(f^{(1)} \otimes \cdots \otimes f^{(n)}, K\big)_{\otimes_{k=1}^n H_{0,k}}, \quad K \in \otimes_{k=1}^n H_{0,k}.$$

Therefore, (6.9) can be continued as follows:

$$
\begin{aligned}
a(f^{(1)},u^{(2)},\dots,u^{(n)})
&= \big(f^{(1)} \otimes J_2^{-1}u^{(2)} \otimes \cdots \otimes J_n^{-1}u^{(n)}, K\big)_{\otimes_{k=1}^n H_{0,k}} \\
&= \Big(\Big(\mathbb{1} \otimes J_2^{-1} \otimes \cdots \otimes J_n^{-1}\Big)\Big(f^{(1)} \otimes u^{(2)} \otimes \cdots \otimes u^{(n)}\Big), K\Big)_{\otimes_{k=1}^n H_{0,k}} \\
&= \Big(f^{(1)} \otimes u^{(2)} \otimes \cdots \otimes u^{(n)}, \Big(\mathbb{1} \otimes J_2^{-1} \otimes \cdots \otimes J_n^{-1}\Big)^+ K\Big)_{\otimes_{k=1}^n H_{0,k}} \\
&= \Big(f^{(1)} \otimes u^{(2)} \otimes \cdots \otimes u^{(n)}, \Big(\mathbb{1} \otimes \mathbf{J}_2^{-1} \otimes \cdots \otimes \mathbf{J}_n^{-1}\Big) K\Big)_{\otimes_{k=1}^n H_{0,k}}. \quad (6.11)
\end{aligned}
$$

Here, we have used the relation $\big(\otimes_{k=1}^n J_k^{-1}\big)^+ = \otimes_{k=1}^n \mathbf{J}_k^{-1}$, where $+$ is the operation of conjugation with respect to chain (6.3). By setting $\mathrm{A} = \Big(\mathbb{1}\otimes\mathbf{J}_2^{-1}\otimes\cdots\otimes\mathbf{J}_n^{-1}\Big)K \in H_{0,1} \otimes H_{-,2} \otimes \cdots \otimes H_{-,n}$ in (6.11), we arrive at the required representation (6.8). The fact that the kernel A is uniquely determined for a given form a follows from the denseness of the linear span of the vectors $f^{(1)} \otimes u^{(2)} \otimes \cdots \otimes u^{(n)}\big(f^{(1)} \in H_{0,1}, u^{(k)} \in H_{+,k}; k = 2, \dots, n\big)$ in $H_{0,1} \otimes H_{-,2} \otimes \cdots \otimes H_{-,n}$.

Let us prove the last statement of the theorem. Assume that the form a admits representation (6.8). Then, just as above, we arrive at the following representation for the form b introduced by (6.10):

$$
\begin{aligned}
b(f^{(1)},\dots,f^{(n)}) &= \big(f^{(1)} \otimes J_2 f^{(2)} \otimes \cdots \otimes J_n f^{(n)}, \mathrm{A}\big)_{\otimes_{k=1}^n H_{0,k}} \\
&= \big(f^{(1)} \otimes \cdots \otimes f^{(n)}, (\mathbb{1} \otimes \mathbf{J}_2 \otimes \cdots \otimes \mathbf{J}_n)\mathrm{A}\big)_{\otimes_{k=1}^n H_{0,k}} \\
&\quad \big(f^{(k)} \in H_{0,k}, k = 1, \dots, n\big).
\end{aligned}
$$

At the same time, $\big(\mathbb{1}\otimes\mathbf{J}_2\cdots\otimes\mathbf{J}_n\big)\mathrm{A} \in \otimes_{k=1}^n H_{0,k}$ and, therefore, in view of Lemma 6.2, we conclude that condition (6.7) is satisfied for the coordinates b_α of the form b. Since a is an arbitrary form and $A_k = O_k J_k \neq 0$, it follows from Lemma 6.1 that $A_k < \infty$, i.e., $O_k < \infty$ $(k = 2, \dots, n)$. □

Corollary 6.1. *The statement of Theorem 6.1 can be made somewhat "more symmetric" (at the expense of making the result slightly less precise):*

Assume that each embedding $H_{+,k} \to H_{0,k}$ $(k = 1, \dots, n)$ in (6.2) is quasi-nuclear. Then every n-linear continuous form (6.4) admits the representation

$$
\begin{aligned}
a(u^{(1)},\dots,u^{(n)}) &= \big(u^{(1)} \otimes \cdots \otimes u^{(n)}, \mathrm{A}\big)_{\otimes_{k=1}^n H_{0,k}} \\
&\quad \big(u^{(k)} \in H_{+,k}, k = 1, \dots, n\big). \quad (6.12)
\end{aligned}
$$

Moreover, the kernel $\mathrm{A} \in \otimes_{k=1}^n H_{-,k}$ in this representation is determined uniquely.

As a rule, Theorem 6.1 is used just in the form of Corollary 6.1.

6.2 Nuclear Riggings

Let us now modify Theorem 6.1 for the case of nuclear riggings and forms defined in nuclear spaces. Consider a collection of n nuclear riggings given by (2.11) and (2.14)

$$\Phi'_k \supseteq H_{0,k} \supseteq \Phi_k = \text{pr} \lim_{\tau_k \in T_k} H_{+,\tau_k} \quad (k = 1, \dots, n). \tag{6.13}$$

According to the scheme presented in Section 5 (see (5.20)), we construct the nuclear chain

$$\bigotimes_{k=1}^{n} \Phi'_k \supseteq \bigotimes_{k=1}^{n} H_{0,k} \supseteq \bigotimes_{k=1}^{n} \Phi_k \tag{6.14}$$

and consider, as in (6.4), the n-linear forms

$$\bigoplus_{k=1}^{n} \Phi_k \ni \langle \varphi^{(1)}, \dots, \varphi^{(n)} \rangle \mapsto a\big(\varphi^{(1)}, \dots, \varphi^{(n)}\big) \in \mathbb{C} \tag{6.15}$$

continuous in the direct product $\otimes_{k=1}^{n} \Phi_k$ of linear topological spaces Φ_k. Since

$$\bigotimes_{k=1}^{n} \Phi_k = \text{pr} \lim_{\tau = (\tau_1, \dots, \tau_n) \in T} \bigotimes_{k=1}^{n} H_{+,\tau_k} \quad (T = \mathop{\times}_{k=1}^{n} T_k),$$

one can easily prove that the continuity of the form (6.15) is equivalent to its continuity in the norm of the space $\otimes_{k=1}^{n} H_{+,\tau_k}$ with certain $\tau = (\tau_1, \dots, \tau_n)$ and, hence, to the validity of the estimate

$$\big|a(\varphi^{(1)}, \dots, \varphi^{(n)})\big| \le c_\tau \prod_{k=1}^{n} \|\varphi^{(k)}\|_{H_{+,\tau_k}} \tag{6.16}$$
$$\big(c_\tau > 0, \varphi^{(k)} \in H_{+,\tau_k}; k = 1, \dots, n\big).$$

Theorem 6.2. *For the nuclear riggings (6.13), every continuous n-linear form (6.15) admits the representation*

$$a\big(\varphi^{(1)}, \dots, \varphi^{(n)}\big) = \big(\varphi^{(1)} \otimes \cdots \otimes \varphi^{(n)}, \mathrm{A}\big)_{\otimes_{k=1}^{n} H_{0,k}} \tag{6.17}$$
$$\big(\varphi^{(k)} \in \Phi_k, k = 1, \dots, n\big);$$

moreover, the generalized kernel $\mathrm{A} \in \otimes_{k=1}^{n} \Phi'_k$ in this representation is uniquely determined for given a.

Proof. In view of the continuity of a, we can choose $\tau = (\tau_1, \dots, \tau_n) \in T$ such that inequality (6.16) is satisfied. For every τ_k, in view of the nuclearity of Φ_k, we find $\tau'_k \in T_k$ such that $H_{+,\tau'_k} \subseteq H_{+,\tau_k}$ and the embedding operator $O_{\tau'_k,\tau_k} : H_{+,\tau'_k} \to H_{+,\tau_k}$ is quasinuclear. Consider the collection of chains

$$H_{-,\tau'_k} \supseteq H_{-,\tau_k} \supseteq H_{0,k} \supseteq H_{+,\tau_k} \supseteq H_{+,\tau'_k} \quad (k = 1, \dots, n)$$

and their tensor product

$$\bigotimes_{k=1}^{n} H_{-,\tau'_k} \supseteq \bigotimes_{k=1}^{n} H_{-,\tau_k} \supseteq \bigotimes_{k=1}^{n} H_{0,k} \supseteq \bigotimes_{k=1}^{n} H_{+,\tau_k} \supseteq \bigotimes_{k=1}^{n} H_{+,\tau'_k}.$$

The form $a(\varphi^{(1)},\dots,\varphi^{(n)})$ $(\varphi^{(k)} \in \Phi_k)$ is continuous in $\otimes_{k=1}^{n} H_{+,\tau_k}$ and the embedding $H_{+,\tau'_k} \to H_{+,\tau_k}$ is quasinuclear $(k = 1,\dots,n)$ Therefore, by virtue of Corollary 6.1, we get the following representation:

$$a(\varphi^{(1)},\dots,\varphi^{(n)}) = (\varphi^{(1)} \otimes \cdots \otimes \varphi^{(n)}, \mathrm{A})_{\otimes_{k=1}^{n} H_{+,\tau_k}} \tag{6.18}$$
$$(\varphi^{(k)} \in \Phi_k, \quad k = 1,\dots,n),$$

where the kernel A lies in $\otimes_{k=1}^{n} H_{-,\tau'_k}^{(\tau_k)}$ and the space $H_{-,\tau'_k}^{(\tau_k)}$ is dual to the space H_{+,τ'_k} with respect to H_{+,τ_k}, i.e., it is an element of the chain $H_{-,\tau'_k}^{(\tau_k)} \supseteq H_{+,\tau_k} \supseteq H_{+,\tau'_k} (k = 1,\dots,n)$. Denote by $\mathbf{I}_{\tau'_k}^{(\tau_k)} : H_{-,\tau'_k}^{(\tau_k)} \to H_{+,\tau'_k}$ the isometries connected with the last chains. Then (6.18) can be rewritten in the form

$$a(\varphi^{(1)},\dots,\varphi^{(n)}) = (\varphi^{(1)} \otimes \cdots \otimes \varphi^{(n)}, \mathrm{B})_{\otimes_{k=1}^{n} H_{+,\tau'_k}}, \tag{6.19}$$

where $\mathrm{B} = (\otimes_{k=1}^{n} \mathbf{I}_{\tau'_k}^{(\tau_k)})\mathrm{A} \in \otimes_{k=1}^{n} H_{+,\tau'_k}$.

Finally, by using the isometries $\mathbf{I}_{\tau'_k} : H_{-,\tau'_k} \to H_{+,\tau'_k}$, we can transform (6.19) into the required representation (6.17) with the kernel

$$\mathrm{A} = \Big(\bigotimes_{k=1}^{n} \mathbf{I}_{\tau'_k}^{-1}\Big)\mathrm{B} \in \bigotimes_{k=1}^{n} H_{-,\tau'_k} \subseteq \bigotimes_{k=1}^{n} \Phi'_k.$$

The fact that the kernel A is uniquely determined for given a follows, as above, from the denseness of the linear span of the vectors

$$\varphi^{(1)} \otimes \cdots \otimes \varphi^{(n)} \qquad (\varphi^{(k)} \in \Phi_k)$$

in $\otimes_{k=1}^{n} \Phi_k$. □

REMARK 6.1. In the case where each Φ_k in (6.15) is a countably Hilbert space, every separately continuous (i.e., continuous in each variable when all other variables are fixed) polylinear form (6.15) is also continuous with respect to the collection of its variables, i.e., in $\times_{k=1}^{n} \Phi_k$. Therefore, for countably Hilbert Φ_k, Theorem 6.2 remains true for separately continuous forms. A similar situation is also observed for some other Φ_k, e.g., for the space $\mathcal{D}(\mathbb{R}^N)$.

6.3 Bilinear Forms

By using Theorem 6.1, one can also establish the kernel theorem for bilinear (sesquilinear) forms. Thus, consider the chain

$$H_- \supseteq H_0 \supseteq H_+ \tag{6.20}$$

and assume that H_+ is equipped with an involution, which is, at the same time, an involution in H_0. This means that one can indicate an antilinear mapping $H_+ \ni u \mapsto u^* \in H_+$ acting on H_+ such that $(u^*)^* = u, (u^*, v^*)_{H_+} = \overline{(u, v)_{H_+}}$ and $(u^*, v^*)_{H_0} = \overline{(u, v)_{H_0}}$ $(u, v \in H_+$; cf. Section 8.4).

It is easy to see that, in this case, $(u^*, v^*)_{H_-} = \overline{(u, v)_{H_-}}$, i.e., the operation $*$ is also an involution in H_- and, consequently, can be extended by continuity to the involution in the whole H_-: $H_- \ni \alpha \mapsto \alpha^* \in H_-$. The restriction of this mapping $H_- \supseteq H_0 \ni f \mapsto f^* \in H_0$ is an involution in H_0. Clearly, $(\mathbf{I}\alpha)^* = \mathbf{I}\alpha^*$ $(\alpha \in H_-)$. If the spaces of chain (6.20) are equipped with this involution, then we say that (6.20) is a chain with the involution $*$.

Let $a(f, g)$ be a bilinear form defined in H_0, i.e., a continuous function $H_0 \oplus H_0 \ni \langle f, g \rangle \mapsto a(f, g) \in \mathbb{C}$ linear in the first variable and antilinear in the second variable. The following theorem is true:

Theorem 6.3. *Assume that the embedding $H_+ \to H_0$ in chain (6.20) is quasinuclear. Then, for every continuous bilinear form $a(f, g)$ $(f, g \in H_0)$, one can construct a unique generalized kernel $\mathrm{A}_a \in H_0 \otimes H_-$ such that*

$$a(u, g) = (\mathrm{A}_a, g \otimes u^*)_{H_0 \otimes H_0} \quad (u \in H_+; g \in H_0). \tag{6.21}$$

Conversely, if every form a of the indicated type admits representation (6.21) with $\mathrm{A}_a \in H_0 \otimes H_-$, then the embedding $H_+ \to H_0$ is quasinuclear.

Proof. Let us construct a continuous bilinear form b by setting $b(f, g) = a(g^*, f)$ $(f, g \in H_0)$. According to Theorem 6.1, there exists a generalized kernel $\mathrm{A} \in H_0 \otimes H_-$ such that $b(f, u) = (f \otimes u, \mathrm{A})_{H_0 \otimes H_0}$ $(f \in H_0, u \in H_+)$. Therefore, for $u \in H_+$, $g \in H_0$,

$$a(u, g) = \overline{b(g, u^*)} = \overline{(g \otimes u^*, \mathrm{A})}_{H_0 \otimes H_0} = (\mathrm{A}, g \otimes u^*)_{H_0 \otimes H_0},$$

i.e., representation (6.21) holds with $\mathrm{A}_a = \mathrm{A}$. It is obvious that A_a is uniquely determined for given a.

The converse statement immediately follows from the similar assertion in Theorem 6.1, namely, representation (6.21) implies representation (6.8) for the relevant b which is arbitrary in view of the arbitrariness of a. □

REMARK 6.2. In Theorem 6.1, one can clearly fix any other variable instead of the first one. Hence, in addition to (6.21), a admits the following representation:

$$a(f,v) = \left(\mathrm{A}'_a, v \otimes f^*\right)_{H_0 \otimes H_0} \quad (f \in H_0, v \in H_+), \tag{6.22}$$

where the kernel $\mathrm{A}'_a \in H_- \otimes H_0$ is uniquely determined for given a. It follows from (6.21) and (6.22) that

$$a(v^*, u) = \left(\mathrm{A}_a, u \otimes v\right)_{H_0 \otimes H_0} = \left(\mathrm{A}'_a, u \otimes v\right)_{H_0 \otimes H_0}$$

for $u, v \in H_+$, i.e., the restrictions of the functionals A_a and A'_a to $H_+ \otimes H_+$ coincide. Denote this common restriction by A. Thus, under the conditions of Theorem 6.3, we have the representation

$$a(u,v) = \left(\mathrm{A}, v \otimes u^*\right)_{H_0 \otimes H_0} \quad (u, v \in H_+) \tag{6.23}$$

with the kernel $\mathrm{A} \in H_- \otimes H_-$.

REMARK 6.3. The assertions of Theorem 6.2 for the case of nuclear riggings can also be easily generalized to the case of bilinear forms. Thus, let Φ be a nuclear space with natural involution $*\colon \Phi \to \Phi$. Consider a bilinear form defined in Φ, i.e., a continuous function $\Phi \times \Phi \ni \langle \varphi, \psi \rangle \mapsto a(\varphi, \psi) \in \mathbb{C}$ linear in the first variable and antilinear in the second variable. Then $a(\varphi, \psi) = \left(\mathrm{A}, \psi \otimes \varphi^*\right)_{H_0 \otimes H_0}$, where $\mathrm{A} \in \Phi' \otimes \Phi'$.

Let $A \in \mathcal{L}(H_0)$. Given A, we construct the continuous bilinear form $a(f,g) = (Af, g)_{H_0}$ $(f, g \in H_0)$. Then, by using this form, we construct, according to (6.23), the kernel $\mathcal{A}_A = \mathcal{A} \in H_- \otimes H_-$. It is called the kernel of the operator A,

$$(Au, v)_{H_0} = \left(\mathcal{A}, v \otimes u^*\right)_{H_0 \otimes H_0} \quad (u, v \in H_+). \tag{6.24}$$

Example

6.1. Let $H_0 = L_2(R, \mathfrak{R}, d\mu) = L_2$, where R is a space with a measure μ given on a σ-algebra $\mathfrak{R}$, and let A be a Hilbert-Schmidt operator in H_0, i.e.,

$$(Af)(x) = \int_R K(x,y) f(y) d\mu(y),$$
$$K \in L_2\left(R \times R, \mathfrak{R} \times \mathfrak{R}, d(\mu \times \mu)\right) = H_0 \otimes H_0. \tag{6.25}$$

Then

$$(Af, g)_{H_0} = \int_{R \times R} K(x,y) f(y) \overline{g(x)} d(\mu \times \mu)(x,y)$$
$$= \left(K, g \otimes f^*\right)_{H_0 \otimes H_0}, \tag{6.26}$$

where $f^*(\cdot) = \overline{f(\cdot)}$. By comparing (6.26) with (6.24), we note that, in this case, $\mathcal{A} = K$. Relation (6.24) demonstrates that an arbitrary bounded operator A in L_2 admits, in a certain sense, representation (6.25) with $K = \mathcal{A}$, i.e., it is an "integral operator with generalized kernel". Relation (6.24) also clarifies the arrangement of the variables on the right-hand sides of (6.21)–(6.23).

6.4 One More Kernel Theorem

In order to formulate the kernel theorem for the case of polylinear (bilinear) forms defined in the spaces $L_2(G)$ $(G \subseteq \mathbb{R}^N, N \in \mathbb{N})$ with respect to the Lebesgue measure, one can use the Sobolev space quasinuclearly embedded in $L_2(G)$ (see Theorems 3.2, 4.3, and 4.4) or, in the case of nuclear riggings, the spaces $C^\infty(G)$, $\mathcal{S}(\mathbb{R}^N)$, and $\mathcal{D}(\mathbb{R}^N)$.

The following "elementary" kernel theorem is also true:

Theorem 6.4. *Let $L_2(G) \oplus L_2(G) \ni \langle f, g\rangle \mapsto a(f,g) \in \mathbb{C}$ be a continuous bilinear form. Then there exists a kernel $T \in C(\mathbb{R}^N \times \mathbb{R}^N)$ such that*

$$a(u,v) = \int_{G\times G} T(x,y)(\mathcal{D}u)(y)\overline{(\mathcal{D}v)(x)}dxdy \tag{6.27}$$
$$(\mathcal{D} = D_1 \dots D_N; \quad u, v \in C_0^N(G))$$

for the smooth functions finite with respect to G.

Proof. Consider the indicator of an open parallelepiped in $\mathbb{R}^N$ formed by the coordinate hyperplanes and the hyperplanes parallel to them and passing through a point $x = (x_1, \dots, x_N) \in \mathbb{R}^N$. Let $\omega(x, \cdot)$ denote the product of this function by the indicator of the domain G and by $(-1)^N \mathrm{sign} x_1 \dots \mathrm{sign} x_N$. First, we establish the equality

$$\int_G (f, \omega(x,\cdot))_{L_2(G)} \overline{(\mathcal{D}u)(x)}dx = (f,u)_{L_2(G)}$$
$$(f \in L_2(G), u \in C_0^N(G)). \tag{6.28}$$

Indeed, by extending the functions f and u to the outside of G as zero and integrating by parts, we obtain

$$\int_G (f, \omega(x,\cdot))_{L_2(G)} \overline{(\mathcal{D}u)}(x)dx$$
$$= (-1)^N \int_{\mathbb{R}^N} \Big(\int_0^{x_1} \cdots \int_0^{x_N} f(\xi)d\xi_1 \dots d\xi_N\Big)\overline{(\mathcal{D}u)(x)}dx$$
$$= \int_{\mathbb{R}^N} \mathcal{D}_x\Big(\int_0^{x_1} \cdots \int_0^{x_N} f(\xi)d\xi_1 \dots d\xi_N\Big)\overline{u(x)}dx = (f,u)_{L_2(G)}.$$

The required equality (6.27) is established by using (6.28). Thus, let $A \in \mathcal{L}(L_2(G))$ be the operator corresponding to a, i.e., $a(f,g) = (Af, g)_{L_2(G)}$ $\big(f, g \in L_2(G)\big)$ (see Section 8.5). On the right-hand side of (6.27), we set $T(x,y) =$

$\left(A\omega(y,\cdot),\, \omega(x,\cdot)\right)_{L_2(G)}$ $(x, y \in \mathbb{R}^N)$ and apply (6.28). This gives

$$\begin{aligned}
\int_{G\times G} & T(x,y)(\mathcal{D}u)(y)\overline{(\mathcal{D}u)(x)}dxdy \\
&= \int_G \int_G \left(A\omega(y,\cdot),\omega(x,\cdot)\right)_{L_2(G)} (\mathcal{D}u)(y)\overline{(\mathcal{D}u)(x)}dxdy \\
&= \int_G \{ \int_G \left(A\omega(y,\cdot),\omega(x,\cdot)\right)_{L_2(G)} \overline{(\mathcal{D}v)(x)}dx \} (\mathcal{D}u)(y)dy \\
&= \int_G \left(A\omega(y,\cdot),v\right)_{L_2(G)} (\mathcal{D}u)(y)dy \\
&= \overline{\int_G \left(v, A\omega(y,\cdot)\right)_{L_2(G)} \overline{(\mathcal{D}u)(y)}dy} \\
&= \overline{\int_G \left(A^*v, \omega(y,\cdot)\right)_{L_2(G)} \overline{(\mathcal{D}u)(y)}dy} = (\overline{A^*v,u})_{L_2(G)} \\
&= (Au,v)_{L_2(G)} = a(u,v) \quad (u, v \in C_0^N(G)).
\end{aligned}$$

The continuity of the kernel T follows from the continuity of the vector function $\mathbb{R}^N \ni x \mapsto \omega(x,\cdot) \in L_2(G)$. □

Note that if the kernel T is smooth, then (6.27) has the form (6.26) with $K = \mathcal{A}_A = \mathcal{D}_x\mathcal{D}_y T$. Generally speaking, these derivatives should be understood in the sense of generalized functions (see Section 11.2) and the last kernel is generalized. Note that one can indicate a quasinuclear rigging (6.20) of the space $H_0 = L_2(G)$, which transforms (6.23) into (6.27). Also note that a relation similar to (6.27) can be written for the case of polylinear forms.

Formula (6.27) is connected with the following expression for the elements of the matrix $\left(a_{jk}\right)_{j,k=1}^m$ of an operator A in the m-dimensional space $\mathbb{C}^m$:

$$a_{jk} = (A\delta_k, \delta_j)_{\mathbb{C}^N} \quad (j, k = 1, \dots, m). \tag{6.29}$$

Here, $\delta_j = \left(\delta_{jn}\right)_{n=1}^m$ are vectors of an orthonormal basis in $\mathbb{C}^m$. Let us clarify this assertion. On passing from $\mathbb{C}^m$ to $L_2(G)$, the role of δ_j must be played by the δ-functions $\delta_x(\xi)(x \in G)$ but they do not belong to $L_2(G)$ and, therefore, relation (6.29) becomes meaningless. At the same time, one can transform the basis in (6.29) by setting $\omega_j = \sum_{l=1}^j \delta_l$ $(j = 1, \dots, m)$ and construct the matrix $t_{jk} = (A\omega_k, \omega_j)_{\mathbb{C}^m}$ $(j, k = 1, \dots, m)$. This matrix enables us to present the action of the operator in a simple form. The kernel T is an analogue of the matrix $\left(t_{jk}\right)_{j,k=1}^m$, e.g., if $G = (0, \infty) \subset \mathbb{R}$, then we can formally write $\omega_x(\xi) = \int_0^x \delta_z(\xi)dz$ $(x \in (0,\infty))$. The generalized kernel $\mathcal{D}_x\mathcal{D}_y T$ is an analogue of matrix (6.29) and relation (6.27) is similar to the formula that reconstructs the action of the operator A in terms of the matrix $\left(t_{jk}\right)_{j,k=1}^m$.

7 Completions of a Space with Respect to Two Different Norms

In this section, we study a problem which, in fact, frequently arised above. Indeed, consider two norms given in a linear space, one of which is greater than or equal to the other. We find the completions of this space with respect to each of these norms and, as a result, obtain two different Banach spaces. It is interesting to check the validity of the natural assertion that the space constructed with respect to the greater norm is a subset of the space constructed with respect to the smaller norm.

In the general case, this is not true but, at the same time, in most typical situations, the indicated embedding takes place. This is why we have not paid much attention to this problem yet in order not to make our presentation too complicated. However, this problem is important for the next section dealing with bilinear forms.

7.1 Completions with Respect to Two Different Norms

Let L be a linear space, let $L \ni f \mapsto \|f\|_E \geq 0$ be a norm in this space, and let E be the completion of L with respect to this norm. Recall that E consists of the classes f_E of equivalent fundamental sequences $(f_n)_{n=1}^{\infty}$ ($f_n \in L$). The equivalence relation $(f_n)_{n=1}^{\infty} \sim (g_n)_{n=1}^{\infty}$ means that $\|f_n - g_n\|_E \to 0$ as $n \to \infty$. The linear structure in the space E is induced by the linear operations over the sequences. If $\|\cdot\|_E$ is a Hilbert norm, i.e., if L is equipped with the scalar product $(f, g)_E$ ($f, g \in L$) and $\|\cdot\|_E = (\cdot,\cdot)_E^{1/2}$, then E is a Hilbert space. The space L is embedded in E by the identification of $f \in L$ with a class that contains a stationary sequence $(f, f, \dots)$.

Now assume that L is a linear set with two different norms, namely, $L \ni f \mapsto \|f\|_{E_1} \geq 0$ and $L \ni f \mapsto \|f\|_{E_2} \geq 0$. Let E_1 and E_2 be the corresponding completions of L. Assume that these norms are comparable in the following sense:

$$\|f\|_{E_1} \leq \|f\|_{E_2} \quad (f \in L) \tag{7.1}$$

(it is clear that, instead of (7.1), one can write the inequality $(\exists c > 0)\ (\forall f \in L)$: $\|f\|_{E_1} \leq c\,\|f\|_{E_2}$).

Arguing somewhat inaccurately, one can conclude that, as a result of the completion, inequality (7.1) yields the inclusion $E_1 \supseteq E_2$ and the inequality $\|f\|_{E_1} \leq \|f\|_{E_2}$ ($f \in E_2$). However, it has been already mentioned that this is not true. Let us clarify the situation.

Assume that $(f_n)_{n=1}^{\infty} (f_n \in L)$ is a fundamental sequence with respect to the norm $\|\cdot\|_{E_2}$. Then, by virtue of (7.1), it is also fundamental with respect to the norm $\|\cdot\|_{E_1}$. Let $(f_n)_{n=1}^{\infty} \in f_{E_2} \in E_2$ and $(f_n)_{n=1}^{\infty} \in f_{E_1} \in E_1$. Let us associate the vector f_{E_2} with the vector f_{E_1}. This mapping is well-defined. Indeed, if $(g_n)_{n=1}^{\infty} \in f_{E_2}$, i.e., $(f_n)_{n=1}^{\infty} \sim (g_n)_{n=1}^{\infty}$ with respect to $\|\cdot\|_{E_2}$, then, by virtue of (6.1), the same equivalence relation can be written for $\|\cdot\|_{E_1}$.

Thus, we have constructed the mapping $E_2 \ni f_{E_2} \mapsto Qf_{E_2} = f_{E_1} \in E_1$. It follows from the method, according to which the completions are equipped with linear structure, that Q is linear. Indeed, by virtue of (7.1), it is continuous,

$$\begin{aligned}\|Qf_{E_2}\|_{E_1} &= \|f_{E_1}\|_{E_1} = \lim_{n\to\infty} \|f_n\|_{E_1} \\ &\le \lim_{n\to\infty} \|f_n\|_{E_2} = \|f_{E_2}\|_{E_2} \quad \big((f_n)_{n=1}^{\infty} \in f_{E_2} \in E_2\big).\end{aligned}$$

Note that the restriction $Q \upharpoonright L$ is the embedding operator which embeds $L \subseteq E_2$ in the set L regarded as a subset of the space E_1.

Consider the subspace

$$\operatorname{Ker} Q = \{f \in E_2 |\, Qf = 0\} \subseteq E_2. \tag{7.2}$$

If $\operatorname{Ker} Q = \{0\}$, then E_2 can be identified with the range $\mathcal{R}(Q)$ and one can assume that $E_2 \subseteq E_1$ and $\|f\|_{E_1} \le \|f\|_{E_2}$ $(f \in E_2)$. In the general case, these inclusion and inequality hold for the factor-space $E_2/\mathrm{Ker}Q \subseteq E_1$.

In the case where E_2 is a Hilbert space, instead of the factor-space, we can take the orthogonal complement $E_2 \ominus \operatorname{Ker} Q$. Note that the second ultimate case $(\operatorname{Ker} Q = E_2)$ is impossible. Moreover, $L \cap \operatorname{Ker} Q = \{0\}$ as follows from the fact that $Q \upharpoonright L$ is the indicated embedding.

Let us formulate these results as a theorem.

Theorem 7.1. *Let L be a linear space with two norms $\|\cdot\|_{E_1}$ and $\|\cdot\|_{E_2}$ comparable in the sense of (7.1), let E_1 and E_2 be the corresponding completions of L, and let Q be the operator introduced above. Then*

$$E_1 \supseteq E_2/\operatorname{Ker} Q, \quad \|f\|_{E_1} \le \|f\|_{E_2/\operatorname{Ker} Q} \quad (f \in E_2/\mathrm{Ker}Q). \tag{7.3}$$

If E_2 is a Hilbert space, then $E_2 \ominus \operatorname{Ker} Q$ plays the role of the factor-space in (7.3). If $\operatorname{Ker} Q = \{0\}$, then $E_1 \supseteq E_2$ and $\|f\|_{E_1} \le \|f\|_{E_2}$ $(f \in E_2)$.

The next theorem immediately follows from the construction of the operator Q and relation (7.2).

Theorem 7.2. *The kernel $\operatorname{Ker} Q = \{0\}$ if and only if any sequence $(f_n)_{n=1}^{\infty}$ $(f_n \in L)$ fundamental in the norm $\|\cdot\|_{E_2}$ and approaching zero in the norm $\|\cdot\|_{E_1}$ approaches zero in the norm $\|\cdot\|_{E_2}$.*

7.2 Examples

We consider several examples dealing with the most typical situations.

7.1. Let $G \subset \mathbb{R}^N (N \in \mathbb{N})$ be a bounded domain, $L = C^1(\tilde{G})$, $\|f\|_{E_1} = \|f\|_{L_2(G)}$, and $\|f\|_{E_2} = \|f\|_{W_2^1(G)}$. Here, $L_2(G) = L_2(G, \mathfrak{B}(G), dx)$ is constructed with respect to the Lebesgue measure dx and $W_2^1(G)$ is a Sobolev space. *In this case,* $\operatorname{Ker} Q = \{0\}$.

In fact, let $\left(f_n\right)_{n=1}^{\infty} \subset L$ be a sequence fundamental in the norm of $E_2 = W_2^1(G)$ and, at the same time, $f_n \xrightarrow[n\to\infty]{} 0$ in $E_1 = L_2(G)$. Then, for each derivative D_j, the sequence $\left(D_j f_n\right)_{n=1}^{\infty}$ is fundamental in $L_2(G)$. Let $h_j \in L_2(G)$ be the corresponding limit in $L_2(G)$. For a function $g \in C_0^\infty(G)$ finite with respect to G, we can write the following equality:

$$\left(h_j, g\right)_{L_2(G)} = \lim_{n\to\infty} \left(D_j f_n, g\right)_{L_2(G)} = - \lim_{n\to\infty} \left(f_n, D_j g\right)_{L_2(G)} = 0,$$

whence it follows that $h_j = 0$ $(j = 1, \dots, N)$. Passing to the limit in the expression

$$\left\|f_n\right\|^2_{W_2^1(G)} = \left\|f_n\right\|^2_{L_2(G)} + \sum_{j=1}^{N} \left\|D_j f_n\right\|^2_{L_2(G)}$$

as $n \to \infty$, we conclude that $f_n \xrightarrow[n\to\infty]{} 0$ in $W_2^1(G)$. □

Thus, we have $E_1 = L_2(G)$ and $E_2 = W_2^1(G)$ and it is possible to write (as is usually done in this case) the inclusion $W_2^1(G) \subset L_2(G)$.

7.2. Let $G \subset \mathbb{R}^N$ be a bounded domain, $L = C(\tilde{G})$, and

$$\|f\|_{E_1} = \|f\|_{L_2(G)}, \quad \left(f, g\right)_{E_2} = \left(f, g\right)_{L_2(G)} + f(x_0)\overline{g(x_0)} \quad (f, g \in L), \tag{7.4}$$

where $x_0 \in \tilde{G}$ is a fixed point. Consider the Hilbert space $L_2(G) \oplus \mathbb{C}$ of pairs $\langle f, p \rangle$ $\left(f \in L_2(G),\ p \in \mathbb{C}\right)$. The space $C(\tilde{G})$ can be embedded in $L_2(G) \oplus \mathbb{C}$ by identifying $f \in C(\tilde{G})$ with a pair $\langle f, f(x_0) \rangle$; clearly, $C(\tilde{G})$ is dense in $L_2(G) \oplus \mathbb{C}$. This implies that $E_2 = L_2(G) \oplus \mathbb{C}$ and $E_1 = L_2(G)$. The operator Q has the form

$$Q\langle f, p \rangle = f \ \left(\langle f, p \rangle \in E_2\right), \quad \operatorname{Ker} Q = \left\{\langle 0, p \rangle | p \in \mathbb{C}\right\} \subset E_2,$$

i.e., $\operatorname{Ker} Q$ regarded as a subspace of E_2 coincides with $\mathbb{C}$.

7.3. Let $G \subset \mathbb{R}^N$ be a bounded domain, $L = C^1(\tilde{G})$, $\|f\|_{E_1} = \|f\|_{L_2(G)}$, and

$$(f, g)_{E_2} = (f, g)_{W_2^1(G)} + f(x_0)\overline{g(x_0)} \qquad (f, g \in L) \tag{7.5}$$

($x_0 \in \tilde{G}$ is fixed). If $N = 1$, then, by virtue of the embedding theorem, $W_2^1(G) \subset C(\tilde{G})$ and this embedding is continuous. Thus, the scalar product (7.5) is equivalent to the scalar product in $W_2^1(G)$ and, therefore, $E_2 = W_2^1(G)$ and $E_1 = L_2(G)$. According to Example 7.1, $\operatorname{Ker} Q = \{0\}$.

Let $N \geq 2$. Consider the Hilbert space $W_2^1(G) \oplus \mathbb{C}$ of pairs $\langle \varphi, p \rangle$ $\big(\varphi \in W_2^1(G), p \in \mathbb{C}\big)$. The space $C^1(\tilde{G})$ is embedded in this space if $\varphi \in C^1(\tilde{G})$ is identified with $\langle \varphi, \varphi(x_0) \rangle$. Since $W_2^1(G)$ is not embedded in $C(\tilde{G})$, it is easy to see that the set $C^1(\tilde{G})$ is dense in $W_2^1(G) \oplus \mathbb{C}$. Hence, $E_2 = W_2^1(G) \oplus \mathbb{C}$ and $E_1 = L_2(G)$. The operator Q has the form $Q\langle \varphi, p \rangle = \varphi$ $\big(\langle \varphi, p \rangle \in E_2\big)$ and $\operatorname{Ker} Q = \mathbb{C} \subset E_2$.

7.4. Let $G \subset \mathbb{R}^N$ be a bounded domain,

$$L = C^\infty(\tilde{G}), \quad \|f\|_{E_1} = \|f\|_{W_2^l(G)}, \quad \|f\|_{E_2}^2 = \|f\|_{W_2^l(G)}^2 + \|D^\nu f\|_{W_2^l(G)}^2, \tag{7.6}$$

where $l \in \mathbb{Z}_+$ and a certain derivative are fixed. *In this case,* $\mathrm{Ker}Q = \{0\}$.

We act as in Example 7.1. Assume that a sequence $(f_n)_{n=1}^\infty \subset L$ is fundamental with respect to the norm of E_2 and $f_n \xrightarrow[n\to\infty]{} 0$ in E_1. Then the sequence $(D^\nu f_n)_{n=1}^\infty$ is fundamental in $W_2^l(G)$. Let $h \in W_2^l(G)$ be the limit of the indicated sequence in this space. For a function $g \in C_0^\infty(G)$, we have

$$(h, g)_{W_2^l(G)} = \lim_{n\to\infty} (D^\nu f_n, g)_{W_2^l(G)} = (-1)^{|\nu|} \lim_{n\to\infty} (f_n, D^\nu g)_{W_2^l(G)} = 0.$$

In view of the arbitrariness of g, this implies that $h = 0$. Passing to the limit in the expression $\|f_n\|_{E_2}^2 = \|f_n\|_{W_2^l(G)}^2 + \|D^\nu f_n\|_{W_2^l(G)}^2$ as $n \to \infty$, we conclude that $f_n \xrightarrow[n\to\infty]{} 0$ in E_2. □

By analogy, one can consider the case where expression (7.6) for the squared norm in E_2 contains the sum of several derivatives of f. In particular, this is true for the case where $L = C^\infty(\tilde{G})$, $E_1 = W_2^l(G)$, and $E_2 = W_2^m(G)$, where, in turn, $m \geq l$.

Examples 7.1 -7.4 can be easily reformulated for unbounded domains G and weighted Sobolev spaces.

8 Semibounded Bilinear Forms

In Section 8.5 it was proved that an arbitrary continuous bilinear form a in a Hilbert space H admits the representation $a(f, g) = (Af, g)_H$ $(f, g \in H)$, where A is a bounded operator in H. An important role is played by similar theorems on representations in the case of forms that are not continuous but possess certain additional properties. The problem is that physical objects often appear as forms but, at the same time, for the application of mathematical methods, it is highly desirable that these objects be associated with operators.

In this section, we present the corresponding theory of representations and clarify its relation to the theory of Hilbert riggings. More precisely, we interpret simple facts of the theory of Hilbert riggings as theorems of the theory of bilinear forms.

8.1 Lemma on Hilbert Riggings

Our presentation is based on the data about the completions of a space with respect to two different norms (Section 7) and the following lemma of the theory of Hilbert space riggings:

Lemma 8.1. *Assume that $\mathcal{D}(A) = \{u \in H_+ \mid \mathbf{I}^{-1}u \in H_0\}$. In H_0, we consider the operator $A = \mathbf{I}^{-1} \restriction \mathcal{D}(A)$. It is stated that A is selfadjoint and satisfies the relations*

$$(u, v)_{H_+} = (Au, v)_{H_0} \quad (u \in \mathcal{D}(A), v \in H_+),$$
$$(u, v)_{H_+} = (\sqrt{A}u, \sqrt{A}v)_{H_0} \quad (u, v \in H_+ = \mathcal{D}(\sqrt{A})). \tag{8.1}$$

Proof. Since H_0 is dense in H_-, the space $\mathcal{D}(A)$ is dense in H_+ and, consequently, in H_0. Further, according to (1.13), $(\alpha, v)_{H_0} = (\mathbf{I}\alpha, v)_{H_+}$ $(\alpha \in H_-,\ v \in H_+)$ and, therefore,

$$(u, v)_{H_+} = (\mathbf{I}^{-1}u, v)_{H_0} \quad (u, v \in H_+). \tag{8.2}$$

It follows from (8.2) and the definition of A that $(u, v)_{H_+} = (Au, v)_{H_0}$ $(u \in \mathcal{D}(A), v \in H_+)$, i.e., we obtain the first relation in (8.1). By setting here $v = u$, we obtain $(Au, u)_{H_0} = (u, u)_{H_+} \geq 0$ $\big(u \in \mathcal{D}(A)\big)$, which means that A is an Hermitian operator and satisfies the inequality $A \geq \mathbb{1}$. Further, for every $u \in \mathcal{D}(A)$, we have $\|Au\|_{H_0} = \|\mathbf{I}^{-1}u\|_{H_0} \geq \|\mathbf{I}^{-1}u\|_{H_-} = \|u\|_{H_+}$.

Let us establish the second relation in (8.1). We use the operators mentioned in (1.11). According to the proof of Theorem 1.3, $J = \sqrt{OI}$ if the operator on the right-hand side is regarded as acting from H_0 to H_+. Therefore, $(\sqrt{OI}f, \sqrt{OI}g)_{H_+} = (f, g)_{H_0}$ $(f, g \in H_0)$ and $\mathcal{R}(\sqrt{OI}) = H_+$. In other words,

$$(u, v)_{H_+} = \big((OI)^{-1/2}u, (OI)^{-1/2}v\big)_{H_0} \quad (u, v \in H_+). \tag{8.3}$$

At the same time, $I = \mathbf{I} \restriction H_0$ and, consequently, $(OI)^{-1} = A$. Thus, equality (8.3) turns into the required second relation in (8.1). □

8.2 Positive Forms

Here, we introduce the notion of a prechain, which is closely related to the notion of a chain. In fact, the presence of a prechain is equivalent to the determination of a positive form and the existing close relation between the theories of bilinear forms and rigged Hilbert spaces is largely based on this fact.

Let H_0 be a Hilbert space and let L be a linear set dense in this space with a scalar product $(f, g)_{L_+}$ $(f, g \in L)$ such that $\|f\|_{H_0} \leq \|f\|_{L_+}$ $\big(\|f\|_{L_+} = \big((f, f)_{L_+}\big)^{1/2}, f \in L\big)$. In this case, we say that a *prechain*

$$H_0 \supseteq L \tag{8.4}$$

is defined.

Denote by L_+ the completion of L with respect to the norm $\|\cdot\|_{L_+}$. In this case, all the requirements of the scheme presented in Section 7 are satisfied for $E_1 = H_0$ and $E_2 = L_+$. Let $Q\colon L_+ \to H_0$ be the corresponding operator. According to (7.3), for given prechain (8.4), one can construct the chain

$$H_- \supseteq H_0 \supseteq H_+ = L_+ \ominus \operatorname{Ker} Q. \tag{8.5}$$

We say that prechain (8.4) is *closed* if L is complete with respect to $\|\cdot\|_{L_+}$ and *closable* (admitting a closure) if $\operatorname{Ker} Q = \{0\}$ (it is obvious that closeness yields closability). In view of Theorem 7.2, we can formulate the following criterion of closability:

Prechain (8.4) is closable if and only if every sequence $(f_n)_{n=1}^{\infty} \subset L$ fundamental in the norm $\|\cdot\|_{L_+}$ and convergent to zero in the norm $\|\cdot\|_{H_0}$ converges to zero in the norm $\|\cdot\|_{L_+}$.

The prechain $H_0 \supseteq L_+$ constructed according to the closable prechain (8.1) by completing L with respect to $\|\cdot\|_{L_+}$ is called a closure of prechain (8.1). Throughout this Section, we consider only closed or closable prechains $H_0 \supseteq L$ (in the case of closeness, $L_+ = L$). Every prechain of this sort can be extended to the chain

$$H_- \supseteq H_0 \supseteq H_+ = L_+, \tag{8.6}$$

by constructing the corresponding negative space. After this, it becomes possible to apply the general facts established in Section 1 (in this book, we do not dwell upon the theory of nonclosable prechains and forms).

In our subsequent presentation, we use the language of forms.

A function $\mathcal{D}(a) \times \mathcal{D}(a) \ni \langle f, g\rangle \mapsto a(f,g) \in \mathbb{C}$ linear in the first variable and antilinear in the second variable is called a bilinear (sesquilinear) form a in a Hilbert space H_0 (here, $\mathcal{D}(a)$ — a linear set dense in H_0 — is the domain of the form a). The diagonal values of a represent the quadratic form $a[\cdot]$ associated with the bilinear form under consideration, i.e., $\mathcal{D}(a) \ni f \mapsto a[f] = a(f,f) \in \mathbb{C}$.

Given a quadratic form, one can uniquely reconstruct the corresponding bilinear form by using the polarization identity (see Section 8.5)

$$a(f,g) = \frac{1}{4}\big(a[f+g] - a[f-g] + ia[f+ig] - ia[f-ig]\big) \quad (f, g \in \mathcal{D}(a)). \tag{8.7}$$

The linear operations are introduced on bilinear forms in a natural way. Thus, if a and b are two bilinear forms and, at the same time, $\mathcal{D}(a) \cap \mathcal{D}(b)$ is dense in H_0, then the bilinear form $a + b$ is defined by the equality

$$(a+b)(f,g) = a(f,g) + b(f,g) \quad \big(f, g \in \mathcal{D}(a+b) = \mathcal{D}(a) \cap \mathcal{D}(b)\big).$$

The product λa, where $\lambda \in \mathbb{C}$, is always defined. Indeed,

$$(\lambda a)(f,g) = \lambda a(f,g) \quad \big(f, g \in \mathcal{D}(\lambda a) = \mathcal{D}(a)\big).$$

Given a bilinear form a, one can always construct the adjoint bilinear form a^* according to the equality

$$a^*(f,g) = \overline{a(g,f)} \quad (f,g)(f,g \in \mathcal{D}(a^*) = \mathcal{D}(a)).$$

A bilinear form a is called Hermitian if $a^* = a$. It follows from (8.7) that in order for a to be Hermitian, it is necessary and sufficient that the quadratic form $a[\cdot]$ take only real values. Every bilinear form a can be expressed as a linear combination of two Hermitian forms $\operatorname{Re} a$ and $\operatorname{Im} a$, namely,

$$a = \operatorname{Re} a + i\operatorname{Im} a, \quad Re\, a = \frac{1}{2}(a + a^*), \quad \operatorname{Im} a = \frac{1}{2i}(a - a^*).$$

A bilinear form a is called positive with vertex $\alpha > 0$ if

$$a(f,f) \geq \alpha \|f\|_{H_0}^2 \quad (f \in \mathcal{D}(a)). \tag{8.8}$$

In this subsection, it is convenient to assume that $\alpha = 1$. Positive forms are always Hermitian because $a[\cdot]$ is real-valued.

Given a positive form a in H_0, one can naturally construct prechain (8.4) by setting $L = \mathcal{D}(a)$, $(f,g)_{L_+} = a(f,g)$ $(f,g \in L)$. Conversely, prechain (8.4) determines the positive form $a(f,g) = (f,g)_{L_+}$ $(f,g \in \mathcal{D}(a) = L)$.

The definitions introduced above for prechains can be easily reformulated for positive forms.

A positive form a is called closed if the corresponding prechain is closed. The closure $\tilde{a}$ of a closable form a is defined by the equality $\tilde{a}(f,g) = (f,g)_{L_+}$, where $f,g \in \mathcal{D}(\tilde{a}) = L_+$ and L and $(\cdot,\cdot)_{L_+}$ are constructed according to a.

Thus, to calculate $\tilde{a}(f,g)$ for $f,g \in \mathcal{D}(\tilde{a}) \subseteq H_0$, we must construct sequences $(f_n)_{n=1}^{\infty}$, $(g_m)_{m=1}^{\infty} \subset \mathcal{D}(a)$ fundamental in the norm $(a[\cdot])^{1/2}$ and convergent in H_0 to f and g, respectively. Then $\tilde{a}(f,g) = \lim_{n,m\to\infty} a(f_n, g_m)$.

Lemma 8.1 yields the following theorem on the representation of a positive form:

Theorem 8.1. *Let a be a closed positive bilinear form with vertex $\alpha = 1$. It is stated that there exists a selfadjoint operator $A \geq \mathbb{1}$ acting on the space H_0 such that*

$$a(f,g) = (Af,g)_{H_0} \quad (f \in \mathcal{D}(A) \subseteq \mathcal{D}(a), g \in \mathcal{D}(a)). \tag{8.9}$$

Its domain $\mathcal{D}(A)$ is dense in $\mathcal{D}(a)$ with respect to the norm $(a[\cdot])^{1/2}$ and, moreover, $\|Af\|_{H_0} \geq (a[f])^{1/2}$ $(f \in \mathcal{D}(a))$. In addition to (8.9), the form a admits the following representation in terms of the operator $\sqrt{A}$:

$$a(f,g) = (\sqrt{A}f, \sqrt{A}g)_{H_0} \quad (f,g \in \mathcal{D}(\sqrt{A}) = \mathcal{D}(a)). \tag{8.10}$$

Proof. By using the procedure described above, we pass from the form a to the closed prechain (8.4) and then to chain (8.5). By applying Lemma 8.1 to (8.5), we obtain the required statement. □

If the form a is nonclosed but admits a closure $\tilde{a}$, then relations (8.9) and (8.10) also hold for $\tilde{a}$. By setting in these formulas $f \in \mathcal{D}(A) \cap \mathcal{D}(a)$, $g \in \mathcal{D}(a)$, or $f, g \in \mathcal{D}(\sqrt{A}) \cap \mathcal{D}(a)$, respectively, we obtain the representation of the original form.

Note that if a and b are two positive forms such that $\mathcal{D}(a) \cap \mathcal{D}(b)$ is dense in H_0, then $a + b$ is also a positive form. It is easy to see that if a and b are closed (closable), then $a + b$ is also closed (closable). This immediately follows from the definitions and the criterion of closability if we note that the fact that a sequence is fundamental in the norm $(a[\cdot] + b[\cdot])^{1/2}$ is equivalent to the fact that it is fundamental both in the norm $(a[\cdot])^{1/2}$ and in the norm $(b[\cdot])^{1/2}$.

Example

8.1. Examples 7.1 and 7.2 can be interpreted as standard examples of closable and nonclosable forms. Thus, in both cases, $H_0 = L_2(G)$. The form $a(f, g) = (f, g)_{W_2^1(G)}$ $(f, g \in C^1(\tilde{G}))$ is positive and admits a closure. The form $a(f, g) = (f, g)_{L_2(G)} + f(x_0)\overline{g(x_0)}$ $(f, g \in C(\tilde{G}))$ is positive but does not admit a closure. The same situation takes place in Examples 7.3 and 7.4.

8.3 Semibounded Forms

The theory of representations is, as a rule, constructed for semibounded forms. Below, we present the fundamentals of this theory.

A bilinear form a in the space H_0 is called semibounded (from below) with vertex $\alpha \in \mathbb{R}$ if

$$a(f, f) \geq \alpha \, \|f\|_{H_0}^2 \quad (f \in \mathcal{D}(a)). \tag{8.11}$$

If $\alpha = 0$, the form a is called nonnegative. For $\alpha > 0$, it is called a positive form and has been already introduced above.

Semibounded forms are clearly Hermitian.

A semibounded form a can be associated with a positive form a_p (whose vertex is equal to one) by setting

$$a_p(f, g) = a(f, g) + (1 - \alpha)(f, g)_{H_0} \quad (f, g \in \mathcal{D}(a_p) = \mathcal{D}(a)). \tag{8.12}$$

The definitions, related to the form a are formulated in terms of the form a_p, namely, *a is closed (closable) if a_p is closed (closable); the closure $\tilde{a}$ of a closable form a is determined, according to (8.12), by the formula*

$$\tilde{a}(f, g) = \tilde{a}_p(f, g) - (1 - \alpha)(f, g)_{H_0} \quad (f, g \in \mathcal{D}(\tilde{a}) = \mathcal{D}(\tilde{a}_p)); \tag{8.13}$$

this closure is a semibounded form with the same vertex α.

In the case of positive forms, we can act in a somewhat different manner: If a is a positive form with vertex α, then $\frac{1}{\alpha}a$ is a positive form with vertex one and the application of the definitions introduced in Subsection 2 to $\frac{1}{\alpha}a$ leads us to the corresponding definitions for a (since the norms $(a_p[\cdot])^{1/2}$ and $(\frac{1}{\alpha}a[\cdot])^{1/2}$ are equivalent). This implies that, in the case of semibounded forms, a_p can also be defined by relation (8.12) with 1 replaced by $\varepsilon > 0$.

For semibounded forms, Theorem 8.1 takes the following form:

Theorem 8.2. *Let a be a closed semibounded bilinear form with vertex $\alpha \in \mathbb{R}$. There exists a selfadjoint operator $A \geq \alpha\mathbb{I}$ acting on the space H_0 such that representation (8.9) is true. Its domain $\mathcal{D}(A)$ is dense in $\mathcal{D}(a)$ with respect to the norm $(a_p[\cdot])^{1/2}$. If a is nonnegative, then it is also representable in the form (8.10).*

Proof. Let us write representation (8.9) for the form a_p. Let $A_p \geq \mathbb{I}$ be the corresponding operator. In view of relation (8.12), we obtain (8.9) for a with the operator $A = A_p - (1-\alpha)\mathbb{I} \geq \alpha\mathbb{I}$. To obtain representation (8.10) in the case where a is positive, it suffices to write this representation for $\frac{1}{\alpha}a$. Now assume that the form a is nonnegative and A is the corresponding nonnegative selfadjoint operator. For any $\varepsilon > 0$, the form $a(f,g) + \varepsilon(f,g)_{H_0}$ $(f,g \in \mathcal{D}(a))$ is positive and, therefore,

$$a(f,g) + \varepsilon(f,g)_{H_0} = (\sqrt{A+\varepsilon\mathbb{I}}f, \sqrt{A+\varepsilon\mathbb{I}}g)_{H_0} \quad (f,g \in \mathcal{D}(a)).$$

It follows from the spectral decomposition of the operator A (Theorem 13.6.1) that $\sqrt{A+\varepsilon\mathbb{I}}f \to \sqrt{A}f$ as $\varepsilon \to 0$ in H_0 for any $f \in \mathcal{D}(\sqrt{A}) = \mathcal{D}(\sqrt{A+\varepsilon\mathbb{I}})$. Hence, by passing to the limit, we conclude that a is representable in the form (8.10). □

If the form a admits a closure $\tilde{a}$, then the latter is representable in the form (8.9) and (8.10), whence we get the required representations for the form a.

Let us dwell upon an important procedure for constructing extensions of semibounded operators to selfadjoint operators, which was introduced by Friedrichs. Consider an Hermitian semibounded operator $A \geq \alpha\mathbb{I}$ $(\alpha \in \mathbb{R})$ acting on H_0 with dense domain of definition. It generates, in a standard way, a semibounded form a with vertex α

$$a(f,g) = (Af,g)_{H_0} \quad (f,g \in \mathcal{D}(a) = \mathcal{D}(A)). \tag{8.14}$$

The properties of this form are described by the following theorem:

Theorem 8.3 (Friedrichs). *The bilinear form (8.14) admits the closure $\tilde{a}$ representable in the form $\tilde{a}(f,g) = (A_F f,g)_{H_0}$ $(f \in \mathcal{D}(A_F) \subseteq \mathcal{D}(\tilde{a}), g \in \mathcal{D}(\tilde{a}))$, where $A_F \geq \alpha\mathbb{I}$ is a selfadjoint operator in H_0 — an extension of the operator A (the so-called Friedrichs extension). The operator A_F is the unique selfadjoint extension of A whose domain lies in $\mathcal{D}(\tilde{a})$.*

Proof. Let $A_p = A + (1-\alpha)\mathbb{I} \geq \mathbb{I}$. Then $a_p(f,g) = (A_p f,g)_{H_0}$ $(f,g \in \mathcal{D}(A_p) = \mathcal{D}(A))$. The closability of a means that a_p is closable. We prove the last statement. Assume that $(f_n)_{n=1}^{\infty}$ $(f_n \in \mathcal{D}(A_p))$ is such that

$$(A_p(f_n - f_m), f_n - f_m)_{H_0} = a_p[f_n - f_m] \to 0, \quad n,m \to \infty$$

and $\|f_n\|_{H_0} \to 0$ as $n \to \infty$. It is necessary to show that $(A_p f_n, f_n)_{H_0} = a_p[f_n] \to 0$ as $n \to \infty$.

For any $\varepsilon > 0$, one can find $N(\varepsilon)$ such that $(A_p(f_n - f_m), f_n - f_m)_{H_0} < \varepsilon$ for $n, m > N(\varepsilon)$. Since fundamental sequences are bounded, by using the Cauchy-Buniakowski inequality, we obtain

$$\begin{aligned}(A_p f_n, f_n)_{H_0} &\le \left|(A_p f_n, f_n - f_m)_{H_0}\right| + \left|(A_p f_n, f_m)_{H_0}\right| \\ &\le (A_p f_n, f_n)_{H_0}^{1/2} (A_p(f_n - f_m), f_n - f_m)_{H_0}^{1/2} + \left|(A_p f_n, f_m)_{H_0}\right| \\ &\le c\varepsilon^{1/2} + \left|(A_p f_n, f_m)_{H_0}\right|\end{aligned}$$

for $n, m > N(\varepsilon)$. Passing to the limit in the last inequality as $m \to \infty$, we obtain $(A_p f_n, f_n)_{H_0} < c\varepsilon^{1/2}$ for $n > N(\varepsilon)$ as required.

By applying Theorem 8.2 to a closed semibounded form a with vertex α, we arrive at the representation

$$\tilde{a}(f, g) = (A_F f, g)_{H_0} \quad \big(f \in \mathcal{D}(A_F) \subset \mathcal{D}(\tilde{a}), g \in \mathcal{D}(\tilde{a})\big),$$

where $A_F \ge \alpha \mathbb{I}$ is selfadjoint.

Let us now show that if an Hermitian operator $B \supseteq A$ is such that $\mathcal{D}(B) \subseteq \mathcal{D}(\tilde{a})$, then $B \subseteq A_F$. To do this, first, it is necessary to establish the equality

$$(f, Bg)_{H_0} = \tilde{a}(f, g) \quad \big(f \in \mathcal{D}(\tilde{a}), g \in \mathcal{D}(B)\big). \tag{8.15}$$

Assume that $f \in \mathcal{D}(A) \subseteq \mathcal{D}(B)$. Then $(f, Bg)_{H_0} = (Bf, g)_{H_0} = (Af, g)_{H_0}$. Since $\mathcal{D}(B) \subseteq \mathcal{D}(\tilde{a})$, one can indicate a sequence $(g_n)_{n=1}^{\infty}$ ($g_n \in \mathcal{D}(a) = \mathcal{D}(A)$) such that $g_n \to g$ in the norm $(a_p[\cdot])^{1/2}$ as $n \to \infty$. This implies that

$$(f, Bg)_{H_0} = (Af, g)_{H_0} = \lim_{n\to\infty} (Af, g_n)_{H_0} = \lim_{n\to\infty} a(f, g_n) = \tilde{a}(f, g).$$

Thus, for $f \in \mathcal{D}(A)$, relation (8.15) is established. Now let $f \in \mathcal{D}(\tilde{a})$ and let a sequence $(f_n)_{n=1}^{\infty}$ ($f_n \in \mathcal{D}(a) = \mathcal{D}(A)$) be such that $f_n \to f$ in the norm $(a_p[\cdot])^{1/2}$ as $n \to \infty$. Thus, to establish (8.15) in the general case, it remains to write (8.15) for $f = f_n \in \mathcal{D}(A)$ and pass to the limit in the equality obtained.

By using the definition of A_F and (8.15), we get

$$(A_F f, g)_{H_0} = \tilde{a}(f, g) = (f, Bg)_{H_0} \quad (f \in \mathcal{D}(A_F) \subseteq \mathcal{D}(\tilde{a}), \quad g \in \mathcal{D}(B)).$$

This and the selfadjointness of A_F imply that $g \in \mathcal{D}(A_F)$ and $A_F g = Bg$, i.e., $A_F \supseteq B$.

By using this inclusion, one can easily complete the proof of the theorem. Indeed, by setting $B = A$, we obtain $A_F \supseteq A$. If B is selfadjoint, then it follows from the inclusion $A_F \supseteq B$ that $B = A_F$. □

If A is selfadjoint, then, clearly, $A_F = A$. Let A be positive and selfadjoint. Then its spectral decomposition (Theorem 13.6.1) implies the equalities $\mathcal{D}(\tilde{a}) = \mathcal{D}(\sqrt{A})$ and $\tilde{a}(f,g) = (\sqrt{A}f, \sqrt{A}g)_{H_0}$ $(f, g \in \mathcal{D}(\tilde{a}))$. This enables us to conclude that, for unbounded A, form (8.14) is necessarily nonclosed.

In conclusion, we give a brief presentation of the theory of sectorial forms, which can be regarded as a generalization of semibounded forms.

A bilinear form a in the space H_0 is called a sectorial form with vertex $\alpha \in \mathbb{R}$ if one can indicate an angle $S(\alpha, k)$ $(k \in [0, \infty))$ in the complex plane (less than π) with vertex at the point α having the form

$$S(\alpha, k) = \{z \in \mathbb{C} \mid \operatorname{Re} z \geq \alpha, \quad |\operatorname{Im} z| \leq k(\operatorname{Re} z - \alpha)\}$$

and satisfying the condition

$$a(f,f)/\|f\|_{H_0}^2 \in S(\alpha, k) \quad (f \in \mathcal{D}(a)).$$

For $k = 0$, this is, in fact, the definition of a semibounded form with vertex α. If a is a sectorial form with vertex α, then $\operatorname{Re} a$ is a semibounded form with the same vertex.

Given a sectorial form a, we can define a sectorial form a_p by formula (8.12) (a_p is related to the angle $S(1, k)$). The form $\operatorname{Re} a_p$ is positive with $\alpha = 1$. By using the same procedure as above, for given $\operatorname{Re} a_p$, we now introduce a scalar product $(f, g)_{L_+} = (\operatorname{Re} a_p)(f, g)$ $(f, g \in L)$ in $L = \mathcal{D}(\operatorname{Re} a_p) = \mathcal{D}(a_p) = \mathcal{D}(a)$.

The investigation of sectorial forms is based on the following simple assertion:

Assume that a form a is sectorial. Then the form a_p satisfies the inequality

$$|a_p(f,g)| \leq (k+1)\|f\|_{L_+}\|g\|_{L_+} \quad (f, g \in L). \tag{8.16}$$

Indeed, since $a_p(f,f)\,\|f\|_{H_0}^{-2} \in S(1, k)$, we have

$$\begin{aligned} |\operatorname{Im}(a_p(f,f)\|f\|_{H_0}^{-2})| &\leq k\,(\operatorname{Re}(a_p(f,f)\,\|f\|_{H_0}^{-2}) - 1) \\ &\leq k \operatorname{Re}(a_p(f,f)\,\|f\|_{H_0}^{-2}) \quad (f \in L). \end{aligned}$$

In other words, $|(\operatorname{Im} a_p)(f,f)| \leq k\,\|f\|_{L_+}^2$ $(f \in L)$. Since the form $\operatorname{Im} a_p$ is Hermitian, it follows from the last estimate that

$$|(\operatorname{Im} a_p)(f,g)| \leq k\,\|f\|_{L_+}\|g\|_{L_+} \quad (f, g \in L).$$

Hence, in view of the decomposition $a_p = \operatorname{Re} a_p + i \operatorname{Im} a_p$ and the form of $(\cdot,\cdot)_{L_+}$, we arrive at (8.16). □

A sectorial form a is called closed (closable) if $\operatorname{Re} a_p$ *is closed (closable).*

For a closable form a, its closure $\tilde{a}$ can be found in the following way: We complete L to the space L_+; then, by virtue of (8.16), we extend a_p by continuity to the form $\tilde{a}_p$ defined in $\mathcal{D}(\tilde{a}_p) = L_+$ and, finally, construct the required form $\tilde{a}$ by using (8.13).

Formula (8.9) can be generalized as follows:

Any closed sectorial form a admits representation (8.9) with a closed operator A acting on H_0 whose domain is dense in $\mathcal{D}(a)$ with respect to the norm $((\operatorname{Re} a_p)[\,\cdot\,])^{1/2}$.

Indeed, in this case, the space L is complete and inequality (8.16) yields the representation

$$a_p(f,g) = (Bf,g)_{L_+} = (\operatorname{Re} a_p)(Bf,g) \quad (f,g \in L), \tag{8.17}$$

where B is an operator bounded in L. Since

$$\|Bf\|_{L_+}\|f\|_{L_+} \geq |(Bf,f)_{L_+}| = |a_p(f,f)| \geq (\operatorname{Re} a_p)(f,f) = \|f\|^2_{L_+},$$

we can write the inequality $\|Bf\|_{L_+} \geq \|f\|_{L_+}$ $(f \in L)$, which ensures the existence of the inverse operator B^{-1} in L.

Let us apply Theorem 8.1 to the form $\operatorname{Re} a_p$. By using (8.9), we obtain

$$(\operatorname{Re} a_p)(f,g) = (Cf,g)_{H_0} \quad (f \in \mathcal{D}(C) \subseteq L, g \in L),$$

where $C \geq \mathbb{1}$ is selfadjoint. The domain $\mathcal{D}(C)$ is dense in L and, moreover, $\|Cf\|_{H_0} \geq \|f\|_{L_+}$ $(f \in L)$. It follows from (8.17) and the indicated representation that

$$a(f,g) = (\operatorname{Re} a_p)(Bf,g) = (CBf,g)_{H_0} \quad (f \in L \cap \mathcal{D}(CB), g \in L). \tag{8.18}$$

The set $\mathcal{D}(A_p) = \{f \in L \mid Bf \in \mathcal{D}(C)\}$ is dense in L (and, hence, in H_0) because $\mathcal{D}(C)$ is dense in L and B is invertible in L. The operator A_p acting on H_0 is defined by the formula

$$A_p f = CBf \quad (f \in \mathcal{D}(A_p)).$$

Thus, by using (8.18), we get $a(f,g) = (A_p f, g)_{H_0}$ $(f \in \mathcal{D}(A_p),\ g \in L)$.

The invertibility of the operators C and B in the corresponding spaces enables us to conclude that $\mathcal{R}(A_p) = H_0$. Further,

$$\|A_p f\|_{H_0} = \|CBf\|_{H_0} \geq \|Bf\|_{L_+} \geq \|B^{-1}\|^{-1}\|f\|_{L_+} \geq \|B^{-1}\|^{-1}\|f\|_{H_0}$$
$$(f \in \mathcal{D}(A_p)).$$

Therefore, the inverse operator A_p^{-1} exists in the space H_0. This, in particular, means that A_p is closed.

To obtain representation (8.9), one must use the operator $A = A_p - (1-\alpha)\mathbb{1}$. □

An operator A acting on H_0 is called sectorial (with vertex $\alpha \in \mathbb{R}$) if the corresponding bilinear form (8.14) is sectorial (this is a generalization of the notion of the Hermitian semibounded operator).

Formula (8.9) just proved demonstrates that every closed sectorial form is generated by a certain sectorial operator which, unlike arbitrary sectorial operators, possesses a property of maximality induced by the representation $A = CB - (1-\alpha)\mathbb{I}$, where B is a bounded invertible operator in L and C is a selfadjoint invertible operator in H_0 (this is, in fact, the definition of a maximal sectorial operator, which is a generalization of the notion of selfadjoint semibounded operator).

Here, we do not study the properties of maximal sectorial operators and do not try to give a self-consistent description of this class of operators. We only note that the required properties can be relatively simply obtained from the formula for the operator A presented above, and Theorem 8.3 can be easily generalized to the case of the extension of an arbitrary sectorial operator to the maximal sectorial operator.

8.4 Form Sums of Operators

Let us now return to the problem considered in Section 13.10. Let A and B be, respectively, selfadjoint and Hermitian operators in the space H_0. It is necessary to study the operator $A+B$ $\big(\mathcal{D}(A+B) = \mathcal{D}(A) \cap \mathcal{D}(B)\big)$ and indicate conditions under which it is selfadjoint or essentially selfadjoint. Here, we consider several approaches to this problem based on the use of forms.

First, we present the well-known result for forms which allows one, under certain restrictions, to make the operator $A+B$ meaningful even in the case where $\mathcal{D}(A) \cap \mathcal{D}(B) = \{0\}$.

Theorem 8.4 (KLMN[1]). *Let a be a closed positive form and let b be an Hermitian form on $\mathcal{D}(b) = \mathcal{D}(a)$ such that*

$$|b(f,f)| \le pa(f,f) + q(f,f)_{H_0} \quad (f \in \mathcal{D}(a)) \tag{8.19}$$

for some $p \in [0,1)$ and $q \in \mathbb{R}$. Then the form $a+b$ $(\mathcal{D}(a+b) = \mathcal{D}(a))$ is semibounded and closed.

Proof. It follows from the positivity of a that $a(f,f) \ge \varepsilon \|f\|^2_{H_0}$ $(f \in \mathcal{D}(a))$ for some $\varepsilon \in (0,1)$. By using (8.9) and this inequality, we get

$$\begin{aligned} \alpha\|f\|^2_{H_0} &\le (1-p)a(f,f)_{H_0} \le a(f,f) + b(f,f) \\ &\le (1+p)a(f,f) + (1-\alpha+q)(f,f)_{H_0} \quad (f \in \mathcal{D}(a)), \end{aligned} \tag{8.20}$$

[1]This theorem belongs to Kato (1955), Lax and Milgram (1954), Lions (1961), and Nelson (1964).

which ensures, in particular, the semiboundedness of the form $a+b$. According to (8.12) and (8.20), we obtain

$$\begin{aligned}(1-p)a(f,f) &\le (1-p)a(f,f)+(1-\alpha-q)(f,f)_{H_0}\\ &\le (a+b)_p(f,f) \le (1+p)a(f,f)+(1-\alpha+q)(f,f)_{H_0}\\ &\qquad (f\in\mathcal{D}(a)).\end{aligned}$$

This estimate and the inequality $a(f,f)\ge\varepsilon\,\|f\|^2_{H_0}$ imply that the norms $(a\,[\,\cdot\,])^{1/2}$ and $((a+b)_p[\,\cdot\,])^{1/2}$ are equivalent in $\mathcal{D}(a)$. In view of the closeness and positivity of a, we conclude that $\mathcal{D}(a)$ is complete with respect to $(a\,[\,\cdot\,])^{1/2}$ and, hence, with respect to $((a+b)_p[\,\cdot\,])^{1/2}$, but this means that $a+b$ is closed. □

Let us also present another scheme which, in particular, enables us to formulate the theorem just proved in the operator form. Let $A\ge\mathbb{I}$ be a selfadjoint operator in the space H_0. Assume that $D=A^{1/2}$ and construct, following procedure (1.19), a chain

$$\begin{gathered}H_-\supseteq H_0\supseteq H_+\supseteq\mathcal{D}(A^{1/2}),\\ (u,v)_{H_+}=\left(A^{1/2}u,A^{1/2}v\right)_{H_0}\quad (u,v\in H_+).\end{gathered}\tag{8.21}$$

In this case, $(f,g)_{H_-}=(A^{-1/2}f,A^{-1/2}g)_{H_0}$ $(f,g\in H_0)$ and H_- coincides with the complement of H_0 with respect to this scalar product. The operator A can be regarded as acting isometrically from $\mathcal{D}(A)\subseteq H_+$ into H_-, i.e.,

$$\|Au\|_{H_-}=\|A^{1/2}u\|_{H_0}=\|u\|_{H_+}\quad (u\in\mathcal{D}(A))$$

and, therefore, it can be extended by continuity to the operator $\mathcal{A}\colon H_+\to H_-$ that realizes an isometry between H_+ and H_- (and coincides with $\mathbf{I}^{-1}$). Clearly, $A=\mathcal{A}\restriction\mathcal{D}(A)$ and, moreover, $\mathcal{D}(A)=\{u\in H_+\mid\mathcal{A}u\in H_0\}$. The operator $\mathcal{A}$ is selfadjoint with respect to H_0, i.e., $(\mathcal{A}u,v)_{H_0}=(u,\mathcal{A}v)_{H_0}$.

For given A, we define, according to (8.14), the form

$$a(f,g)=(Af,g)_{H_0}=\int_1^\infty \lambda d\big(E(\lambda)f,g\big)_{H_0}\quad (f,g\in\mathcal{D}(A)),$$

where E is the resolution of the identity that corresponds to A. It follows from this integral representation that, for the closure $\tilde{a}$ of the form a, we have $\mathcal{D}(\tilde{a})=H_+$ and

$$\tilde{a}(f,g)=(A^{1/2}f,A^{1/2}g)_{H_0}=(f,g)_{H_+}=(\mathcal{A}f,g)_{H_0}\quad (f,g\in H_+).$$

Consider a perturbation of the operator A. Assume that $\mathcal{B}$ is a continuous operator acting from the whole space H_+ into H_- and selfadjoint with respect to the zero space in chain (8.21). Therefore, the operator $B=\mathcal{B}\restriction\mathcal{D}(B)$, where

$\mathcal{D}(B) = \{u \in H_+ \mid \mathcal{B}u \in H_0\}$, is Hermitian, although it may be nondensely defined or equal to zero on $\mathcal{D}(B)$. The formal sum $A + B$ is defined in $\mathcal{D}(A) \cap \mathcal{D}(B)$ and may coincide with A or be defined only at 0. Therefore, in the last case, generally speaking, we cannot discuss its selfadjointness. However, we can also act in the following way:

For given $\mathcal{B}$, we introduce an Hermitian bilinear form $b(f,g) = (\mathcal{B}f, g)_{H_0}$ $(f, g \in \mathcal{D}(b) = H_+ = \mathcal{D}(\tilde{a}))$ in H_0 and assume that it satisfies condition (8.19), which can now be written in the form

$$\left|(\mathcal{B}f, f)_{H_0}\right| \le p(A^{1/2}f, A^{1/2}f)_{H_0} + q(f,f)_{H_0} \tag{8.22}$$
$$\big(f \in H_+;\ p \in [0,1),\ q \in \mathbb{R}\big).$$

Then, according to Theorem 8.4, the form $\tilde{a} + b$ $(\mathcal{D}(\tilde{a} + b) = H_+)$ is semibounded and closed. By virtue of Theorem 8.2 on representation, there exists a selfadjoint semibounded operator C in H_0 such that

$$\tilde{a}(f,g) + b(f,g) = (Cf, g)_{H_0} \quad (f \in \mathcal{D}(C) \subseteq H_+, g \in H_+).$$

Therefore, A, B, and C are connected by the following relation:

$$(A^{1/2}f, A^{1/2}g)_{H_0} + (\mathcal{B}f, g)_{H_0} = (Cf, g)_{H_0} \tag{8.23}$$
$$(f \in \mathcal{D}(C) \subseteq H_+ = \mathcal{D}(A^{1/2}), \quad g \in H_+).$$

The operator C defined as indicated above is called the form sum of the operators A and B and denoted by $A \dot{+} B$ (it would be more accurate to write $A \dot{+} \mathcal{B}$). The form sum can also be defined in the following natural way:

$$A \dot{+} B = (\mathcal{A} + \mathcal{B}) \restriction \mathcal{D}(A \dot{+} B),$$
$$\mathcal{D}(A \dot{+} B) = \{u \in H_+ \mid (\mathcal{A} + \mathcal{B})u \in H_0\} \tag{8.24}$$

(the restriction is understood as an operator in H_0).

Indeed, denote by F the operator constructed by using (8.24). It is easy to see that $C \subseteq F$. In fact, since

$$(A^{1/2}f, A^{1/2}g)_{H_0} = (\mathcal{A}f, g)_{H_0} \quad (f, g \in H_+),$$

the equality $H_0 \ni Cf = (\mathcal{A} + \mathcal{B})f = Ff$ for $f \in \mathcal{D}(C)$ follows from (8.23). Further, F is clearly an Hermitian operator in H_0. Hence, due to the selfadjointness of C, we arrive at the equality $F = C$. □

Thus, if condition (8.22) is satisfied, then the operator $A\dot{+}B$ defined by using (8.23) or (8.24) is selfadjoint in H_0. Note that the form b was constructed above by using the operator B. It is obvious that, for the form b appearing in Theorem 8.4, such operator always exists because condition (8.19) ensures its continuity on $H_+ = \mathcal{D}(\tilde{a})$. Therefore, one can say that the operator A is "perturbed" by the form b.

REMARK 8.1. The form $\tilde{a}+b$ can be closed even if we do not assume the validity of (8.22); the form sum $A\dot{+}B$ is clearly defined in this case.

REMARK 8.2. The notion of a form sum admits the following generalization:

Consider an operator $H_+ \supseteq \mathcal{D}(\mathcal{B}) \ni f \mapsto \mathcal{B}f \in H_-$ whose domain is dense in H_+ and assume that it is nonnegative with respect to H_0, i.e., $(\mathcal{B}f, f)_{H_0} \geq 0$ $(f \in \mathcal{D}(\mathcal{B}))$. (One can also consider the case of semiboundedness: $(\exists\alpha \in \mathbb{R})$: $(\mathcal{B}f, f)_{H_0} \geq \alpha(f, f)_{H_0}$ $(f \in \mathcal{D}(\mathcal{B})$.) In the space H_0, we now construct the form

$$b(f, g) = (\mathcal{B}f, g)_{H_0} \quad (f, g \in \mathcal{D}(b) = \mathcal{D}(\mathcal{B}) \subseteq H_+ \subseteq H_0).$$

It is not difficult to show that the form $\tilde{a}+b$ $(\mathcal{D}(\tilde{a}+b) = \mathcal{D}(b))$ admits a closure. The selfadjoint nonnegative operator C in H_0 associated with this closure by equality (8.9) is just the required generalization of the form sum $A\dot{+}B$.

Let us show that *the form* $\tilde{a}+b$ *is closable.*

Indeed, it admits the representation

$$(\tilde{a}+b)(f, g) = \big((\mathcal{A}+\mathcal{B})f, g\big)_{H_0} = \big(\mathcal{A}^{-1}(\mathcal{A}+\mathcal{B})f, g\big)_{H_+} \quad (f, g \in \mathcal{D}(\mathcal{B}))$$

and, therefore, according to the first part of the proof of Theorem 8.3, it is closable as a form in H_+, i.e., the corresponding completion of $\mathcal{D}(\mathcal{B})$ belongs to H_+ and, consequently, to $H_0 \supseteq H_+$. □

Chapter 15
Expansion in Generalized Eigenvectors

It was shown in the introduction to Chapter 13 that expressions of type (13.0.1) for the expansion of a vector in eigenvectors of an operator A in a finite-dimensional Hilbert space H cannot be directly generalized to the case of an infinite-dimensional space because the operator A may have no eigenvectors.

A simple example of a selfadjoint operator A that has no eigenvectors was presented in Section 8.8. Let us recall it in brief. Assume that $H = L_2((a,b))$ with respect to Lebesgue measure. Consider the operator $(Af)(x) = xf(x)$ $(f \in L_2(a,b),\ x \in (a,b))$, which is bounded and selfadjoint and has the spectrum $S(A) = [a,b]$. The equation for the eigenvector $\varphi \in L_2((a,b))$ that corresponds to a point $\lambda \in [a,b]$ of the spectrum has the following form:

$$(x - \lambda)\varphi(x) = 0. \tag{0.1}$$

On the one hand, this implies that $\varphi(x) = 0$ almost everywhere, i.e., $\varphi(x) = 0$ in $L_2((a,b))$ and, therefore, it is not an eigenvector in H. On the other hand, the δ-function at the point λ: $\varphi = \delta_\lambda$ is also a formal solution of equation (0.1). As an element of a corresponding space, it differs from zero and, therefore, can be regarded as an eigenvector.

Thus, the operator A has no ordinary eigenvectors but, at the same time, it has eigenvectors which are generalized functions. It turns out that this is a general property of selfadjoint operators in a separable space H; below, we present the corresponding results. We show that, under certain restrictions, the spectral theorem for A, i.e., the formulas

$$\mathbb{I} = \int_{-\infty}^{\infty} dE(\lambda), \qquad A = \int_{-\infty}^{\infty} \lambda dE(\lambda), \tag{0.2}$$

can be rewritten in the form similar to that in the case of a discrete spectrum where

$$\mathbb{I} = \sum_{k=1}^{\infty} P(\lambda_k) \qquad A = \sum_{k=1}^{\infty} \lambda_k P(\lambda_k).$$

Namely, formulas (0.2) can be rewritten as follows:

$$\mathbb{I} = \int_{-\infty}^{\infty} P(\lambda)d\rho(\lambda), \qquad A = \int_{-\infty}^{\infty} \lambda P(\lambda)d\rho(\lambda), \tag{0.3}$$

where ρ is a measure and $P(\lambda)$ is an operator of "generalized projection", whose range consists of the generalized eigenvectors of the operator A that correspond to the eigenvalue λ.

1 Differentiation of Operator-Valued Measures and Resolutions of the Identity

Below, we prove a theorem of Radon-Nikodym type on differentiation of an operator-valued measure with respect to its trace and present the corollary concerning the differentiation of a resolution of the identity. This fact will be used in establishing the main result of Chapter 15.

1.1 Differentiation of Operator-Valued Measures

Let us fix a chain

$$H_- \supseteq H_0 \supseteq H_+, \tag{1.1}$$

all spaces in which are separable (clearly, it is sufficient that H_+ be separable). Recall that an operator $A\colon H_+ \to H_-$ is called nonnegative if $(Au, u)_{H_0} \geq 0$ ($u \in H_+$) (see Section 14.1). By definition, the trace of a nonnegative operator is equal to

$$\mathrm{Tr}(A) = \sum_{j=1}^{\infty} (Ae_j, e_j)_{H_0},$$

where $(e_j)_{j=1}^{\infty}$ is an orthonormal basis in H_+. The value $\mathrm{Tr}(A)$ does not depend on the choice of this basis. Indeed, if $\mathbf{I}$ is an isometry associated with (1.1), then, by virtue of the relation $(\alpha, u)_{H_0} = (\mathbf{I}\alpha, u)_{H+}$ ($\alpha \in H_-, u \in H_+$), we can conclude that the nonnegativity of A is equivalent to the ordinary nonnegativity of $\mathbf{I}A\colon H_+ \to H_+$ and $\mathrm{Tr}(A) = \mathrm{Tr}(\mathbf{I}A)$.

The following definition is connected with the definition of the general resolution of the identity presented in Section 13.1. Assume that R is an abstract space not necessarily equipped with a topology and $\mathfrak{R}$ is a σ-algebra of sets from R. We say that a function $\mathfrak{R} \ni \alpha \mapsto \theta(\alpha)$ is an *operator-valued measure with a finite trace* if the following conditions are satisfied:

(a) $\theta(\alpha)$ is a nonnegative operator from H_+ to H_- such that $\theta(\emptyset) = 0$ and $\mathrm{Tr}(\theta(R)) < \infty$;

(b) countable additivity takes place, i.e., if $\alpha_j \in \mathfrak{R}(j \in \mathbb{N})$ do not intersect each other, then

$$\theta\left(\bigcup_{j=1}^{\infty} \alpha_j\right) = \sum_{j=1}^{\infty} \theta(\alpha_j),$$

where the series converges in the weak sense.

It follows from the additivity and nonnegativity of θ that it is monotone. i.e., if $\alpha' \subseteq \alpha''$, then $\theta(\alpha') \leq \theta(\alpha'')$. Therefore, $\theta(\alpha) \leq \theta(R)$ and $\mathrm{Tr}(\theta(\alpha)) \leq \mathrm{Tr}(\theta(R))$ ($\alpha \in \mathfrak{R}$).

Let us introduce a numerical nonnegative set function $\mathfrak{R} \ni \alpha \mapsto \rho(\alpha) = \operatorname{Tr}(\theta(\alpha))$. If $\alpha_j \in \mathfrak{R}$ $(j \in \mathbb{N})$ are disjoint, then, by virtue of condition (b) and nonnegativity of the terms, we have

$$\begin{aligned}\rho\left(\bigcup_{j=1}^{\infty}\alpha_j\right) &= \operatorname{Tr}\left(\theta\left(\bigcup_{j=1}^{\infty}\alpha_j\right)\right) = \operatorname{Tr}\left(\sum_{j=1}^{\infty}\theta(\alpha_j)\right)\\ &= \sum_{k=1}^{\infty}\left(\left(\sum_{j=1}^{\infty}\theta(\alpha_j)\right)e_k, e_k\right)_{H_0} = \sum_{k=1}^{\infty}\sum_{j=1}^{\infty}(\theta(\alpha_j)e_k, e_k)_{H_0}\\ &= \sum_{j=1}^{\infty}\operatorname{Tr}(\theta(\alpha_j)) = \sum_{j=1}^{\infty}\rho(\alpha_j).\end{aligned}$$

Thus, $\mathfrak{R} \ni \alpha \mapsto \rho(\alpha)$ is a numerical nonnegative finite measure. The measure ρ is called a *trace measure* for the measure θ.

Theorem 1.1. *An operator-valued measure θ with a finite trace can be differentiated with respect to its trace measure ρ. This means that there exists an operator-valued function $Q(\lambda)\colon H_+ \to H_-$, $Q(\lambda) \geq 0$, $Q(\lambda) \leq \operatorname{Tr}(Q(\lambda)) = 1$ weakly measurable with respect to $\mathfrak{R}$, defined for ρ-almost all $\lambda \in R$, and such that*

$$\theta(\alpha) = \int_{\alpha} Q(\lambda)d\rho(\lambda) \qquad (\alpha \in \mathfrak{R}) \tag{1.2}$$

(the integral converges in the Hilbert-Schmidt norm). The function $Q(\lambda)$ is uniquely defined up to values on a set of measure ρ zero and is called the Radon-Nikodym derivative $(d\theta/d\rho)(\lambda) = Q(\lambda)$.

Note that the convergence of integral (1.2) in the Hilbert-Schmidt norm means its convergence in the Bochner sense if $Q(\lambda)$ is regarded as a vector function with values in the space of Hilbert-Schmidt operators from H_+ to H_- (see Section 10.3). The strong measurability of $Q(\lambda)$ can be proved as follows: Let $\left(Q_{jk}(\lambda)\right)_{j,k=1}^{\infty}$ be a matrix of $Q(\lambda)$. One can construct a sequence of matrices $\left(Q_{jk}(\lambda)\right)_{j,k=1}^{n}$ $(n \in \mathbb{N})$. The corresponding finite-dimensional operators converge to $Q(\lambda)$ for any $\lambda \in R$ in the norm $\cdot$. By Theorem 2.5.2, the measurable functions $Q_{jk}(\lambda)$ $(j, k = 1, \dots, n)$ can be approximated by simple ones.

Proof. We fix an orthonormal basis $(e_j)_{j=1}^{\infty}$ in the space H_+. The measure θ is absolutely continuous with respect to ρ, i.e., if $\rho(\alpha) = 0$, then $\theta(\alpha) = 0$ $(\alpha \in \mathfrak{R})$; indeed

$$|(\theta(\alpha)e_j, e_k)_{H_0}|^2 \leq (\theta(\alpha)e_j, e_j)_{H_0}(\theta(\alpha)e_k, e_k)_{H_0} \leq \rho^2(\alpha) = 0 \qquad (j, k \in \mathbb{N}).$$

This implies that, for fixed $u, v \in H_+$, the complex-valued measure $\mathfrak{R} \ni \alpha \mapsto (\theta(\alpha)u, v)_{H_0} \in \mathbb{C}$ is also absolutely continuous with respect to ρ and, according to the ordinary Radon-Nikodym theorem (see Section 5.2), we have

$$(\theta(\alpha)u, v)_{H_0} = \int_{\alpha} q(\lambda; u, v)d\rho(\lambda) \qquad (\alpha \in \mathfrak{R};\ u, v \in H_+), \tag{1.3}$$

where the derivative $q(\lambda; u, v)$ is defined on the set $\beta_{u,v} \subseteq R$ of complete measure ρ, measurable with respect to $\mathfrak{R}$, and summable; for $u = v$, it is nonnegative. Denote by L a linear span of the vectors $(e_j)_{j=1}^{\infty}$ with rational complex coefficients; $\tilde{L} = H_+$. Since L is countable, the set $\bigcap_{u,v\in L} \beta_{u,v}$ is also a set of complete measure; for λ from this set, all functions $q(\lambda; u, v)(u, v \in L)$ are defined and $q(\lambda; u, u) \geq 0$ $(u \in L)$.

Since the derivative is uniquely defined to within its values on a set of measure zero, the bilinearity of the left-hand side of (1.3) with respect to u and v yields the bilinearity of $q(\lambda; u, v)$. More exactly, there exists a set of complete measure $\beta \subseteq \bigcap_{u,v\in L} \beta_{u,v}$ such that, for $\lambda \in \beta$, we have

$$\begin{aligned} &q(\lambda; p_1u_1 + p_2u_2,\ r_1v_1 + r_2v_2) \\ &\quad = p_1\bar{r}_1q(\lambda; u_1, v_1) + p_1\bar{r}_2q(\lambda; u_1, v_2) + p_2\bar{r}_1q(\lambda; u_2, v_1) + p_2\bar{r}_2q(\lambda; u_2, v_2) \end{aligned}$$

for any $u_1, u_2, v_1, v_2 \in L$ and complex rational p_1, p_2, r_1, and r_2. To prove this, we use the bilinearity of $(\theta(\alpha)u, v)_{H_0}$ and the fact that $\alpha \in \mathfrak{R}$ in (1.3) and conclude that this equality holds for λ from the set $\beta_{p_1,p_2,r_1,r_2,u_1,u_2,v_1,v_2} \subseteq \bigcap_{u,v\in L} \beta_{u,v}$ of complete measure. Then we take the (countable) intersection of all such sets as β. Furthermore, as was mentioned above, for such λ, we have $q(\lambda; u, u) \geq 0$ $(u \in L)$. The bilinearity and nonnegativity yield, in a standard way, the Cauchy-Buniakowski inequality

$$|q(\lambda; u, v)|^2 \leq q(\lambda; u, u)q(\lambda; v, v) \qquad (\lambda \in \beta; \quad u, v \in L). \tag{1.4}$$

By setting $u = v = e_j$ in (1.3), summing over $j \in \mathbb{N}$, and using the Fubini theorem (see Section 4.3), we get

$$\rho(\alpha) = \int_\alpha \left(\sum_{j=1}^{\infty} q(\lambda; e_j, e_j) \right) d\rho(\lambda) \qquad (\alpha \in \mathfrak{R}).$$

Hence, for almost all $\lambda \in \beta$, we have

$$\sum_{j=1}^{\infty} q(\lambda; e_j, e_j) = 1. \tag{1.5}$$

Reducing, if necessary, the set β, we can assume that (1.5) holds for all $\lambda \in \beta$. Relations (1.4) and (1.5) yield

$$\sum_{j,k=1}^{\infty} |q(\lambda; e_j, e_j)|^2 \leq \sum_{j,k=1}^{\infty} q(\lambda; e_j, e_j)q(\lambda; e_k, e_k) = 1 \qquad (\lambda \in \beta). \tag{1.6}$$

Let us fix $\lambda \in \beta$ and denote by $A(\lambda)$ the operator in H_+ that corresponds to the matrix $(a_{jk}(\lambda))_{j,k=1}^{\infty}$ in the basis $(e_j)_{j=1}^{\infty}$; here, $a_{jk}(\lambda) = q(\lambda; e_k, e_j)$. By virtue of (1.6), $A(\lambda)$ is well-defined and is a Hilbert-Schmidt operator. The measurability of each function $\beta \ni \lambda \mapsto q(\lambda; e_j, e_k)$ $(j, k \in \mathbb{N})$ implies that the operator-valued function $\beta \ni \lambda \mapsto A(\lambda)$ is weakly measurable. Let us introduce a continuous operator $Q(\lambda) = \mathbf{I}^{-1}A(\lambda)\colon H_+ \to H_-$ and show that this operator is the required one.

It follows from the measurability of $A(\lambda)$ that $\beta \ni \lambda \mapsto Q(\lambda)$ is weakly measurable. Further, for $u = \sum_{k=1}^{\infty} p_k e_k$, $v = \sum_{j=1}^{\infty} r_j e_j \in L$, we have

$$\begin{aligned}(Q(\lambda)u, v)_{H_0} &= (\mathbf{I}^{-1}A(\lambda)u, v)_{H_0} = (A(\lambda)u, v)_{H_+} \\ &= \sum_{j,k=1}^{\infty} q(\lambda; e_k, e_j) p_k \bar{r}_j = q(\lambda; u, v). \end{aligned} \tag{1.7}$$

In particular, $(Q(\lambda)u, u)_{H_0} = q(\lambda; u, u) \geq 0$. Passing to the limit, we find that the inequality remains valid for arbitrary $u \in H_+$, i.e., $Q(\lambda) \geq 0$. According to (1.5), $\mathrm{Tr}(Q(\lambda)) = \mathrm{Tr}(A(\lambda)) = 1$ $(\lambda \in \beta)$. Thus, $Q(\lambda) \leq 1$ and $Q(\lambda)$ $(\lambda \in \beta)$ is weakly measurable. Hence, there exists the integral

$$\int_\alpha Q(\lambda) d\rho(\lambda) \qquad (\alpha \in \mathfrak{R})$$

convergent in the Hilbert-Schmidt norm. According to (1.7) and (1.3), for $u, v \in L$ and $\alpha \in \mathfrak{R}$, we have

$$\begin{aligned}\left(\left(\int_\alpha Q(\lambda) d\rho(\lambda)\right) u, v\right)_{H_0} &= \int_\alpha (Q(\lambda)u, v)_{H_0} d\rho(\lambda) \\ &= \int_\alpha q(\lambda; u, v) d\rho(\lambda) = (\theta(\alpha)u, v)_{H_0}, \end{aligned} \tag{1.8}$$

i.e., (1.2) is true.

Finally, let us establish the uniqueness of $Q(\lambda)$. Assume that, parallel with $Q(\lambda)$, there is an operator-valued function $Q_1(\lambda)$ of the same type satisfying the equality

$$\int_\alpha Q(\lambda) d\rho(\lambda) = \int_\alpha Q_1(\lambda) d\rho(\lambda) \quad (\alpha \in \mathfrak{R}).$$

Then, for every $u, v \in L$, we have $(Q(\lambda)u, v)_{H_0} = (Q_1(\lambda)u, v)_{H_0}$ for λ from the set of complete measure $\beta_{1;u,v} \subseteq R$. But then, for λ from the set of complete measure $\beta_1 = \bigcap_{u,v \in L} \beta_{1;u,v}$, we have

$$(Q(\lambda)u, v)_{H_0} = (Q_1(\lambda)u, v)_{H_0}$$

for all $u, v \in L$. This and the continuity of the operators $Q(\lambda)$ and $Q_1(\lambda)$ for each fixed $\lambda \in \beta_1$ imply that $Q(\lambda) = Q_1(\lambda)$. □

REMARK 1.1. We can also consider an operator-valued measure with σ-finite trace. This means that there exists a sequence $(R_k)_{k=1}^{\infty} \subset \mathfrak{R}$ such that $R_1 \subseteq R_2 \subseteq \ldots, \bigcup_{k=1}^{\infty} R_k = R$ and $\mathrm{Tr}(\theta(R_k)) < \infty$ $(k \in \mathbb{N})$. In this case, Theorem 1.1 is modified as follows: One must consider a σ-finite trace measure instead of a finite one and state that representation (1.2) holds for every $\mathfrak{R} \ni \alpha \subseteq R_k$ for some $k \in \mathbb{N}$. The proof remains the same.

REMARK 1.2. The formulation of Theorem 1.1 can be made similar to the Radon-Nikodym theorem. Namely, assume that an operator-valued measure θ with σ-finite trace and a σ-finite nonnegative numerical measure $\mathfrak{R} \ni \alpha \mapsto \rho(\alpha) \in [0, \infty]$ with respect to which θ is absolutely continuous (i.e., if $\rho(\alpha) = 0$ for some $\alpha \in \mathfrak{R}$, the $\theta(\alpha) = 0$) are given. In this case, representation (1.2), in which $\mathfrak{R} \ni \alpha \subseteq R_k$ ($k \in \mathbb{N}$) and $Q(\lambda)$ is a weakly measurable operator-valued function defined for ρ-almost all $\lambda \in R$, is true. The values of the function $Q(\lambda)$ are nonnegative operators from H_+ to H_- each having a finite trace summable with respect to ρ over R_k ($k \in \mathbb{N}$) (one should write representation (1.2) with a trace measure and differentiate this measure with respect to ρ).

1.2 Differentiation of a Resolution of the Identity

Let us consider the differentiation of a resolution of the identity and the concept of a spectral measure.

Assume that R is an abstract space, $\mathfrak{R}$ is a σ-algebra of its sets, and $\mathfrak{R} \ni \alpha \mapsto E(\alpha)$ is a general resolution of the identity acting on the space H_0. As a rule, the measure E has no finite or σ-finite trace and, therefore, Theorem 1.1 cannot be directly applied. It is convenient for us to act as follows:

Assume that the rigging (1.1) of the space H_0 is given. Let $O\colon H_+ \to H_0$, $O^+\colon H_0 \to H_-$ be the corresponding embedding operators (it follows from the equality $(f, Ou)_{H_0} = (f, u)_{H_0} = (O^+ f, u)_{H_0}$ ($f \in H_0, u \in H_+$) that O^+ is, in fact, adjoint to O with respect to H_0). The function

$$\mathfrak{R} \ni \alpha \mapsto \theta(\alpha) = O^+ E(\alpha) O, \tag{1.9}$$

whose values are continuous operators from H_+ to H_-, is an operator-valued measure ($O^+E(\alpha)O \geq 0$ because $(O^+E(\alpha)Ou, u)_{H_0} = (E(\alpha)Ou, Ou)_{H_0} \geq 0$ for $u \in H_+$). Recall that rigging (1.1) is called quasinuclear if the embedding operator O is quasinuclear.

Lemma 1.1. *If rigging (1.1) is quasinuclear, then the operator-valued measure (1.9) has a finite trace.*

Before proving the lemma, we note the following: *If $A\colon H_0 \to H_0$ is nonnegative, then $O^+AO\colon H_+ \to H_-$ is also nonnegative and*

$$\operatorname{Tr}(O^+AO) \leq \|A\| \; O^2. \tag{1.10}$$

Indeed, the inequality $O^+AO \geq 0$ has already been explained by the example of $A = E(\alpha)$. Further, let $(e_j)_{j=1}^{\infty}$ be an orthonormal basis in H^+. Then

$$\begin{aligned}\operatorname{Tr}(O^+AO) &= \sum_{j=1}^{\infty} (O^+AOe_j, e_j)_{H_0} = \sum_{j=1}^{\infty} (AOe_j, Oe_j)_{H_0} \\ &\leq \|A\| \sum_{j=1}^{\infty} \|(Oe_j)\|_{H_0}^2 = \|A\| \; O^2.\end{aligned}$$

□

Proof of Lemma 1.1. According to (1.10), we have

$$\mathrm{Tr}(\theta(R)) = \mathrm{Tr}(O^+E(R)O) \leq O^2 < \infty. \qquad \square$$

Let us fix the quasinuclear rigging (1.1). The nonnegative finite measure $\mathfrak{R} \ni \alpha \mapsto \rho(\alpha) = \mathrm{Tr}(O^+E(\alpha)O) \in [0,\infty)$ is called the *spectral measure of the resolution of the identity* E. Clearly, E and ρ are mutually absolutely continuous: for some $\alpha \in \mathfrak{R}$, the equalities $E(\alpha) = 0$ and $\rho(\alpha) = 0$ are equivalent. By applying Theorem 1.1 to (1.10) and ρ, we obtain the following assertion:

Theorem 1.2. *Suppose that* $\mathfrak{R} \ni \alpha \mapsto E(\alpha)$ *is a resolution of the identity acting on the space* H_0, *(1.1) is a fixed quasinuclear rigging, and* $\mathfrak{R} \ni \alpha \mapsto \rho(\alpha) \in [0,\infty)$ *is the corresponding spectral measure. Then the following representation in the form of an integral convergent in the Hilbert-Schmidt norm is true:*

$$O^+E(\alpha)O = \int_\alpha P(\lambda)d\rho(\lambda) \qquad (\alpha \in \mathfrak{R}). \tag{1.11}$$

Here, $P(\lambda)\colon H_+ \to H_-$ *is an operator-valued function weakly measurable with respect to* $\mathfrak{R}$, *defined for* ρ*-almost all* $\lambda \in R$, *and such that* $P(\lambda) \geq 0$ *and* $P(\lambda) \leq \mathrm{Tr}(P(\lambda)) = 1$. $P(\lambda)$ *is called a generalized projector.*

As mentioned above, in the case of resolution of the identity E of a selfadjoint operator in H_0 with discrete spectrum $(\lambda_j)_{j=1}^\infty$, the equality

$$E(\alpha) = \sum_{\lambda_j \in \alpha} P(\lambda_j) \ (\alpha \in \mathfrak{B}(\mathbb{R}))$$

holds, in which $P(\lambda_j)$ is the projector onto the eigensubspace A corresponding to the eigenvalue λ_j. The comparison of this formula with (1.11) has determined the choice of the term "generalized projector".

REMARK 1.3. According to Remark 1.2, we can also introduce the concept of the general spectral measure corresponding to the resolution of the identity E. This measure is defined as a σ-finite nonnegative measure $\mathfrak{R} \ni \alpha \mapsto \rho(\alpha) \in [0,\infty]$ such that ρ and E are mutually absolutely continuous. Representation (1.11) remains valid for the general spectral measure except that the operator $P(\lambda)$ takes a scalar multiplier.

1.3 The Case of a Nuclear Rigging

Similar results can be obtained if we use the nuclear rigging

$$\Phi' \supseteq H_0 \supseteq \Phi \tag{1.12}$$

of the Hilbert space H_0 instead of its quasinuclear rigging (1.1) (recall, that a rigging is called nuclear if $\Phi = \mathrm{pr}\lim_{\tau \in T} H_\tau$ is a nuclear space; see Section 14.2). More exactly, the case of chain (1.12) is reduced to chain (1.1). Let us show this.

Consider rigging (1.12) and denote by O and O^+ the embedding operators $\Phi \subseteq H_0$ and $H_0 \subseteq \Phi'$, respectively. We also consider continuous operators $A\colon \Phi \to$

Φ'. These operators are called nonnegative ($A \geq 0$) if $(A\varphi, \varphi)_{H_0} \geq 0$ ($\varphi \in \Phi$). In particular, if $\mathfrak{R} \ni \alpha \mapsto E(\alpha)$ is the resolution of the identity mentioned above, then the operator $O^+E(\alpha)O\colon \Phi \to \Phi'$ is nonnegative. The function $\mathfrak{R} \ni \alpha \mapsto \theta(\alpha) = O^+E(\alpha)O$ is a measure similar to those studied in Subsections 1 and 2 but with values in $\mathcal{L}(\Phi, \Phi')$.

Theorem 1.3. *Let $\mathfrak{R} \ni \alpha \mapsto E(\alpha)$ and $\rho(\alpha)$ be a resolution of the identity acting on the space H_0 and some spectral measure corresponding to it, respectively. Suppose that* (1.12) *is a fixed nuclear rigging. Then representation* (1.11) *in the form of a weakly convergent integral is true, in which $0 \leq P(\lambda)\colon \Phi \mapsto \Phi'$ (a generalized projector) is an operator-valued function weakly measurable with respect to $\mathfrak{R}$ and defined for ρ-almost all $\lambda \in R$.*

Proof. As was shown in Section 14.2, every H_τ ($\tau \in T$) is embedded densely and continuously in H_0 and $0 \in T$. We choose τ so that the embedding $H_\tau \subseteq H_0$ is quasinuclear; this is possible because Φ is nuclear. As a result, we obtain the chain

$$\Phi' \supseteq H_{-\tau} \supseteq H_0 \supseteq H_\tau \supseteq \Phi \tag{1.13}$$

with dense and continuous embeddings. Let us take the central part of chain (1.12) as (1.1) and introduce the embedding operators $O_1\colon H_\tau \to H_0$ and $O_1^+\colon H_0 \to H_{-\tau}$. Then, by virtue of Theorem 1.2 and Remark 1.3, we get

$$O_1^+E(\alpha)O_1 = \int_\alpha P_1(\lambda)d\rho(\lambda) \qquad (\alpha \in \mathfrak{R}), \tag{1.14}$$

where $P_1(\lambda)\colon H_\tau \to H_{-\tau}$ is a corresponding generalized projector.

Consider the embeddings $O_2\colon \Phi \to H_\tau$ and $O_3\colon H_{-\tau} \to \Phi'$. Multiplying (1.14) from the right and from the left by O_2 and O_3, respectively, we obtain the required equality (1.11) with $P(\lambda) = O_3P_1(\lambda)O_2$. The convergence of integral (1.14) in the Hilbert-Schmidt norm (from H_τ to $H_{-\tau}$) obviously yields the weak convergence of integral (1.11). □

The results presented in this section are valid, e.g., for the resolution of the identity corresponding to a given selfadjoint or normal operator. In the next section, we examine this situation in detail.

2 Generalized Eigenvectors and the Projection Spectral Theorem

2.1 The Case of a Selfadjoint Operator

Consider the simplest classic case of a selfadjoint operator A acting on a rigged Hilbert space H_0. Assume that there exists a linear topological space D densely and continuously embedded in H_+ and such that $D \subseteq \mathcal{D}(A)$ and the restriction $A \upharpoonright D$ acts continuously from D to H_+. In this case, instead of (1.1), we have the rigging (chain)

$$H_- \supseteq H_0 \supseteq H_+ \supseteq D. \tag{2.1}$$

We say that the operator A with the mentioned properties and rigging (2.1) are standardly connected (or A admits (2.1)). Chain (2.1) is called an extension of (1.1). As before, (2.1) is quasinuclear by definition if the embedding of the (separable) space H_+ in H_0 is quasinuclear.

A nonzero vector $\varphi \in H_-$ is called a *generalized eigenvector* of the operator A corresponding to an eigenvalue $\lambda \in \mathbb{C}$ if

$$(\varphi, Au)_{H_0} = \lambda(\varphi, u)_{H_0} \qquad (u \in D). \tag{2.2}$$

If $\varphi \in \mathcal{D}(A)$, then the operator A in (2.2) can be transferred to φ and, since $u \in D$ is arbitrary, we get $A\varphi = \lambda\varphi$, i.e., φ is an ordinary eigenvector of A corresponding to the eigenvalue λ. Thus, the definition introduced above generalizes the classic concept.

The collection of eigenvalues corresponding to all possible generalized eigenvectors is called the *generalized spectrum* $g(A)$ of the operator A. Clearly, this spectrum depends on the choice of rigging (2.1) and, generally speaking, differs from the spectrum $S(A)$ of the operator A.

Example

2.1. Let $H_0 = L_2(\mathbb{R}, dx)$ with respect to the Lebesque measure dx and let A be the minimal operator generated by the expression $(\mathcal{L}u)(x) = -iu'(x)$, i.e., the closure of the operator $H_0 \supseteq C_0^\infty(\mathbb{R}) \ni u \mapsto iu'(x) \in H_0$ (see Section 12.2); this operator is selfadjoint and $S(A) = \mathbb{R}$ (see Section 16.4). Assume that $H_+ = L_2(\mathbb{R}, e^{x^2} dx)$ and $D = \mathcal{D}(\mathbb{R})$. Thus, the conditions concerning rigging (2.1) are satisfied and $H_- = L_2(\mathbb{R}, e^{-x^2} dx)$. For every $\lambda \in \mathbb{C}$, the function $\varphi(x) = e^{i\lambda x}$ ($x \in \mathbb{R}$) belongs to H_- and satisfies (2.2). Thus, $g(A) = \mathbb{C} \neq \mathbb{R} = S(A)$; $g(A) \supset S(A)$.

Consider the same operator A and assume that $H_+ = H_0$. In this case, $H_- = H_0$. There are no vectors $H_- \ni \varphi \neq 0$ satisfying (2.2). Indeed, each $\varphi \in \mathcal{D}'(\mathbb{R}) \supset H_0$ satisfying (2.2) is a generalized solution of the equation $-\varphi' = \lambda\varphi$ and, therefore, $\varphi(x) = ce^{i\lambda x}$ ($x \in \mathbb{R}, \mathbb{C} \ni c \neq 0$) (see Section 16.4). But such function φ cannot belong to $H_0 = H_-$ for any $\lambda \in \mathbb{C}$. Thus, $g(A) = \emptyset \neq \mathbb{R} = S(A)$; $g(A) \subset S(A)$.

Let us proceed to the spectral theorem. Assume that A is a selfadjoint operator acting on a separable Hilbert space H_0 and standardly connected with quasinuclear rigging (2.1). According to Theorem 1.2, its resolution of the identity E defined on the σ-algebra of Borel sets $\mathfrak{B}(\mathbb{R})$ admits representation (1.11), i.e.,

$$O^+ E(\alpha) O = \int_\alpha P(\lambda) d\rho(\lambda) \quad (\alpha \in \mathfrak{B}(\mathbb{R})). \tag{2.3}$$

Here, by definition, $\mathfrak{B}(\mathbb{R}) \ni \alpha \mapsto \rho(\alpha) = \operatorname{Tr}(O^+ E(\alpha) O) \in [0, \infty)$ is the spectral measure of A and $P(\lambda) \colon H_+ \to H_-$ is an operator-valued function weakly measurable with respect to $\mathfrak{B}(\mathbb{R})$, defined for ρ-almost all $\lambda \in \mathbb{R}$, and such that $P(\lambda) \geq 0$ and $P(\lambda) \leq \operatorname{Tr}(P(\lambda)) = 1$. Since, for $B \in \mathcal{L}(H_+, H_-)$, $B \geq 0$, the equalities $B = 0$

and $\operatorname{Tr}(B) = 0$ are equivalent, we have $S(A) = \operatorname{supp} E = \operatorname{supp} \rho$. Therefore, we can assume that λ takes values not in $\mathbb{R}$ but in $S(A)$. Naturally, such a measure connected with the resolution of the identity of the operator A is called a *general spectral measure* of this operator.

Theorem 2.1. (BGKM[1]-decomposition). *Let A be a selfadjoint operator acting on a separable Hilbert space H_0 and standardly connected with quasinuclear chain (2.1) in which D is separable. Then there exists an operator-valued function $P(\lambda)$ weakly measurable, defined for almost all λ from the spectrum $S(A)$ in the sense of the spectral measure ρ, and such that its values are nonnegative operators from H_+ to H_- and $P(\lambda) \leq \operatorname{Tr}(P(\lambda)) = 1$, which realizes the following representation of the operator A and its resolution of the identity E:*

$$\begin{aligned} E(\alpha)u &= \Big(\int_\alpha P(\lambda)d\rho(\lambda)\Big)u \qquad (\alpha \in \mathfrak{B}(\mathbb{R}),\ u \in H_+), \\ Au &= \Big(\int_{S(A)} \lambda P(\lambda)d\rho(\lambda)\Big)u \qquad (u \in \mathcal{D}(A) \cap H_+). \end{aligned} \tag{2.4}$$

The range $\mathfrak{R}(P(\lambda)) \subseteq H_-$ consists of generalized eigenvectors φ of A corresponding to the eigenvalue λ, i.e.,

$$(\varphi, Au)_{H_0} = \lambda(\varphi, u)_{H_0} \qquad (u \in D). \tag{2.5}$$

Thus, formulas (2.4), in fact, have the form of representations (0.3) with selected "projectors on the eigensubspaces" consisting of generalized eigenvectors. In this connection, *the spectral theorem for A in the form (2.4) is called the projection theorem.*

Proof. By comparing Theorem 2.1 with Theorem 1.2 in the particular case of the resolution of the identity E corresponding to A (see (2.3)), we conclude that it suffices to prove only the last statement of the theorem. Note that the second equality in (2.4) follows from representation (13.6.1) and the equality $dE(\lambda)u = P(\lambda)d\rho(\lambda)u$ for $u \in H_+$, which, in fact, is the first equality in (2.4) (i.e., (2.3)) rewritten in a different form.

Let us prove that $\varphi \in \mathfrak{R}(P(\lambda))$ satisfies (2.5). In other words, we must establish the following relation: There exists the set $\beta \in \mathfrak{B}(\mathbb{R})$ of complete spectral measure ρ such that, for any $\lambda \in \beta$,

$$(P(\lambda)u, Av)_{H_0} = \lambda(P(\lambda)u, v)_{H_0} \qquad (u \in H_+, v \in D). \tag{2.6}$$

First, note that it suffices to establish the existence of a set $\beta \in \mathfrak{B}(\mathbb{R})$ of complete measure ρ such that, for $\lambda \in \beta$, relation (2.6) holds for u and v taking

[1]This result belongs to Yu. Berezansky (1956), L. Gårding (1954), I. Gelfand and A. Kostyuchenko (1955), G. Kats (1958), and K. Maurin (1958).

values, respectively, in the fixed sets $G \subseteq H_+$ and $F \subseteq D$, which are dense in these spaces.

Indeed, assume that $u \in H_+$, $v \in D$, $D \ni u_n \underset{n\to\infty}{\longrightarrow} u$ in H_+, and $F \ni v_m \underset{m\to\infty}{\longrightarrow} v$ in D. By assumption, we have

$$(P(\lambda)u_n,\ Av_m)_{H_0} = \lambda(P(\lambda)u_n,\ v_m)_{H_0} \qquad (\lambda \in \beta;\ n, m \in \mathbb{N}). \tag{2.7}$$

Let us pass to the limit in (2.7), first, as $n \to \infty$ and then as $m \to \infty$. This is possible because $P(\lambda) \in \mathcal{L}(H_+, H_-)$ and $A \restriction D \in \mathcal{L}(D, H_+)$. As a result, we obtain the required relation (2.6).

We fix $u, v \in D$. By using (2.3), we get

$$\begin{aligned}\int_\alpha (P(\lambda)u, Av)_{H_0} d\rho(\lambda) &= \left(\left(\int_\alpha P(\lambda)d\rho(\lambda)\right) u, Av\right)_{H_0} \\ &= (O^+E(\alpha)Ou, Av)_{H_0} = (AE(\alpha)u, v)_{H_0} \\ &= \left(\left(\int_\alpha \lambda dE(\lambda)\right) u, v\right)_{H_0} \\ &= \int_\alpha \lambda d(O^+E(\lambda)Ou, v)_{H_0} \qquad (\alpha \in \mathfrak{B}(\mathbb{R})).\end{aligned} \tag{2.8}$$

Here, we have used the relation

$$AE(\alpha) = \int_\alpha \lambda dE(\lambda) \qquad (\alpha \in \mathfrak{B}(\mathbb{R})),$$

which follows from equalities (13.6.1) and (13.2.22) (see also Remark 13.2.1). Further, by replacing the differential in the last integral in (2.8) according to (2.3), we obtain

$$\int_\alpha (P(\lambda)u, Av)_{H_0} d\rho(\lambda) = \int_\alpha \lambda(P(\lambda)u, v)_{H_0} d\rho(\lambda) \qquad (\alpha \in \mathfrak{B}(\mathbb{R})).$$

Since α is arbitrary, the last relation implies that there exists a set $\beta_{u,v} \in \mathfrak{B}(\mathbb{R})$ of complete measure ρ such that, for $\lambda \in \beta_{u,v}$, relation (2.6) is true.

Assume now that L is a countable set dense in the space D and, hence, in H_+. Consider the countable intersection $\bigcap_{u,v\in L} \beta_{u,v} = \beta$, which is a set of complete measure ρ. If $\lambda \in \beta$, relation (2.6) holds for all $u, v \in L$. But then, according to the remark made at the beginning of the proof, this relation is also true for all $u \in H_+$ and $v \in D$. □

Under somewhat stronger restrictions on the operator A, this theorem can be reformulated in terms of nuclear riggings of the space H_0. Below, we present necessary definitions.

Consider a rigging of H_0 by linear topological spaces

$$\Phi' \supseteq H_0 \supseteq \Phi. \tag{2.9}$$

We say that a selfadjoint operator A in H_0 and chain (2.9) are *standardly connected* (or A admits (2.9)) if $\Phi \subseteq \mathcal{D}(A)$ and $A \upharpoonright \Phi$ acts continuously on Φ. The definition of generalized eigenvectors remains, in fact, the same, i.e., equality (2.2) must hold for $u \in \Phi$. Furthermore, the generalized spectrum $g(A)$ is, as before, a collection of all eigenvalues corresponding to generalized eigenvectors.

Note that if A is standardly connected with (2.9) and $\Phi = \operatorname{pr}\lim_{\tau \in T} H_\tau$, then A is also standardly connected with any chain

$$H_{-\tau} \supseteq H_0 \supseteq H_\tau \supseteq \Phi = D \tag{2.10}$$

of the form (2.1). Let us apply Theorem 1.3 instead of Theorem 1.2 to the resolution of the identity E of the operator A and use the scheme of its proof. Since A is standardly connected with (2.10), by using Theorem 2.1, we establish the following result:

Theorem 2.2. *Let A be a selfadjoint operator acting on a separable Hilbert space H_0 and standardly connected with the nuclear chain (2.9). Then all statements of Theorem 2.1 remain valid with the only modification that $P(\lambda)$ acts continuously from Φ to Φ' (and one cannot speak about its Hilbert norm and trace). In this case, the measure ρ is a spectral measure of the operator A.*

2.2 The Case of a Normal Operator

Let us show how the results obtained above change if the operator A is normal. First, we must modify the definition of a generalized eigenvector.

Let A be a normal operator in H_0 and let $\varphi \in H_0$ be its eigenvector corresponding to an eigenvalue λ_0, in a certain neighborhood $U \subset \mathbb{C}$ of which there are no other points of the spectrum of A. Then the spectral decomposition of A has the following form (see (13.6.19)):

$$A = \int_{\mathbb{C}} \lambda dE(\lambda) = \lambda_0 P(\lambda_0) + \int_{\mathbb{C}\setminus U} \lambda dE(\lambda), \tag{2.11}$$

where E is the resolution of the identity of the operator A and $P(\lambda_0)$ is the projector to the eigensubspace corresponding to λ_0 and consisting of the vectors φ. For the adjoint operator, relations (13.6.3) and (2.11) yield

$$A^* = \int_{\mathbb{C}} \bar{\lambda} dE(\lambda) = \bar{\lambda}_0 P(\lambda_0) + \int_{\mathbb{C}\setminus U} \bar{\lambda} dE(\lambda). \tag{2.12}$$

It follows from (2.12) that φ is also an eigenvector of the operator A^* corresponding to the eigenvalue $\bar{\lambda}_0$. In view of this, it is convenient to introduce the following definition:

Chain (2.1) considered above and a normal operator A acting on H_0 are called standardly connected if $D \subseteq \mathcal{D}(A)$ and the restrictions $A \restriction D$ and $A^* \restriction D$ act continuously from D to H_+. The generalized eigenvector of the operator A corresponding to an eigenvalue $\lambda \in \mathbb{C}$ is defined as a vector $\varphi \in H_-$ such that

$$(\varphi, A^* u)_{H_0} = \lambda(\varphi, u)_{H_0}, \quad (\varphi, Au)_{H_0} = \bar{\lambda}(\varphi, u)_{H_0} \qquad (u \in D). \tag{2.13}$$

As before, we can conclude from (2.13) that if, in addition, $\varphi \in \mathcal{D}(A)$, then $A\varphi = \lambda\varphi$ and $A^*\varphi = \bar{\lambda}\varphi$. Thus, (2.13) is a generalization of the concept of an eigenvector of a normal operator. Obviously, equality (2.3) remains valid; in this case, however, $\alpha \in \mathfrak{B}(\mathbb{C})$. The concept of the spectral measure ρ of an operator A is introduced analogously. In the case under consideration, the following analogue of Theorem 2.1 is true:

Theorem 2.3. *Let A be a normal operator acting on a separable Hilbert space H_0 and standarly connected with quasinuclear chain (2.1) in which D is separable. Then all statements of Theorem 2.1 remain valid with formulas (2.4) replaced by*

$$\begin{aligned} E(\alpha)u &= \left(\int_\alpha P(\lambda)d\rho(\lambda)\right)u \qquad && (\alpha \in \mathfrak{B}(\mathbb{C}),\ u \in H_+), \\ Au &= \left(\int_{S(A)} \lambda P(\lambda)d\rho(\lambda)\right)u \qquad && (u \in \mathcal{D}(A) \cap H_+), \\ A^*u &= \left(\int_{S(A)} \bar{\lambda} P(\lambda)d\rho(\lambda)\right)u \qquad && (u \in \mathcal{D}(A) \cap H_+) \end{aligned} \tag{2.14}$$

and equality (2.5) replaced by (2.13).

Proof. As in the case of Theorem 2.1, the question is reduced to proving the following assertion: There exists a set $\beta \in \mathfrak{B}(\mathbb{C})$ of complete spectral measure such that, for every $\lambda \in \beta$, we have

$$\begin{aligned} (P(\lambda)u, A^*v)_{H_0} &= \lambda(P(\lambda)u, v)_{H_0}, \\ (P(\lambda)u, Av)_{H_0} &= \bar{\lambda}(P(\lambda)u, v)_{H_0} \end{aligned} \tag{2.15}$$

$$(u \in H_+,\ v \in D).$$

First, we prove the existence of a set $\beta_1 \in \mathfrak{B}(\mathbb{C})$ of complete measure ρ such that, for any $\lambda \in \beta_1$, the first relation in (2.15) holds. This can be done by analogy with the proof of Theorem 2.1; one should only use representation (13.6.32) instead of (13.6.1).

Similarly, one can show that there exists a set $\beta_2 \in \mathfrak{B}(\mathbb{C})$ of complete measure ρ such that, for any $\lambda \in \beta_2$, the second relation in (2.15) is true (by using representation (13.6.19) instead of representation (13.6.32)). Finally, we set $\beta = \beta_1 \cap \beta_2$. $\square$

As in the case of selfadjoint operators, we can use the nuclear chain (2.9) standardly connected with a normal operator. The definitions of standard connection and generalized eigenvector are similar to the corresponding definitions presented above. The corresponding analogue of Theorem 2.2 can be obtained by obvious modification and we do not present it here.

In a special case, the results of this subsection can be applied to unitary operators. We do not present here the corresponding theorems on expansion of unitary operators in generalized eigenvectors; the reader can easily formulated them by himself.

2.3 Families of Commuting Operators

Let us discuss one more problem concerning selfadjoint operators. We assume that unbounded selfadjoint operators $A_1, \dots, A_n$ act on a Hilbert space H_0 and denote their resolutions of the identity by $E_1, \dots, E_n$, respectively. These operators are called *commuting* if their resolutions of the identity commute, namely, $E_j(\alpha_j)E_k(\alpha_k) = E_k(\alpha_k)E_j(\alpha_j)$ $(\alpha_j, \alpha_k \in \mathfrak{B}(\mathbb{R});\ j, k = 1, \dots, n)$. In Section 13.9, we have clarified the conditions which guarantee that the operators $A_1, \dots, A_n$ commute. Recall that if these operators are bounded, then the relations $A_jA_k = A_kA_j$ $(j, k = 1, \dots, n)$ are necessary and sufficient conditions for these operators to commute. If they are unbounded, the situation is much more complicated and, for example, the fact that such conditions are satisfied on certain dense sets does not guarantee that the corresponding resolutions of the identity commute, i.e., that A_j commute.

Assume that the family $A = (A_j)_{j=1}^n$ of commuting selfadjoint operators in H_0 is given. Let us construct the expansion of these operators in generalized joint eigenvectors. *A vector $0 \neq \varphi \in H_0$ is called an (ordinary) joint eigenvector of the family A if $\varphi \in \mathcal{D}(A_j)$ and $A_j\varphi = \lambda_j\varphi$ with some $\lambda_j \in \mathbb{R}$ $(j = 1, \dots, n)$, where $\lambda = (\lambda_1, \dots, \lambda_n) \in \mathbb{R}^n$ is the eigenvalue of the family A corresponding to φ.*

In accordance with this definition, we introduce the concept of a generalized joint eigenvector. Consider chain (2.1) standardly connected with each A_j $(j = 1, \dots, n)$. Then, by definition, $\varphi \in H_-$ is a *generalized joint eigenvector of the family A corresponding to the eigenvalue $\lambda = (\lambda_1, \dots, \lambda_n) \in \mathbb{C}^n$ if*

$$(\varphi, A_j u)_{H_0} = \lambda_j (\varphi, u)_{H_0} \qquad (j = 1, \dots, n;\ u \in D). \tag{2.16}$$

The collection of all these λ forms the generalized spectrum $g(A)$ of the family.

The family A is naturally associated with so called joint resolution of the identity E, i.e., the resolution of the identity in the space $\mathbb{R}^n$ defined on the σ-algebra $\mathfrak{B}(\mathbb{R}^n)$ as a direct product $E = \times_{j=1}^n E_j$ of resolutions of the identity $E_1, \dots, E_n$ (see Section 13.3). By definition, the support of the measure E is the spectrum of A, $S(A) = \operatorname{supp} E \subseteq \times_{j=1}^n \operatorname{supp} E_j = \times_{j=1}^n S(A_j)$ (see (13.5.23)). As before, by using E and the quasinuclear chain (2.1), we construct the spectral measure of the family A: $\mathfrak{B}(\mathbb{R}^n) \ni \alpha \mapsto \rho(\alpha) = \operatorname{Tr}(O^+ E(\alpha) O)$. Similarly, we introduce the general spectral measure of the family A. Recall that the operators A_j can be reconstructed by using E as follows (see (13.5.16)):

$$A_j = \int_{\mathbb{R}^n} \lambda_j dE(\lambda) \qquad (j = 1, \ldots, n). \tag{2.17}$$

Theorem 2.4. *Let $A = (A_j)_{j=1}^n$ be a family of commuting selfadjoint operators acting on a separable Hilbert space H_0, each of which is standardly connected with the quasinuclear chain (2.1) in which D is separable. Then all statements of Theorem 2.1 remain valid with formulas (2.4) replaced by*

$$\begin{gathered} E(\alpha)u = \left(\int_\alpha P(\lambda)d\rho(\lambda)\right) u \qquad (\alpha \in \mathfrak{B}(\mathbb{R}^n),\ u \in H_+), \\ A_j u = \left(\int_{S(A)} \lambda_j P(\lambda)d\rho(\lambda)\right) u \qquad (j = 1, \ldots, n;\ u \in \mathcal{D}(A_j) \cap H_+) \end{gathered} \tag{2.18}$$

and equality (2.5) replaced by (2.16).

Proof. As in the case of Theorem 2.1, the problem is reduced to proving the following relation: There exists a set $\beta \in \mathfrak{B}(\mathbb{R}^n)$ of complete measure ρ such that, for any $\lambda \in \beta$, we have

$$(P(\lambda)u,\ A_j v)_{H_0} = \lambda_j (P(\lambda)u, v)_{H_0} \qquad (j = 1, \ldots, n;\ u \in H_+,\ v \in D). \tag{2.19}$$

As in the proof of Theorem 2.1, we can conclude that, for any fixed $j = 1, \ldots, n$, there is a set $\beta_j \in \mathfrak{B}(\mathbb{R}^n)$ for which relation (2.19) is satisfied. To prove Theorem 2.4, one must repeat the proof of Theorem 2.1, using representation (2.17) instead of (13.6.1) and setting $\beta = \bigcap_{j=1}^n \beta_j$. $\square$

Theorem 2.4 remains valid for commuting normal operators (i.e., operators whose resolutions of the identity commute). The corresponding results can be also formulated for the case where the nuclear chain (2.9) is used instead of the quasinuclear chain (2.1). For infinite families of commuting selfadjoint (or normal) operators, the facts presented above remain valid as well, but their formulations and proofs are more complicated (especially, in the case of more than countable families).

REMARK 2.1. It is worth noting that the space D from (2.1) may be not dense in H_+. If D is dense in H_0, then one can construct a chain $D' \supseteq H_0 \supseteq D$, where D' is the dual space of antilinear continuous functionals on D. In a certain sense, D' contains H_-; indeed, any $\alpha \in H_-$ is also a continuous antilinear functional in D and, therefore, α can be interpreted as an element $l_\alpha \in D'$ (more exactly, l_α is identified with the class of $\beta \in H_-$ such that $(\beta, u)_{H_0} = (\alpha, u)_{H_0}$, $u \in D$). It is clear that all results obtained above for $\varphi \in H_-$ remain valid for generalized eigenvectors φ from D'.

2.4 Cyclic Vectors

Recall that the spectral measure of a selfadjoint operator A in H_0 was defined as the trace measure $\mathfrak{B}(\mathbb{R}) \ni \alpha \mapsto \rho(\alpha) = \mathrm{Tr}(O^+ E(\alpha) O) \in [0, \infty)$ constructed by using the quasinuclear chain (2.1). The general spectral measure was introduced as a scalar measure ρ such that ρ and E are absolutely continuous with respect to each other (see Remark 1.3).

Here, we consider an important case where the role of the spectral measure ρ can be played by a measure different from a trace one.

A unit vector $\Omega \in H_0$ is called a *cyclic vector* of an operator A (or a *vacuum*) if $\Omega \in \bigcap_{m=1}^{\infty} \mathcal{D}(A^m)$ and the set of vectors $\{A^m \Omega \mid m \in \mathbb{Z}_+\}$ is total.

Theorem 2.5. *Assume that a selfadjoint operator A with the resolution of the identity E has a cyclic vector Ω. Then the finite nonnegative measure $\mathfrak{B}(\mathbb{R}) \ni \alpha \mapsto \rho(\alpha) = (E(\alpha)\Omega, \Omega)_{H_0} \in [0, \infty)$ is a spectral measure of this operator.*

Proof. If $E(\alpha) = 0$ for some $\alpha \in \mathfrak{B}(\mathbb{R})$, then $\rho(\alpha) = 0$. Let us prove the inverse implication. Assume that $0 = \rho(\alpha) = (E(\alpha)\Omega, \Omega)_{H_0} = \|E(\alpha)\Omega\|^2_{H_0}$, i.e., $E(\alpha)\Omega = 0$. Then, for any $m \in \mathbb{Z}_+$, we have $0 = A^m E(\alpha)\Omega = E(\alpha) A^m \Omega$, and, hence, $E(\alpha) f = 0$, where f belongs to the linear span of the vectors $A^m \Omega$. By assumption, this span is dense in H_0 and, therefore, $E(\alpha) = 0$. □

In the case of normal operators, the formulation of the theorem remains the same. For a family of commuting operators $A = (A_j)_{j=1}^n$, instead of A^m, one must take the products $A_1^{m_1} \ldots A_n^{m_n}$, where $m_1, \ldots, m_n \in \mathbb{Z}_+$.

3 Fourier Transformation in Generalized Eigenvectors and the Direct Integral of Hilbert Spaces

Note that the initial formulas (13.0.1) and (13.0.2) of expansion in eigenvectors of a selfadjoint operator in a finite-dimensional space differ from expressions (2.4) proved above. Namely, the initial formulas contain the eigenvectors while expressions (2.4) contain the projectors $P(\lambda)$. Let us show that analogous expansions can be also obtained in the general case.

Below, we consider the case of one selfadjoint operator, but the results obtained can be easily generalized to normal operators and families of commuting operators considered above. We leave it for the reader to formulate and prove the corresponding results.

3.1 Fourier Transformation

Assume that a selfadjoint operator A satisfies the conditions of the projection spectral theorem (Theorem 2.1). By virtue of the first relation in (2.4), we have

$$(E(\alpha)u, v)_{H_0} = \int_{\alpha} (P(\lambda)u, v)_{H_0} \, d\rho(\lambda) \qquad (u, v \in H_+; \quad \alpha \in \mathfrak{B}(\mathbb{R})). \tag{3.1}$$

In particular, for $\alpha = \mathbb{R}$, equality (3.1) gives the decomposition of $(u,v)_{H_0}$ into the integral of $(P(\lambda)u,v)_{H_0}$. The last expression determines a scalar product and (3.1) turns into the decomposition of H_0, which is called the *direct integral* of the corresponding Hilbert spaces. Let us consider this case in detail.

We fix $\lambda \in \mathbb{R}$ so that $\operatorname{Tr}(P(\lambda)) = 1$. The points λ possessing this property form a set of complete spectral measure. Below, we assume that λ belongs to this set. By virtue of the relations $P(\lambda) \geq 0$ and $P(\lambda) \leq \operatorname{Tr}(P(\lambda)) = 1$, the operator $\mathbf{J}P(\lambda)J\colon H_0 \to H_0$ is a nonnegative Hilbert-Schmidt operator (see (1.1) and (1.2)). Let $h_\gamma(\lambda) \in H_0$ $(\gamma = 1, 2, \dots, N(\lambda) \leq \infty)$ be an orthonormal sequence of the eigenvectors of the operator $\mathbf{J}P(\lambda)J$ corresponding to the eigenvalues $\nu_\gamma(\lambda)$. Since the dependence of $\mathbf{J}P(\lambda)J$ on the parameter λ is weakly measurable with respect to $\mathfrak{B}(\mathbb{R})$, we can assume (this can be easily proved) that $h_\gamma(\lambda)$ is weakly measurable and $\nu_\gamma(\lambda)$ is measurable $(\gamma = 1, 2, \dots, N(\lambda))$. Then

$$(P(\lambda)Jf, Jg)_{H_0} = (\mathbf{J}P(\lambda)Jf, g)_{H_0} = \sum_{\gamma=1}^{N(\lambda)} \nu_\gamma(\lambda)(f, h_\gamma(\lambda))_{H_0}\overline{(g, h_\gamma(\lambda))}_{H_0}$$

$$= \sum_{\gamma=1}^{N(\lambda)} (Jf, \varphi_\gamma(\lambda))_{H_0}\overline{(Jg, \varphi_\gamma(\lambda))}_{H_0} \qquad (f, g \in H_0), \tag{3.2}$$

where $\varphi_\gamma(\lambda) = \sqrt{\nu_\gamma(\lambda)}\mathbf{J}^{-1}h_\gamma(\lambda) = P(\lambda)((\nu_\gamma(\lambda))^{-1/2}Jh_\gamma(\lambda)) \in \mathcal{R}(P(\lambda)) \subseteq H_-$.

The vectors $\varphi_\gamma(\lambda)$ $(\lambda \in \mathbb{R};\ \gamma = 1, 2, \dots, N(\lambda) \leq \infty)$ are individual generalized eigenvectors of the operator A.

It follows from (3.2) that

$$(P(\lambda)u, v)_{H_0} = \sum_{\gamma=1}^{N(\lambda)} (u, \varphi_\gamma(\lambda))_{H_0}\overline{(v, \varphi_\gamma(\lambda))}_{H_0} \qquad (u, v \in H_+) \tag{3.3}$$

for ρ-almost all $\lambda \in \mathbb{R}$.

Denote $l_2(\infty) = l_2$ and $l_2(N) = \mathbb{C}^N$ $(N < \infty)$, assuming that the last space is embedded in l_2 (all the coordinates of the vector, beginning from $N+1$, are equal to zero). *The mapping*

$$H_+ \ni u \mapsto \hat{u}(\lambda) = (\hat{u}_1(\lambda), \hat{u}_2(\lambda), \dots) \in l_2(N(\lambda)),$$
$$\hat{u}_\gamma(\lambda) = (u, \varphi_\gamma(\lambda))_{H_0} \qquad (\gamma = 1, 2, \dots, N(\lambda)) \tag{3.4}$$

is called the Fourier transformation corresponding to the operator A (the inclusion $\hat{u}(\lambda) \in l_2(N(\lambda))$ follows from (3.3) for $v = u$). The Fourier transform $\hat{u}(\lambda)$ of the vector u is defined for ρ-almost all $\lambda \in \mathbb{R}$ and each coordinate is measurable with respect to $\mathfrak{B}(\mathbb{R})$. By inserting (3.3) in (3.1) and using (3.4), we obtain the Parseval equality for Fourier transforms

$$(E(\alpha)u, v)_{H_0} = \int_\alpha (\hat{u}(\lambda), \hat{v}(\lambda))_{l_2(N(\lambda))}\, d\rho(\lambda) \tag{3.5}$$
$$(\alpha \in \mathfrak{B}(\mathbb{R});\ u, v \in H_+).$$

Note the following important fact:

Lemma 3.1. *For fixed* λ, *the set* $\{\hat{u}(\lambda) \mid u \in H_+\}$ *coincides with* $l_2(N(\lambda))$ *for* $N(\lambda) < \infty$ *and contains all finite vectors from* l_2 *for* $N(\lambda) = \infty$.

Proof. It suffices to show that every vector $(0, \dots, 0, 1, 0, \dots)$ with the unit on the kth place belongs to this set. We set $u = (\nu_k(\lambda))^{-1/2} J h_k(\lambda) \in H_+$. Then

$$\begin{aligned}(u, \varphi_\gamma(\lambda)_{H_0}) &= ((\nu_k(\lambda))^{-1/2}(J h_k(\lambda), \varphi_\gamma(\lambda))_{H_0} \\ &= (\nu_k(\lambda))^{-1/2}(h_k(\lambda), \mathbf{J}\varphi_\gamma(\lambda))_{H_0} = (h_k(\lambda), h_\gamma(\lambda))_{H_0} = \delta_{k\gamma} \\ &\quad (\gamma = 1, 2, \dots, N(\lambda)).\end{aligned}$$ □

3.2 The Direct Integral of Hilbert Spaces

Consider the direct integral of the Hilbert spaces $l_2(N(\lambda))$ over $\mathbb{R}$ with measure ρ

$$L_2 = \int_{\mathbb{R}} \oplus l_2(N(\lambda))\, d\rho(\lambda).$$

This integral is defined as the collection of all vector functions $\mathbb{R} \ni \lambda \mapsto F(\lambda) \in l_2(N(\lambda))$ given for ρ-almost all λ, measurable with respect to $\mathfrak{B}(\mathbb{R})$ in the sense that each coordinate $F_\gamma(\lambda)$ $(\gamma = 1, 2, \dots, N(\lambda))$ is measurable, and such that

$$\int_{\mathbb{R}} \|F(\lambda)\|^2_{l_2(N(\lambda))}\, d\rho(\lambda) < \infty.$$

One can easily prove (cf. Sections 6.8, 7.9) that the direct integral is a Hilbert space with the scalar product

$$(F(\cdot), G(\cdot))_{L_2} = \int_{\mathbb{R}} (F(\lambda), G(\lambda))_{l_2(N(\lambda))}\, d\rho(\lambda) \qquad (3.6)$$
$$(F(\cdot), G(\cdot) \in L_2).$$

By comparing (3.5) and (3.6), we conclude that, for $\alpha = \mathbb{R}$, the expression on the right-hand side of (3.5) determines a scalar product in the direct integral and, therefore, the Parseval equality can be rewritten in the form $(u, v)_{H_0} = (\hat{u}(\cdot), \hat{v}(\cdot))_{L_2}$ $(u, v \in H_+)$. Extending this equality by continuity to the whole H_0, we obtain

$$(f, g)_{H_0} = \int_{\mathbb{R}} (\hat{f}(\lambda), \hat{g}(\lambda))_{l_2(N(\lambda))}\, d\rho(\lambda) \qquad (f, g \in H_0), \qquad (3.7)$$

where the Fourier transform $\hat{f}(\lambda)$ of the vector f is understood as the limit of the Fourier transforms (3.4) in the norm of the direct integral (clearly, it is no longer possible to use the last formula in (3.4) for $\hat{f}_\gamma(\lambda)$).

Theorem 3.1. *If D is a base of the operator A, i.e., if the closure of $A \upharpoonright D$ in H_0 coincides with A, then the Fourier transforms $\hat{u}(\lambda)$ $(u \in H_+)$ are dense in the direct integral and, hence, the Fourier transformation $H_0 \ni f \mapsto \hat{f}(\lambda) \in L_2$ realizes an isomorphism between the spaces H_0 and L_2.*

Thus, in the indicated sense, H_0 can be regarded as decomposed into the direct integral

$$H_0 = \int_{\mathbb{R}} \oplus l_2(N(\lambda))\, d\rho(\lambda). \tag{3.8}$$

Lemma 3.2. *Under mapping (3.4), the operator $A \upharpoonright D$ turns into the operator of multiplication by λ, i.e., for ρ-almost all $\lambda \in \mathbb{R}$, the following relation holds:*

$$(Au)\widehat{\ }(\lambda) = \lambda \hat{u}(\lambda) \qquad (u \in D). \tag{3.9}$$

Proof. Denote by $\beta \in \mathfrak{B}(\mathbb{R})$ a set of complete spectral measure ρ such that, for $\lambda \in \beta$, $\mathcal{R}(P(\lambda))$ consists of the generalized eigenvectors of the operator A corresponding to the eigenvalue λ. The existence of this set is guaranteed by Theorem 2.1. Let $\lambda \in \beta$. For all $\gamma = 1, 2, \dots, N(\lambda)$, we have $\varphi_\gamma(\lambda) \in \mathcal{R}(P(\lambda))$. Therefore, according to (2.5) and (3.4), the following equality holds for all $u \in D$:

$$\begin{aligned}(Au)\widehat{\ }(\lambda) &= (Au, \varphi_\gamma(\lambda))_{H_0} = \overline{(\varphi_\gamma(\lambda), Au)_{H_0}} \\ &= \lambda\overline{(\varphi_\gamma(\lambda), u)_{H_0}} = \lambda(u, \varphi_\gamma(\lambda))_{H_0} = \lambda \hat{u}_\gamma(\lambda).\end{aligned}$$

This relation is the required equality (3.9) in the coordinate form. □

Proof of Theorem 3.1. Let us fix a nonreal $z \in \mathbb{C}$. It should be proved that, for every $u \in H_+$, we have $(\lambda - z)^{-1}\hat{u}(\lambda) \in \hat{H}_0 \subset L_2$. Indeed, since $(\lambda - z)^{-1}$ regarded as a function of $\lambda \in R$ is bounded, the operator of multiplication by $(\lambda - z)^{-1}$ is continuous in L_2 and, therefore, for $v \in D$, according to (3.9) and (3.7), we have

$$\begin{aligned}\|(\lambda - z)^{-1}\hat{u}(\lambda) - \hat{v}(\lambda)\|_{L_2} &\le c\|\hat{u}(\lambda) - (\lambda - z)\hat{v}(\lambda)\|_{L_2} \\ &= c\|\hat{u}(\lambda) - ((A - z\mathbb{1})v)\widehat{\ }(\lambda)\|_{L_2} = c\|u - (A - z\mathbb{1})v\|_{H_0}.\end{aligned}$$

Since D forms a base of the selfadjoint operator A, the right-hand side of this estimate can be made as small as desired, which proves the assertion made above.

Let $F(\lambda) \in L_2$ be orthogonal in L_2 to $\hat{H}_0$. Then, in particular, for nonreal $z \in \mathbb{C}$ and $u \in H_+$ chosen so that $\hat{u}(\lambda) = (1, 0, 0, \dots)$ (this is possible, see Lemma 3.1), we get

$$0 = \int_{\mathbb{R}} \left(F(\lambda), \frac{1}{\lambda - z}\hat{u}(\lambda)\right)_{l_2(N(\lambda))} d\rho(\lambda) = \int_{\mathbb{R}} \frac{1}{\lambda - \bar{z}} F_1(\lambda)\, d\rho(\lambda) = \int_{\mathbb{R}} \frac{1}{\lambda - \bar{z}}\, d\omega(\lambda);$$

$$\omega(\delta) = \int_\delta F_1(\lambda)\, d\rho(\lambda) \qquad (\delta \in \mathfrak{B}(\mathbb{R})).$$

This implies that $\omega = 0$ and, hence, $F_1(\lambda) = 0$ for ρ-almost all $\lambda \in \mathbb{R}$ (this follows from the following well-known fact: If a charge ω of bounded variation given on

$\mathfrak{B}(\mathbb{R})$ is such that $\int_{\mathbb{R}}(\lambda - z)^{-1}\, d\omega(\lambda) = 0$ for all nonreal $z \in \mathbb{C}$, then $\omega = 0$; see, e.g., [Ban, Shi1]). The same reasoning can be applied to $F_2(\lambda)$ and so on. As a result, we find that $F(\lambda) = 0$ in L_2, which implies that $\hat{H}_0 = L_2$. □

The results obtained above yield $N(\lambda) = \dim(\mathcal{R}(P(\lambda)))$, i.e., $N(\lambda)$ is the "multiplicity" of the eigenvalue λ. In the case where $N(\lambda) = 1$ for ρ-almost all λ, i.e., if the "spectrum is simple", we have $\hat{u}(\lambda) = \hat{u}_1(\lambda) = (u, \varphi_1(\lambda))_{H_0} \in \mathbb{C}$ ($u \in H_+$) and (3.8) gives a decomposition of H_0 into the direct integral of the complex planes $\mathbb{C} = l_2(1)$.

Theorem 3.2. *Suppose that a selfadjoint operator A satisfying the conditions of the projection spectral theorem (Theorem 2.1) has a cyclic vector $\Omega \in D$ such that, for any $m \in \mathbb{N}$, $A^m\Omega \in D$ and the linear span of these vectors is dense not only in H_0, but also in H_+. Then the spectrum of the operator A is simple.*

Proof. Assume that $N(\lambda) > 1$ for a set of λ of positive measure ρ. Then there exists $\lambda \in \mathbb{R}$ such that $N(\lambda) > 1$, $\mathrm{Tr}(P(\lambda)) = 1$, and $\mathcal{R}(P(\lambda))$ consists of generalized eigenvectors corresponding to λ. For $f \in H_0$ and $g = J^{-1}\Omega$, relation (3.2) yields

$$(P(\lambda)Jf, \Omega)_{H_0} = \sum_{\gamma=1}^{N(\lambda)} \nu_\gamma(\lambda)(f, h_\gamma(\lambda))_{H_0}\overline{(J^{-1}\Omega, h_\gamma(\lambda))_{H_0}} = \sum_{\gamma=1}^{N(\lambda)} (f, h_\gamma(\lambda))_{H_0}\bar{a}_\gamma. \tag{3.10}$$

Since $(h_\gamma(\lambda))_{\gamma=1}^{N(\lambda)}$ is an orthonormal sequence in H_0, the vectors $b(f) = ((f, h_1(\lambda))_{H_0}, (f, h_2(\lambda))_{H_0}, \dots) \in l_2(N(\lambda))$ run over $l_2(N(\lambda))$ as f runs over H_0. This implies that, for $N(\lambda) > 1$, there exists $f_0 \in H_0$ such that $b(f_0)$ is not equal to zero and is orthogonal in $l_2(N(\lambda))$ to the vector $a = (a_1, a_2, \dots)$ introduced in (3.10) (the vector a belongs to $l_2(N(\lambda))$ because $J^{-1}\Omega \in H_0$ and the factors $\nu_\gamma(\lambda)$ are bounded). By setting $f = f_0$ in (3.10), we get $(P(\lambda)Jf_0, \Omega)_{H_0} = 0$.

Since $P(\lambda)Jf_0$ is a generalized eigenvector of the operator A corresponding to λ, successive application of equality (2.2) yields

$$(P(\lambda)Jf_0, A^m\Omega)_{H_0} = \lambda^m(P(\lambda)Jf_0, \Omega)_{H_0} = 0 \qquad (m \in \mathbb{Z}_+).$$

In view of the fact that the linear span of the vectors $A^m\Omega$ is dense in H_+, this implies $P(\lambda)Jf_0 = 0$.

By setting $f = f_0$ in (3.2), we get

$$0 = \sum_{\gamma=1}^{N(\lambda)} \nu_\gamma(\lambda)b_\gamma(f_0)\overline{b_\gamma(g)} \qquad (g \in H_0).$$

The vectors $b(g) \in l_2(N(\lambda))$ run over the whole $l_2(N(\lambda))$ as g runs over H_0. Therefore, the last equality and the relations $0 < \nu_\gamma(\lambda) \le c < \infty$ yield $b(f_0) = 0$. We arrived at a contradiction. □

REMARK 3.1. In all formulas of this section, we can, clearly, replace $\mathbb{R}$ and $\mathfrak{B}(\mathbb{R})$ by the spectrum $S(A)$ of the operator A and $\mathfrak{B}(S(A))$, respectively.

4 Expansion in Eigenfunctions of Càrleman Operators

The expansion in generalized eigenvectors of a self-adjoint operator A was constructed under the assumption that the embedding $H_+ \subseteq H_0$ in chain (2.1) is quasinuclear. In this section, we establish that the quasinuclearity of this embedding is also a necessary condition for the possibility of expansion of an arbitrary operator A. At the same time, for some types of operators A, the choice of chain (2.1) may be less restricted. An important class of such operators is formed by so called Càrleman operators.

4.1 The Inverse Theorem

Theorem 4.1. *Assume that there exists chain (2.1) such that the resolution of the identity E of an arbitrary self-adjoint operator A in H_0 satisfies relation (1.11) in which $\mathfrak{B}(\mathbb{R}) \ni \alpha \mapsto \rho(\alpha)$ is a finite measure and $\mathbb{R} \ni \lambda \mapsto P(\lambda) \in \mathcal{L}(H_+, H_0)$ is an operator-valued function defined ρ-almost everywhere and such that $\|P(\lambda)\| \leq c < \infty$ for ρ-almost all $\lambda \in \mathbb{R}$. Then the embedding $H_+ \to H_0$ is quasinuclear.*

Lemma 4.1. *Let H be a Hilbert space, let $(e_j)_{j=1}^\infty$ be an orthonormal basis in this space, and let $A \in \mathcal{L}(H)$. If the matrix of this operator has the form $(a_{jk})_{j,k=1}^\infty$, $a_{jk} = (Ae_k, e_j)_H = \alpha_j\bar{\alpha}_k$, where $\alpha = (\alpha_j)_{j=1}^\infty \in l_2$, then A is a Hilbert-Schmidt operator and*

$$A = \|A\| = \sum_{j=1}^{\infty} |\alpha_j|^2. \tag{4.1}$$

Proof. It is clear that

$$A^2 = \sum_{j,k=1}^{\infty} |a_{jk}|^2 = \left(\sum_{j=1}^{\infty} |\alpha_j|^2\right)^2 < \infty;$$

moreover, $\|A\| \leq A$. Therefore, to prove (4.1), it suffices to verify the inequality $\|A\| \geq \|\alpha\|_{l_2}^2$. But this inequality follows from the relation $Af = \|\alpha\|_{l_2}^2 f$, where $f = \sum_{j=1}^\infty \alpha_j e_j \in H$. □

Proof of Theorem *4.1*. Consider the operators O, O^+, J, and $\mathbf{J}$ connected with chain (2.1) and an arbitrary resolution of the identity $\mathfrak{B}(\mathbb{R}) \ni \alpha \mapsto E(\alpha)$ in H_0. We set $C = OJ\colon H_0 \to H_0$. Obviously, $C^* = \mathbf{J}O^+\colon H_0 \to H_0$. For all disjoint $\alpha_l \in \mathfrak{B}(\mathbb{R})$ ($l \in \mathbb{N}$) such that $\bigcup_{l=1}^\infty \alpha_l = \mathbb{R}$, by using (1.11) and the conditions of the theorem, we get

$$\begin{aligned}\sum_{l=1}^{\infty}\|C^*E(\alpha_l)C\| &= \sum_{l=1}^{\infty}\|\mathbf{J}O^+E(\alpha_l)OJ\| \\ &\le \sum_{l=1}^{\infty}\|O^+E(\alpha_l)O\| = \sum_{l=1}^{\infty}\left\|\int_{\alpha_l}P(\lambda)\,d\rho(\lambda)\right\| \\ &\le \sum_{l=1}^{\infty}\int_{\alpha_l}\|P(\lambda)\|\,d\rho(\lambda) \le c\sum_{l=1}^{\infty}\rho(\alpha_l) = c\rho(\mathbb{R}) < \infty. \end{aligned} \tag{4.2}$$

We construct the following one-dimensional resolution of the identity in H_0: Assume that $(e_l)_{l=1}^{\infty}$ is an orthonormal basis in H_0, P_l is a projector onto the one-dimensional subspace spanned by e_l, and $(\lambda_l)_{l=1}^{\infty}$ is a fixed sequence of real numbers monotonically increasing to $+\infty$. We set $\mathfrak{B}(\mathbb{R}) \ni \alpha \mapsto E(\alpha) = \sum_{l\in\alpha} P_l$; the set function thus constructed is the required resolution of the identity. Let us apply estimate (4.2) with $\alpha_1 = (-\infty, \lambda_1]$, $\alpha_2 = (\lambda_1, \lambda_2]$, and $\alpha_3 = (\lambda_2, \lambda_3], \dots$ to this resolution. As a result, we get

$$\sum_{l=1}^{\infty}\|C^*P_lC\| = \sum_{l=1}^{\infty}\|C^*E(\alpha_l)C\| < \infty. \tag{4.3}$$

One can easily verify that the matrix $\left(a_{jk}^{(l)}\right)_{j,k=1}^{\infty}$ of the operator C^*P_lC has the form $a_{jk}^{(l)} = \bar{c}_{lj}c_{lk}$, where $(c_{jk})_{j,k=1}^{\infty}$ is the matrix of the operator C. Therefore, we can use Lemma 4.1. As a result, we obtain $\|C^*P_lC\| = \sum_{j=1}^{\infty}|c_{lj}|^2$. Thus, (4.3) yields the condition $\sum_{l,j=1}^{\infty}|c_{lj}|^2 < \infty$, i.e., $C = OJ$ is a Hilbert-Schmidt operator in H_0. Since J is an isometry between H_0 and H_+, we can conclude that O is a Hilbert-Schmidt operator from H_+ to H_0. □

4.2 Nonquasinuclear Riggings

As already mentioned, the choice of chain (2.1) may be less restricted for fixed A. Let us present some relevant results. Let A be a selfadjoint operator acting on H_0 and let E be its resolution of the identity. If we construct chain (2.1) so that it is standardly connected with A and the operator-valued measure $\mathfrak{B}(\mathbb{R}) \ni \alpha \mapsto \theta(\alpha) = O^+E(\alpha)O$ has a σ-finite trace $\rho(\alpha)$, then, according to Remark 1.3, it will be possible to differentiate θ with respect to ρ and repeat the arguments of Sections 2 and 3. The formulation of Theorem 2.1 remains the same except that, generally speaking, the spectral measure $\mathfrak{B}(\mathbb{R}) \ni \alpha \mapsto \rho(\alpha) = \mathrm{Tr}(O^+E(\alpha)O)$ is not finite in this case.

Theorem 4.2. *Let an operator A and chain (2.1) standardly connected with it be such that there exists a bounded continuous nonzero complex-valued function*

$a(\lambda)$ defined on the spectrum $S(A)$ of this operator and such that the operator $a(A)O\colon H_+ \to H_0$ is quasinuclear. Then the operator-valued measure $\mathfrak{B}(\mathbb{R}) \ni \alpha \mapsto O^+E(\alpha)O$ has a σ-finite trace and, hence, chain (2.1) can be used to construct expansions in generalized eigenvectors of the operator A.

First, we prove the following lemma:

Lemma 4.2. *If there exists a bounded continuous positive function $b(\lambda)$ defined on $S(A)$ and such that $\mathrm{Tr}(\mathbf{J}O^+b(A)OJ) < \infty$, then the measure $\mathfrak{B}(\mathbb{R}) \ni \alpha \mapsto O^+E(\alpha)O$ has a σ-finite trace.*

Proof. We set $S_n(A) = S(A) \cap [-n, n]$ $(n \in \mathbb{N})$. Then, for every n, one can find $\varepsilon_n > 0$ such that $b(\lambda) \geq \varepsilon_n$ $(\lambda \in S_n(A))$. Therefore,

$$0 \leq \varepsilon_n E([-n,n]) = \varepsilon_n E(S_n(A)) \leq \int_{S_n(A)} b(\lambda)\, dE(\lambda) \leq \int_{S(A)} b(\lambda)\, dE(\lambda) = b(A).$$

Hence, for any $C \in \mathcal{L}(H_0)$, we have $0 \leq \varepsilon_n C^*E([-n,n])C \leq C^*b(A)C$. Setting $C = OJ$ and taking into account that $C^* = \mathbf{J}O^+$, we get $0 \leq \mathbf{J}O^+E([-n,n])OJ \leq \varepsilon_n^{-1}\mathbf{J}O^+b(A)OJ$. Therefore,

$$\mathrm{Tr}(\mathbf{J}O^+E([-n,n])OJ) \leq \varepsilon_n^{-1}\mathrm{Tr}(\mathbf{J}O^+b(A)OJ) < \infty \qquad (n \in \mathbb{N}).$$

Furthermore, if $C \in \mathcal{L}(H_+, H_-)$ is nonnegative, then $\mathbf{J}CJ \in \mathcal{L}(H_0)$ is also nonnegative, and

$$\mathrm{Tr}C = \mathrm{Tr}(\mathbf{J}CJ). \tag{4.4}$$

Indeed, let $(e_j)_{j=1}^{\infty}$ be an orthonormal basis in the space H_+. Then $(J^{-1}e_j)_{j=1}^{\infty}$ is an orthonormal basis in H_0 and

$$\mathrm{Tr}(C) = \sum_{j=1}^{\infty}(Ce_j, e_j)_{H_0} = \sum_{j=1}^{\infty}(\mathbf{J}CJJ^{-1}e_j, J^{-1}e_j)_{H_0} = \mathrm{Tr}(\mathbf{J}CJ).$$

Finally, by using (1.11) and (4.4), we obtain

$$\mathrm{Tr}(O^+E([-n,n])O) = \mathrm{Tr}(\mathbf{J}O^+E([-n,n])OJ) < \infty \qquad (n \in \mathbb{N}). \qquad \square$$

Proof of Theorem 4.2. We use the following evident relation: For any $C \in \mathcal{L}(H_0)$, we have $\mathrm{Tr}(C^*C) = C^2 \leq \infty$. By applying it to $C = a(A)OJ \in \mathcal{L}(H_0)$, we obtain the general formula

$$\mathrm{Tr}(\mathbf{J}O^+(|a|^2(A))OJ) = \mathrm{Tr}(\mathbf{J}O^+(a(A))^*a(A)OJ) = a(A)OJ^2 = a(A)O^2 < \infty.$$

To complete the proof, it remains to use Lemma 4.2 with $b(\lambda) = |a(\lambda)|^2$. $\square$

4.3 Càrleman Operators

Let us apply the results obtained in Subsection 2 to certain individual cases. Assume that $H_0 = L_2(R, \mathfrak{R}, d\mu) = L_2(R, d\mu)$, where R is a space with measure μ given on a certain σ-algebra $\mathfrak{R}$ of sets in R and such that $\mu(R) \leq \infty$. A selfadjoint operator A acting on this space is called a Càrleman operator if there exists a bounded continuous nonzero complex-valued function $a(\lambda)$ defined on its spectrum and such that $a(A)$ is an integral Càrleman operator. The last condition means that there exists a kernel $K(x, y)$ measurable with respect to $\mathfrak{R} \times \mathfrak{R}$, defined for $\mu \times \mu$-almost all $(x, y) \in R \times R$, and such that, for some set of functions f dense in $L_2(R, d\mu)$, the representation

$$(a(A)f)(x) = \int_R K(x, y) f(y)\, d\mu(y)$$

is valid and, furthermore,

$$\int_R |K(x, y)|^2\, d\mu(x) < \infty \tag{4.5}$$

for μ-almost all $y \in R$.

Examples of Càrleman operators will be given in Section 16.5.

Theorem 4.3. *Let A be a selfadjoint Càrleman operator acting on the space $L_2(R, d\mu)$. Then there exists an $\mathfrak{R}$-measurable weight $p(x) \geq 1$ $(x \in R)$ such that the chain (see Example 14.1.1)*

$$L_2(R, p^{-1}(x)\, d\mu(x)) \supseteq L_2(R, d\mu(x) \supseteq L_2(R, p(x)\, d\mu(x)), \tag{4.6}$$

extended properly to (2.1) can be used to construct the expansion in eigenvectors of the operator A.

Proof. Consider an arbitrary chain of the form (4.6). For this chain, we have $(If)(x) = p^{-1}(x) f(x)$ $(f \in L_2(R, d\mu))$. Therefore,

$$(Jf)(x) = p^{-1/2}(x) f(x) \qquad (f \in L_2(R, d\mu)). \tag{4.7}$$

Hence, the operator $a(A)OJ$ continuously acting on the space $L_2(R, d\mu)$ is associated with the kernel $K_1(x, y) = K(x, y) p^{-1/2}(y)$ and

$$a(A)O^2 = a(A)OJ^2 = \int_R \int_R |K(x, y)|^2 p^{-1}(y) d\mu(x) d\mu(y) < \infty. \tag{4.8}$$

According to (4.5), the function $k(y) = \int_R |K(x, y)|^2\, d\mu(x)$ is measurable and finite almost everywhere. Therefore, one can choose a measurable weight $p(y) \geq 1$ so that $\int_R k(y) p^{-1}(y)\, d\mu(y) < \infty$. In view of (4.8), the last condition means that $a(A)O < \infty$. Clearly, chain (4.6) and the weight $p(x)$ thus chosen satisfy the conditions of Theorem 4.2. □

We emphasize that $p(x)$ in Theorem 4.3 is an arbitrary measurable weight $p(x) \geq 1$ $(x \in R)$ for which integral (4.8) is convergent.

Thus, for Càrleman operators, eigenfunctions are generalized only in the sense that they belong not to $L_2(R, d\mu(x))$ but to the space $L_2(R, p^{-1}(x)\, d\mu(x))$ with the weight $p(x)$ indicated above. The expansion in its individual eigenfunctions is carried out according to the general scheme presented in Section 3. Below, we present four simple assertions concerning Càrleman operators A.

(1) *The operator* $P(\lambda)\colon L_2(R, p(x)\, d\mu(x)) \to L_2(R, p^{-1}(x)\, d\mu(x))$ *is an integral operator, i.e.,*

$$(P(\lambda)u)(x) = \int_R P(x, y; \lambda) u(y)\, d\mu(y) \qquad (u \in L_2(R, p(x)\, d\mu(x))). \tag{4.9}$$

Here, the kernel $P(x, y; \lambda)$ *(the spectral kernel of* A*) is positive definite, satisfies the estimate*

$$\int_{R\times R} |P(x, y; \lambda)|^2 p^{-1}(x) p^{-1}(y)\, d(\mu \times \mu)(x, y) \leq 1 \tag{4.10}$$

for ρ*-almost all* $\lambda \in R$*, and is measurable with respect to* $\mathfrak{R} \times \mathfrak{R} \times \mathfrak{B}(\mathbb{R})$ *over the collection of its variables.*

Indeed, $P(\lambda) \leq \operatorname{Tr}(P(\lambda)) = 1$ for ρ-almost all λ. Consider the operator $\mathbf{J}P(\lambda)J\colon L_2(R, d\mu) \to L_2(R, d\mu)$. We have $\mathbf{J}P(\lambda)J = P(\lambda) \leq 1$ ρ-almost everywhere. Therefore, for a proper fixed λ, this operator is integral in $L_2(R, d\mu)$ and its kernel $K(x, y)$ satisfies the estimate $\int_{R\times R} |K(x, y)|^2\, d(\mu \times \mu)(x, y) \leq 1$. By virtue of (4.7), $P(x, y; \lambda) = K(x, y) p^{1/2}(x) p^{1/2}(y)$ is the kernel of the operator $P(\lambda)$ and, hence, relations (4.9) and (4.10) hold. One can also prove that P is measurable. □

(2) *Let* $\mathbb{R} \ni \lambda \mapsto c(\lambda) \in \mathbb{C}$ *be a Borel measurable function such that* $|c(\lambda)| \leq |a(\lambda)|^2$ $(\lambda \in S(A))$*. Then* $\int_{\mathbb{R}} |c(\lambda)|\, d\rho(\lambda) < \infty$*.*

Indeed, for $\alpha \in \mathfrak{B}(\mathbb{R})$, we have

$$\begin{aligned} \rho(\alpha) &= \operatorname{Tr}(O^+ E(\alpha) O) = \operatorname{Tr}(\mathbf{J}O^+ E(\alpha) O J) \\ &= \sum_{j=1}^{\infty} (\mathbf{J}O^+ E(\alpha) O J e_j, e_j)_{H_0} = \lim_{n\to\infty} \rho_n(\alpha), \\ \rho_n(\alpha) &= \sum_{j=1}^{n} (\mathbf{J}O^+ E(\alpha) O J e_j, e_j)_{H_0}, \end{aligned}$$

where $(e_j)_{j=1}^{\infty}$ is a certain orthonormal basis in $H_0 = L_2(R, d\mu)$. By using the Helly theorems (see Section 7.7), one can easily justify the following limit procedure:

$$\begin{aligned}\int_{\mathbb{R}} |c(\lambda)| d\rho(\lambda) &= \lim_{n\to\infty} \int_{\mathbb{R}} |c(\lambda)|\, d\rho_n(\lambda)\\ &= \lim_{n\to\infty} \int_{\mathbb{R}} \sum_{j=1}^{n} |c(\lambda)|\, d(\mathbf{J}O^{+}E(\lambda)OJe_j, e_j)_{H_0}\\ &= \lim_{n\to\infty} \sum_{j=1}^{n} (\mathbf{J}O^{+}(|c|(A))OJe_j, e_j)_{H_0}\\ &\le \lim_{n\to\infty} \sum_{j=1}^{n} (\mathbf{J}O^{+}|a|^2(A)OJe_j, e_j)_{H_0}\\ &= \mathrm{Tr}(\mathbf{J}O^{+}|a|^2(A)OJ) = \ a(A)^2 < \infty. \end{aligned}$$ □

(3) *For any function $c(\lambda)$ from assertion (2), the operator $O^{+}c(A)O$: $L_2(R, p(x)\, d\mu(x)) \to L_2(R, p^{-1}(x)\, d\mu(x))$ is integral and its kernel $K(x,y)$ belong to $L_2(R \times R, (p^{-1}(x)\, d\mu(x)) \times (p^{-1}(y)\, d\mu(y)))$.*

Indeed, by virtue of (1.11), we have

$$O^{+}c(A)O = \int_{\mathbb{R}} c(\lambda)\, d(O^{+}E(\lambda)O) = \int_{\mathbb{R}} c(\lambda)P(\lambda)\, d\rho(\lambda).$$

According to assertion (1), $P(\lambda)$ is an integral operator and, therefore, formally, $O^{+}c(A)O$ is also an integral operator with the kernel

$$K(x,y) = \int_{\mathbb{R}} c(\lambda)P(x,y;\lambda)\, d\rho(\lambda) \qquad (x,y \in R). \tag{4.11}$$

To prove assertion (3), it suffices to establish that function (4.11) belongs to $L_2(R \times R, (p^{-1}(x)\, d\mu(x)) \times (p^{-1}(y)\, d\mu(y)))$. The Cauchy-Buniakowski inequality and (4.10) imply that

$$\begin{aligned}&\int_{R\times R} \left| \int_{\mathbb{R}} c(\lambda)P(x,y;\lambda)d\rho(\lambda)\right|^2 p^{-1}(x)p^{-1}(y)d(\mu\times\mu)(x,y)\\ &\le \int_{\mathbb{R}} |c(\lambda)|d\rho(\lambda) \int_{R\times R}\left(\int_{\mathbb{R}} |c(\lambda)||P(x,y;\lambda)|^2 d\rho(\lambda)\right)p^{-1}(x)p^{-1}(y)d(\mu\times\mu)(x,y)\\ &< \left(\int_{\mathbb{R}} |c(\lambda)|\,d\rho(\lambda)\right)^2 < \infty.\end{aligned}$$ □

(4) *If, for a selfadjoint operator A acting on the space $L_2(R, d\mu)$ and a function $a(\lambda)$ from the definition of Càrleman operators, one can choose rigging (4.6) so that $a(A)O^2 = \mathrm{Tr}(\mathbf{J}O^{+}|a|^2(A)OJ) < \infty$, then the operator A is a Càrleman operator.*

Indeed, it is evident that $a(A)OJ < \infty$. Therefore, the operator $a(A)OJ$ acting on $L_2(R, d\mu)$ is an integral operator with the kernel $K_1(x,y)$ square summable over two variables. But then $K(x,y) = K_1(x,y)p^{1/2}(y)$ is the required kernel for $a(A)$. □

Chapter 16
Differential Operators

In this chapter, we consider applications of some results of the theory of unbounded operators presented in Chapters 12–15 to the most important class of operators in mathematical physics, namely, to differential operators. For ordinary differential operators, the corresponding applications are described in many books (see, e.g., [AkG, DuS2, LySt, Nai2]); on the other hand, for partial differential equations such results can rarely be found in the literature. In this chapter, the main attention is paid to partial differential equations, namely, to the most important case of elliptic equations. This study requires a fairly complex technique presented in Sections 1 and 2 in sufficiently consistent form. We consider second-order equations (for the case of general order, see, e.g., [Ber]). Note that we do not present the complete spectral theory of elliptic differential operators (for example, we do not study the structure of the spectrum, etc.). The results presented here should be regarded as an illustration of the general theorems on important examples.

1 Theorem on Isomorphisms for Elliptic Operators

1.1 Preliminary Information

In the region $G \subseteq \mathbb{R}^N$ with sufficiently smooth boundary ∂G, we consider the general linear differential expression of order $r \in \mathbb{N}$, i.e.,

$$(\mathcal{L}u)(x) = \sum_{|\alpha| \leq r} a_\alpha(x)(D^\alpha u)(x) \quad (x \in G). \tag{1.1}$$

We have already considered such expressions in Section 12.2, where it was required that the coefficients a_α belong to the class $C^{|\alpha|}(G)$. In this chapter, in connection with boundary-value problems, we always assume that a_α are sufficiently smooth up to the boundary of the region, i.e., $a_\alpha \in C^k(\tilde{G})$ for some $k = k(\alpha) \geq |\alpha|$ (recall that we say that a function belongs to the class $C^k(\tilde{G})$ if it is the restriction of some function from $C^k(\mathbb{R}^N)$ to $\tilde{G}$). Thus, for sufficiently smooth u, the expression $(\mathcal{L}u)(x)$ is also defined for $x \in \tilde{G}$.

The expression $\mathcal{L}$ is called *elliptic* in $\tilde{G}$ if, for every nonzero vector $\xi = (\xi_1, \dots \xi_N) \in \mathbb{R}^N$, the r-linear form

$$\mathcal{L}_0(x, \xi) = \sum_{|\alpha| = r} a_\alpha(x)\xi^\alpha \quad (\xi^\alpha = \xi_1^{\alpha_1} \dots \xi_N^{\alpha_N}) \tag{1.2}$$

is not zero for $x \in \tilde{G}$. Since the leading coefficients of the expression $\mathcal{L}^+$ formally adjoint to $\mathcal{L}$ are equal to $(-1)^r \overline{a_\alpha(x)}$, the expressions $\mathcal{L}$ and $\mathcal{L}^+$ are either both elliptic or both not elliptic.

We study the operators constructed by extension of the minimal operator L associated with $\mathcal{L}$ to functions satisfying certain boundary conditions. The case of an arbitrary order r is quite complicated and, therefore, we restrict ourselves to the second-order expressions and assume that $a_0(x)$ may be complex-valued while the coefficients $a_\alpha(x)$ of derivatives are real-valued. Furthermore, for simplicity, we consider only zero boundary conditions, which are also called *Dirichlet conditions*. In the case of $r = 2$, it is convenient to rewrite the differential expression (1.1) in the form

$$(\mathcal{L}u)(x) = \sum_{j,k=1}^{N} a_{jk}(x)(D_j D_k u)(x) + \sum_{j=1}^{N} a_j(x)(D_j u)(x) + a(x)u(x) \quad (x \in \tilde{G}). \tag{1.3}$$

Thus, for example, we have $a_{jk} = a_\alpha$, where the multiindex consists of zeros except (for $j \neq k$) the jth and kth coordinates which are equal to one; for $j = k$, the jth coordinate is equal to 2. Form (1.2) can now be rewritten as follows:

$$\mathcal{L}_0(x, \xi) = \sum_{j,k=1}^{N} a_{jk}(x)\xi_j\xi_k.$$

Since $D_j D_k = D_k D_j$, it is convenient to assume that the matrix $\left(a_{jk}(x)\right)_{j,k=1}^{N}$ in (1.3) is symmetric.

In this section, we consider only bounded regions $G \subset \mathbb{R}^N$ with the boundary from the class C^l, where $l \in \mathbb{N}$ is sufficiently large. We shall also need two following two consequences of the embedding theorems for Sobolev spaces (see [KaA, Mikha]):

1. *Let $l \in \mathbb{N}$ be arbitrary. Then the embedding operator $W_2^l(G) \subseteq L_2(G)$ is compact. Note that if $l > N/2$, then this operator is even quasinuclear (Theorem 14.3.2); however, in the general case, one can only establish its compactness* (this result is related to Theorem 9.1.4, according to which the compactness of the operator A^n for some $n \in \mathbb{N}$ implies that A is compact). *Moreover, the embedding operator $W_2^l(G) \subseteq W_2^k(G)$, $k \in \mathbb{Z}_+$, $l > k$, is also compact.*

2. *Let $l \in \mathbb{N}$. We denote by $\overset{\circ}{W}{}_2^l(G)$ the subspace of the space $W_2^l(G)$ which is equal to the closure of the class of smooth functions $C_0^l(G)$ finite with respect to G. It is stated that $\overset{\circ}{W}{}_2^l(G)$ is a proper subspace of $W_2^l(G)$ which coincides with the closure in $W_2^l(G)$ of functions from $C^l(\tilde{G})$ whose derivatives of order $1, \ldots, l-1$ vanish on the boundary ∂G.* Thus, since the metric in $W_2^l(G)$ is integral, the derivatives of order $1, \ldots, l-1$ are "retained" under closure while higher derivatives are lost (this result is similar to the fact that the class $C_0(G)$ is dense in $L_2(G) = W_2^0(G)$; in this connection, see Section 11.1). Nevertheless, we shall use this result only in the case of $l = 1$ where it can be easily proved by the local rectification of the boundary ∂G by changing the variables, i.e., by passing locally to the case of a half space instead of G.

Denote

$$W_2^l(G, \mathrm{b}) = \overset{\circ}{W}{}_2^1(G) \cap W_2^l(G) \quad (l \in \mathbb{N}). \tag{1.4}$$

It follows from the results presented above that $W_2^l(G, \mathrm{b})$ is a proper subspace of $W_2^l(G)$, which coincides with the closure of all functions from $C^l(\tilde{G})$ vanishing on ∂G. We will regard $W_2^l(G, \mathrm{b})$ as a Hilbert space with respect to the scalar product in $W_2^l(G)$ (the "space of smooth solutions of the boundary-value problem with zero boundary conditions").

To solve the boundary-value problem with a spectral parameter for (1.3) with zero boundary conditions means to find a solution of the equation

$$(\mathcal{L}u)(x) - \lambda u(x) = f(x) \quad (x \in \tilde{G}), \tag{1.5}$$

which is equal to zero on the boundary ∂G. Here, f is a given right-hand side and $\lambda \in \mathbb{C}$ is a fixed number (the spectral parameter).

We assume that $f \in W_2^s(G)$, where $s \in \mathbb{Z}_+$ is a fixed number. This problem is associated with the operator $\Lambda_s(\mathrm{b})$ which acts from the space $W_2^{2+s}(G, \mathrm{b})$ to $W_2^s(G)$ according to the law

$$W_2^{2+s}(G, \mathrm{b}) \ni u \mapsto \Lambda_s(\mathrm{b})u = \mathcal{L}u \in W_2^s(G). \tag{1.6}$$

Under the assumption that the coefficients in the expression (1.1) are sufficiently smooth, i.e., $a_\alpha \in C^{\max(|\alpha|, s)}(\tilde{G})$, the operator $\Lambda_s(\mathrm{b})$ is well-defined.

The operator $\Lambda_s(\mathrm{b})$ is a so called *strong operator of the problem*. In the case of $s = 0$, this operator, regarded as an operator $L(\mathrm{b})$ in $L_2(G)$, is the extension of the minimal operator L corresponding to $\mathcal{L}$. More exactly, we set

$$\mathcal{D}(L(\mathrm{b})) = W_2^2(G, \mathrm{b}),$$

$$L_2(G) \supset \mathcal{D}(L(\mathrm{b})) \ni u \mapsto L(\mathrm{b})u = \Lambda_0(\mathrm{b})u = \mathcal{L}u \in L_2(G). \tag{1.7}$$

It is clear that $L(\mathrm{b}) \supset L$. The operator $L(\mathrm{b})$ is the closure of operator (1.7) in $L_2(G)$ on functions from $C^2(\tilde{G})$ vanishing on ∂G. This follows from the fact that $W_2^2(G, \mathrm{b})$ coincides with the closure in $W_2^2(G)$ of the collection of all such functions.

In the subsequent sections, the operator $L(\mathrm{b})$ will be studied in more details; for example, we shall establish that it is a selfadjoint Càrleman operator. Here, we consider the operator $\Lambda_s(\mathrm{b})$ and its extension to generalized functions. For this purpose, besides the standard negative Sobolev space $W_2^{-l}(G)(l \in \mathbb{Z}_+)$, we consider the Sobolev space $W_2^{-l}(G, \mathrm{b})$, where $l \in N$. This space is defined as a negative space in the chain

$$W_2^{-l}(G, \mathrm{b}) \supseteq L_2(G) \supseteq W_2^l(G, \mathrm{b}) \quad (l \in \mathbb{N}). \tag{1.8}$$

Note that the space $W_2^l(G, \mathrm{b})$ equipped with the scalar product from $W_2^l(G)$ is dense in $L_2(G)$ and can be taken as a positive space. It is also convenient to set $W_2^0(G, \mathrm{b}) = L_2(G)$.

1.2 The Principal Result

Theorem 1.1 (on isomorphisms). *Let $\mathcal{L}$ be the second-order ($r = 2$) elliptic expression (1.1) with real-valued coefficients of derivatives and assume that zero is not an eigenvalue of the corresponding problem with zero boundary conditions (i.e., the equation $\mathcal{L}\varphi = 0$, $\varphi \in W_2^2(G, \mathrm{b})$, has only the trivial solution)). Then the operator $\Lambda_s(\mathrm{b})$ is a one-to-one bicontinuous mapping between the following spaces:*

$$\Lambda_s(\mathrm{b})\colon W_2^{2+s}(G, \mathrm{b}) \to W_2^s(G) \quad (s = 0, 1, 2, \dots). \tag{1.9}$$

The operator $\Lambda_0(\mathrm{b})$ can be extended by continuity to a similar mapping between the spaces

$$\begin{aligned} &W_2^1(G, \mathrm{b}) \to W_2^{-1}(G, \mathrm{b}) \quad (s = -1), \\ &W_2^{2+s}(G) \to W_2^s(G, \mathrm{b}) \quad (s = -2, -3, \dots). \end{aligned} \tag{1.10}$$

Here, we assume that the coefficients a_α in the expression $\mathcal{L}$ and the boundary ∂G satisfy the following smoothness conditions: $a_\alpha \in C^{\max(|\alpha|, s)}(\tilde{G})$ and $\partial G \in C^{2+|s|}$ for $s = -1, 0, 1, 2, \dots$ and $a_\alpha \in C^{|\alpha|+|2+s|}(\tilde{G})$ and $\partial G \in C^{2+|2+s|}$ for $s = -2, 3, \dots$.

Thus, an operator of type $u \mapsto \mathcal{L}u$ realizes a topological isomorphism (homeomorphism) between the pairs of spaces in (1.9) and (1.10). In the case of (1.9), u satisfies the zero boundary condition; in the case of (1.10), these conditions are lost after closure, but the class of right-hand sides is extended, namely, it includes generalized functions. Note that the index s that characterizes the degree of smoothness of the right-hand side f in (1.5) is an arbitrary integer, i.e., f can be a generalized function. The action of $\mathcal{L}$ is quite exact — roughly speaking, it decreases the smoothness by the order of $\mathcal{L}$, i.e., by two.

The proof of this theorem is based on the coercivity inequality, which, in our case, has the following form: Assume that $\partial G \in C^{2+s}$ and $a_\alpha \in C^{\max(|\alpha|, s)}(\tilde{G})$ for $\mathcal{L}$ and $s \in \mathbb{Z}_+$ under consideration. Then there exist $p \geq 0$ and $c > 0$ such that

$$\|\mathcal{L}u\|^2_{W_2^s(G)} + p\,\|u\|^2_{L_2(G)} \geq c\,\|u\|^2_{W_2^{2+s}(G)} \qquad \left(u \in W_2^{2+s}(G, \mathrm{b})\right) \tag{1.11}$$

(for the proof of inequality (1.11), see, e.g., [Lad, pp. 116–125] for $s = 0$ and [Ber, pp. 463–474] in the general case; note that inequality (1.11) with $s = 0$ proves Theorem 1.1 in the case of $s = 0, -1, -2$).

Let us establish several lemmas.

Lemma 1.1. *Let the conditions of Theorem 1.1 be satisfied for fixed $s \in \mathbb{Z}_+$. Then there exist $c_1, c_2 > 0$ such that*

$$c_1\|u\|_{W_2^{2+s}(G)} \leq \|\mathcal{L}u\|_{W_2^s(G)} \leq c_2\|u\|_{W_2^{2+s}(G)} \tag{1.12}$$
$$\left(u \in W_2^{2+s}(G, \mathrm{b})\right).$$

Proof. The right inequality in (1.12) is quite evident. Indeed, by using the triangle inequality and the Leibniz formula, we find that the norm $\|\cdot\|_{W_2^s(G)}$ of the expression $\mathcal{L}u$ with $u \in W_2^{2+s}(G, \mathrm{b})$ is bounded from above by $\|u\|_{W_2^{2+s}(G)}$. The coefficients a_α are sufficiently smooth for this calculation (they belong to $C^s(\tilde{G})$).

To prove the left inequality in (1.12), we assume the contrary. Then $(\forall n \in \mathbb{N})(\exists u_n \in W_2^{2+s}(G, \mathrm{b}))$: $\|\mathcal{L}u_n\|_{W_2^s} < n^{-1}\|u_n\|_{W_2^{2+s}}$. We normalize u_n so that $\|u_n\|_{W_2^{2+s}} = 1 (n \in \mathbb{N})$. Since the embedding operator $W_2^{2+s}(G) \subseteq L_2(G)$ is compact (see assertion 1 in Subsection 1.1), one can select a subsequence $(u_{n_k})_{k=1}^\infty$ from the sequence $(u_n)_{n=1}^\infty$ so that $u_{n_k} \to \varphi$ in $L_2(G)$ as $k \to \infty$, where φ is a function from $L_2(G)$.

On the other hand, by virtue of inequality (1.11) and the relation

$$\|\mathcal{L}u_{n_k}\|_{W_2^s(G)} \leq n_k^{-1} \xrightarrow[k\to\infty]{} 0,$$

we have

$$\|u_{n_k} - u_{n_l}\|^2_{W_2^{2+s}(G)} \leq c^{-1}\left(\|\mathcal{L}u_{n_k} - \mathcal{L}u_{n_l}\|^2_{W_2^s(G)} + p\,\|u_{n_k} - u_{n_l}\|^2_{L_2(G)}\right) \to 0.$$

as $k, l \to \infty$. Thus, $(u_{n_k})_{k=1}^\infty$ is fundamental in $W_2^{2+s}(G)$ and converges in $L_2(G)$ to $\varphi \in L_2(G)$. This means that $\varphi \in W_2^{2+s}(G)$ and $u_{n_k} \xrightarrow[k\to\infty]{} \varphi$ in the sense of this space. Since u_{n_k} belongs to the set $W_2^{2+s}(G, \mathrm{b})$, which is closed in $W_2^{2+s}(G)$, we have $\varphi \in W_2^{2+s}(G, \mathrm{b})$ and $\|\varphi\|_{W_2^{2+s}} = 1$.

Further, we have $\mathcal{L}\varphi = \lim_{k\to\infty} \mathcal{L}u_{n_k} = 0$, $0 \neq \varphi \in W_2^{2+s}(G, \mathrm{b}) \subset W_2^2(G, \mathrm{b})$, in the sense of convergence in $W_2^s(G)$. Thus, zero is an eigenvalue of problem (1.5), which contradicts the conditions of the lemma. The left inequality in (1.12) is proved. □

The ellipticity condition applied to (1.3) means that

$$(\forall \xi = (\xi_1, \ldots, \xi_N) \in \mathbb{R}^N : \xi \neq 0) \quad (\forall x \in \tilde{G}) : \mathcal{L}_0(x, \xi) = \sum_{j,k=1}^{N} a_{jk}(x)\xi_j\xi_k \neq 0.$$

Since, by assumption, $a_{jk}(x)$ are real-valued, $\mathcal{L}_0(x, \xi)$ preserves its sign for indicated ξ and x. Below, we always assume that, for example, $\mathcal{L}_0(x, \xi) \leq 0$.

Lemma 1.2. *Let the conditions of Theorem 1.1 be satisfied for $s = 0$. Then one can choose $q \geq 0$ so large that, for some $c_3 > 0$, the following inequality is true:*

$$\operatorname{Re}\left((\mathcal{L} + q\mathbb{1})u, u\right)_{L_2(G)} \geq c_3\|u\|^2_{W_2^1(G)} \quad (u \in W_2^2(G, \mathrm{b})). \tag{1.13}$$

Proof. By using (1.3) and integrating by parts, for $u \in W_2^2(G,\mathrm{b})$, we get

$$
\begin{aligned}
\operatorname{Re}\big((\mathcal{L}+q\mathbb{I})u,u\big)_{L_2(G)}
&= \operatorname{Re}\sum_{j,k=1}^{N}\int_G a_{jk}(x)(D_jD_ku)(x)\overline{u(x)}dx\\
&\quad+\operatorname{Re}\sum_{j=1}^{N}\int_G a_j(x)(D_ju)(x)\overline{u(x)}dx+\int_G(\operatorname{Re}a(x)+q)|u(x)|^2dx\\
&=-\sum_{j,k=1}^{N}\int_G a_{jk}(x)(D_ju)(x)\overline{(D_ku)(x)}dx\\
&\quad+\operatorname{Re}\sum_{j=1}^{N}\int_G\left[-\sum_{k=1}^{N}(D_ka_{jk})(x)+a_j(x)\right](D_ju)(x)\overline{u(x)}dx\\
&\quad+\int_G(\operatorname{Re}a(x)+q)|u(x)|^2dx\\
&\geq\varepsilon\sum_{j=1}^{N}\int_G|(D_ju)(x)|^2dx\\
&\quad+\operatorname{Re}\sum_{j=1}^{N}\int_G b_j(x)(D_ju)(x)\overline{u(x)}dx+\int_G(\operatorname{Re}a(x)+q)|u(x)|^2dx. \qquad (1.14)
\end{aligned}
$$

Here, $\varepsilon > 0$ is such that, for any $\xi = (\xi_1, \dots, \xi_N) \in \mathbb{C}^N$ and $x \in \tilde{G}$, the following inequality holds:

$$
-\sum_{j,k=1}^{N} a_{jk}(x)\xi_j\bar{\xi}_k \geq \varepsilon\sum_{j=1}^{N}|\xi_j|^2 = \varepsilon|\xi|^2. \qquad (1.15)
$$

The existence of such $\varepsilon > 0$ follows from the fact that the quadratic form $-\mathcal{L}_0(x,\xi)$ is positive definite ($\forall x \in \tilde{G}$) and nondegenerate (with respect to $\xi \in \mathbb{R}^N$) and its coefficients depend continuously on $x \in \tilde{G}$. Note also that $b_j(x)$ in (1.14) denotes the corresponding expression in square brackets.

Integration by parts yields

$$
\operatorname{Re}\int_G b_j(x)(D_ju)(x)\overline{u(x)}dx = -\frac{1}{2}\int_G(D_jb_j)(x)|u(x)|^2dx \quad (u \in W_2^2(G,\mathrm{b}))
$$

and, therefore, estimate (1.14) can be extended as follows:

$$
\begin{aligned}
\operatorname{Re}\big((\mathcal{L}+q\mathbb{I})u,u\big)_{L_2(G)} &\geq \varepsilon\sum_{j=1}^{N}\int_G|(D_ju)(x)|^2dx\\
&\quad+\int_G\left[-\frac{1}{2}\sum_{j=1}^{N}(D_jb_j)(x)+\operatorname{Re}a(x)+q\right]|u(x)|^2dx \geq \varepsilon\|u\|^2_{W_2^1(G)}
\end{aligned}
$$

for sufficiently large $q \geq 0$. □

REMARK 1.1. In estimate (1.13), $c_3 = \varepsilon$, where the number $\varepsilon > 0$ is the same as in (1.15), and $q \geq 0$ should be taken so large that the expression in the last square brackets is greater than or equal to ε for $x \in G$.

Corollary 1.1. *Zero is not an eigenvalue of problem (1.5) with $\mathcal{L}$ replaced by $\mathcal{L} + q\mathbb{I}$ with sufficiently large $q \geq 0$ (inequality (1.3) denies the existence of a function $\varphi \in W_2^2(G, \mathrm{b}), \varphi \neq 0$ for which $(\mathcal{L} + q\mathbb{I})\varphi = 0$).* □

Note that the proof of Lemma 1.2 also implies the validity of inequality (1.13) with the right-hand side replaced by $c_4\|u\|^2_{L_2(G)}$, where $c_4 > 0$ can be made as large as desired by choosing a sufficiently large $q \geq 0$.

Consider the differential expression $(\forall t \in [0,1])$

$$\mathcal{L}(t) = t\mathcal{L} - (1-t)\Delta, \quad \Delta = \sum_{j=1}^{N} D_j^2. \tag{1.16}$$

It is also elliptic because

$$-(\mathcal{L}(t))_0(x,\xi) = -t\mathcal{L}_0(x,\xi) + (1-t)|\xi|^2 \geq (1-(1-\varepsilon)t)|\xi|^2,$$

where ε is the same as in (1.15). It follows from the proof of Lemma 1.2 and Corollary 1.1 that the constant $q \geq 0$ can be chosen so large that zero is not an eigenvalue of the problem (1.5) for the expression $\mathcal{L}(t) + q\mathbb{I}(t \in [0,1])$.

Lemma 1.3. *Let the conditions of Theorem 1.1 be satisfied for fixed $s \in \mathbb{Z}_+$. Then there exist a sufficiently large $q \geq 0$ and some $c_5 > 0$ such that the following inequality holds for any $t \in [0,1]$):*

$$\left\|\left(\mathcal{L}(t) + q\mathbb{I}\right)u\right\|_{W_2^s(G)} \geq c_5\|u\|_{W_2^{2+s}(G)} \quad \left(u \in W_2^{2+s}(G,\mathrm{b})\right). \tag{1.17}$$

Proof. It has been already established that $q \geq 0$ can be chosen so large that zero is not an eigenvalue of the problem (1.5) for $\mathcal{L}(t) + q\mathbb{I} \quad (t \in [0,1])$. Let us fix this choice. Assume the contrary, namely, assume that there is no $c_5 > 0$ satisfying (1.17). Then, for any $n \in \mathbb{N}$, there exist $u_n \in W_2^{2+s}(G, \mathrm{b})$ and $t_n \in [0,1]$ such that

$$\left\|\left(\mathcal{L}(t_n) + q\mathbb{I}\right)u_n\right\|_{W_2^s} < n^{-1}\|u_n\|_{W_2^{2+s}}.$$

We may assume that $\|u_n\|_{W_2^{2+s}(G)} = 1 (n \in \mathbb{N})$. By using the compactness of the embedding operator $W_2^{2+s}(G) \subseteq L_2(G)$ and the interval [0,1], we choose a subsequence $(u_{n_k})_{k=1}^{\infty}$ such that $u_{n_k} \to \varphi \in L_2(G)$ in $L_2(G)$ and $t_{n_k} \to \tau \in [0,1]$ as $k \to \infty$. It is easy to see that

$$\|(\mathcal{L}(\tau) + q\mathbb{I})u_{n_k}\|_{W_2^s(G)} \xrightarrow[k\to\infty]{} 0.$$

Indeed, according to (1.16) and the second inequality in (1.12), we have

$$\begin{aligned}\Big\|\big(\mathcal{L}(\tau)+q\mathbb{I}\big)u_{n_k}-\big(\mathcal{L}(t_{n_k})+q\mathbb{I}\big)_{u_{n_k}}\Big\|_{W_2^s}&\\ \le \|(\tau-t_{n_k})\mathcal{L}u_{n_k}\|_{W_2^s}+\|(\tau-t_{n_k})\Delta u_{n_k}\|_{W_2^s}&\\ \le c_6|\tau-t_{n_k}|\,\|u_{n_k}\|_{W_2^{2+s}}=c_6|\tau-t_{n_k}|\to 0,\qquad & k\to\infty.\end{aligned}$$

This and the relation

$$\big\|\big(\mathcal{L}(t_{n_k})+q\mathbb{I}\big)u_{n_k}\big\|_{W_2^s(G)}<n_k^{-1}\xrightarrow[k\to\infty]{}0$$

imply the required assertion.

We now proceed by analogy with the proof of Lemma 1.1. By virtue of inequality (1.11) rewritten for $\mathcal{L}(\tau)+q\mathbb{I}$, we have

$$\begin{aligned}\|u_{n_k}-u_{n_l}\|^2_{W_2^{2+s}(G)}\le c^{-1}\Big(\|(\mathcal{L}(\tau)+q\mathbb{I})u_{n_k}-(\mathcal{L}(\tau)+q\mathbb{I})u_{n_l}\|^2_{W_2^s(G)}&\\ +p\,\|u_{n_k}-u_{n_l}\|^2_{L_2(G)}\Big)\to 0&\end{aligned}$$

as $k,l\to\infty$. Consequently, the sequence $(u_{n_k})_{k=1}^{\infty}$ is fundamental in $W_2^{2+s}(G)$ and, therefore, $\varphi\in W_2^{2+s}(G,\mathrm{b})$, $\|\varphi\|_{W_2^{2+s}(G)}=1$, and this sequence converges to φ in $W_2^{2+s}(G)$. But then

$$(\mathcal{L}(\tau)+q\mathbb{I})\varphi=\lim_{k\to\infty}\big(\mathcal{L}(\tau)+q\mathbb{I}\big)u_{n_k}=0,$$

i.e., zero is an eigenvalue of the problem (1.5) for the expression $\mathcal{L}(\tau)+q\mathbb{I}$, which contradicts the choice of q. □

Let us prove the following general assertion ("the lemma on extension with respect to a parameter"):

Lemma 1.4. *Suppose that E' and E'' are Banach spaces, A_0 and A_1 are linear operators which act continuously from E' to E'', and A_0 realizes a homeomorphism between E' and E''. Assume also that there exists a family of operators B_t which act continuously from E' to E'', depend continuously (in the operator norm) on $t\in[0,1]$, connect A_0 and A_1 (i.e., $B_0=A_0$ and $B_1=A_1$), and are such that*

$$\|B_tu\|_{E''}\ge\varepsilon\|u\|_{E'}\qquad(u\in E'),$$

where $\varepsilon>0$ does not depend on t. Then A_1 also realizes a homeomorphism between E' and E''.

Proof. By virtue of the uniform continuity of B_t in t, one can find $\delta>0$ such that, for $|t'-t''|<\delta$, we have $\|B_{t'}-B_{t''}\|<\varepsilon$. Let us show that if B_{t_0} is a homeomorphism between E' and E'', then B_t, $|t-t_0|<\delta$, is also a homeomorphism. We have

$B_t = B_{t_0} - (B_{t_0} - B_t)$, whence $B_{t_0}^{-1}B_t = \mathbb{I} - B_{t_0}^{-1}(B_{t_0} - B_t)$. According to (1.18), the norm of the operator $B_{t_0}^{-1}$ does not exceed ε^{-1} and, hence, the norm of the operator $B_{t_0}^{-1}(B_{t_0} - B_t)$ acting on E' does not exceed $\|B_{t_0}^{-1}\|\,\|B_{t_0} - B_t\| < \varepsilon^{-1}\cdot\varepsilon = 1$. Therefore, the operator $B_{t_0}^{-1}B_t$ has a continuous inverse operator $(B_{t_0}^{-1}B_t)^{-1}$ in E'. Consequently, $(B_{t_0}^{-1}B_t)^{-1}B_{t_0}^{-1}$ is a continuous operator inverse to B_t. The existence of B_t^{-1} means that B_t realizes a homeomorphism.

It is now obvious how to complete the proof of the Lemma. One must partition the segment [0,1] by the points $0 = t_0, t_1, \dots t_{n-1}, t_n = 1$ with step less than δ. Taking into account that the operator $B_{t_0} = A_0$ is a homeomorphism, we can establish step by step that $B_{t_1}, B_{t_2}, \dots$ are also homeomorphisms. □

Proof of Theorem 1.1. We divide the proof in several steps.

I. Let us prove that the operator $\Lambda_s(\mathrm{b})$ ($s \in \mathbb{Z}_+$) realizes a homeomorphism between spaces (1.9). By virtue of inequalities (1.12), it suffices to verify that

$$\mathcal{R}(\Lambda_s(\mathrm{b})) = W_2^s(G). \tag{1.19}$$

First, we show that the range of the operator

$$W_2^{2+s}(G,\mathrm{b}) \ni u \mapsto (\mathcal{L} + q\mathbb{I})u \in W_2^s(G) \tag{1.20}$$

fills the whole space $W_2^s(G)$; here, the nonnegative number q is chosen in accordance with Lemma 1.3.

We use Lemma 1.4, setting $E' = W_2^{2+s}(G,\mathrm{b})$ and $E'' = W_2^s(G)$. The operators are introduced by the relation $W_2^{s+2}(G,\mathrm{b}) \ni u \mapsto B_t u = \big(\mathcal{L}(t) + q\mathbb{I}\big)u \in W_2^s(G)$ ($\mathcal{L}(t)$ has the form (1.16), $t \in [0,1]$); evidently, they continuously depend on t.

The operator $A_0 = B_0$ has the form $W_2^{2+s}(G,\mathrm{b}) \ni u \mapsto A_0 u = -\Delta u + qu \in W_2^s(G)$. It follows from the classical results on solvability of boundary-value problems with the Laplace operator that the equation $-(\Delta u)(x) + qu(x) = f(x)\,(x \in G)$, where $f \in C^\infty(\tilde{G})$, has a solution $u \in C^{2+s}(\tilde{G})$ satisfying the boundary condition $u(x) = 0$, $x \in \partial G$ (see, e.g., [Vl1, Lad, Mikha]). This implies that $\mathcal{R}(A_0)$ is dense in $W_2^s(G)$ and, hence, by virtue of estimates (1.12) (for $\mathcal{L} = -\Delta + q\mathbb{I}$), we can conclude that $\mathcal{R}(A_0) = W_2^s(G)$.

Thus, A_0 realizes a homeomorphism between $W_2^{2+s}(G,\mathrm{b})$ and $W_2^s(G)$. By virtue of Lemma 1.4, $A_1 = B_1$ also realizes such homeomorphism and, therefore, the range of operator (1.2) coincides with $W_2^s(G)$.

Let us prove (1.19). Denote by A operator (1.20) regarded as an operator in the space $W_2^s(G)$ with the domain $\mathcal{D}(A) = W_2^{2+s}(G,\mathrm{b})$ (nondense for $s > 0$). The operator A^{-1} is compact by virtue of estimates (1.12) (for $\mathcal{L} + q\mathbb{I}$) and the embedding theorems (see assertion 1 in Subsection 1.1). Therefore, the operator $(A^{-1} - \lambda\mathbb{I})^{-1}$ exists if and only if λ is not equal to zero and is not an eigenvalue of the operator A^{-1} (see Section 9.4). Assume that $(A^{-1} - \lambda\mathbb{I})^{-1}$ exists. Then the operator $(A - \lambda^{-1}\mathbb{I})^{-1}$ also exists (it is easy to verify that the last operator is equal to $-\lambda(A^{-1} - \lambda\mathbb{I})^{-1}A^{-1}$). Therefore, $\mathcal{R}(A - \lambda^{-1}\mathbb{I}) = W_2^s(G)$. Since $(A - \lambda^{-1}\mathbb{I})u = \mathcal{L}u = \Lambda_s(\mathrm{b})u$ for $\lambda = q^{-1}$, in order to prove that $\mathcal{R}(\Lambda_s(\mathrm{b})) = W_2^s(G)$ it suffices to show that $\lambda = q^{-1}$ is not an eigenvalue of A^{-1}. Assume that $\varphi \in W_2^s(G)(\varphi \neq 0)$ is such that $A^{-1}\varphi = q^{-1}\varphi$. Then $\varphi \in W_2^{2+s}(G, \mathrm{b})$ and, consequently, $\varphi = q^{-1}A\varphi$, i.e., $0 = (A - q\mathbb{I})\varphi = \mathcal{L}\varphi$, which contradicts the conditions of the theorem. Thus, (1.19) is true and the theorem is proved for $s = 0, 1, 2, \ldots$.

II. Let us prove Theorem 1.1 for $s = -2, -3, \ldots$. This case is treated as "adjoint" to the previous case. The expression $\mathcal{L}^+$ formally adjoint to $\mathcal{L}$ satisfies the equality

$$(\mathcal{L}u, v)_{L_2(G)} = (u, \mathcal{L}^+v)_{L_2(G)} \quad (u, v \in W_2^2(G, \mathrm{b})). \tag{1.21}$$

Note that, unlike equality (12.2.7) defining $\mathcal{L}^+$, generally speaking, none of the functions u and v in (1.21) is finite with respect to G. This, however, makes no importance because integration by parts does not give integrals over ∂G because both the functions u and v vanish on ∂G and the derivatives transferred have at most second order.

As indicated above, $\mathcal{L}^+$ is also an elliptic expression. Let us show that zero is not an eigenvalue of the problem (1.5) for the expression $\mathcal{L}^+$, i.e., that if $\psi \in W_2^2(G, \mathrm{b})$ is such that $\mathcal{L}^+\psi = 0$, then $\psi = 0$. Indeed, let $u \in W_2^2(G, \mathrm{b})$, then, by virtue of (1.21), we have

$$(\mathcal{L}u, \psi)_{L_2(G)} = (u, \mathcal{L}^+\psi)_{L_2(G)} = 0. \tag{1.22}$$

As follows from step I of the proof for $s = 0$, the function $\mathcal{L}u$ runs through the whole space $L_2(G)$ as u runs through $W_2^2(G, \mathrm{b})$. Therefore, (1.22) yields $\psi = 0$.

We set $\sigma = -s = 2, 3, \ldots$. Let us prove an estimate of type (1.12) but in negative norms. We have

$$\begin{aligned} \left|(u, \mathcal{L}^+v)_{L_2(G)}\right| = \left|(\mathcal{L}u, v)_{L_2(G)}\right| &\leq \|\mathcal{L}u\|_{W_2^{-\sigma}(G,\mathrm{b})}\|v\|_{W_2^{\sigma}(G,\mathrm{b})} \\ &\leq c_6^{-1}\|\mathcal{L}u\|_{W_2^{-\sigma}(G,\mathrm{b})}\|\mathcal{L}^+v\|_{W_2^{\sigma-2}(G)}. \end{aligned} \tag{1.23}$$

Here, we have used relation (1.21), the Cauchy-Buniakowski inequality for chain (1.8) (with $l = \sigma$), and the first inequality in (1.12) for $\mathcal{L}^+$: $c_6\|v\|_{W_2^{\sigma}(G,\mathrm{b})} \leq \|\mathcal{L}^+v\|_{W_2^{\sigma-2}(G)}$. According to results of step I applied to $\mathcal{L}^+$, the functions $\mathcal{L}^+v$ in (1.23) run through the whole space $W_2^{\sigma-2}(G)$. Therefore, inequality (1.23) yields

$$\|u\|_{W_2^{2-\sigma}(G)} \leq c_6^{-1}\|\mathcal{L}u\|_{W_2^{-\sigma}(G,\mathrm{b})}(u \in W_2^2(G, \mathrm{b})).$$

Let us estimate $\|\mathcal{L}u\|_{W_2^{-\sigma}(G,\mathrm{b})}$ from above. As before, we assume that $u \in W_2^2(G, \mathrm{b})$ and $v \in W_2^{\sigma}(G, \mathrm{b})$. By using (1.21), the Cauchy-Buniakowski inequality for the chain of ordinary Sobolev spaces, and the second inequality in (1.12) for $\mathcal{L}^+(\|\mathcal{L}^+v\|_{W_2^{\sigma-2}(G)} \le c_7\|v\|_{W_2^{\sigma}(G)})$, we get

$$\begin{aligned} |(\mathcal{L}u, v)_{L_2(G)}| &= |(u, \mathcal{L}^+v)_{L_2(G)}| \\ &\le \|u\|_{W_2^{2-\sigma}(G)}\|\mathcal{L}^+v\|_{W_2^{\sigma-2}(G)} \le c_7\|u\|_{W_2^{2-\sigma}(G)}\|v\|_{W_2^{\sigma}(G)}. \end{aligned} \tag{1.24}$$

The function v in (1.24) runs through the whole space $W_2^{\sigma}(G, \mathrm{b})$. Therefore, (1.24) yields

$$\|\mathcal{L}u\|_{W_2^{\sigma}(G,\mathrm{b})} \le c_7\|u\|_{W_2^{2-\sigma}(G)} \qquad (u \in W_2^2(G, \mathrm{b})).$$

Thus, we have proved the following estimate of the form (1.12):

$$c_6\|u\|_{W_2^{2-\sigma}(G)} \le \|\mathcal{L}u\|_{W_2^{-\sigma}(G,\mathrm{b})} \le c_7\|u\|_{W_2^{2-\sigma}(G)} \qquad (u \in W_2^2(G, \mathrm{b})). \tag{1.25}$$

Let us consider $\Lambda_0(\mathrm{b})$ as an operator acting from the space $W_2^{2-\sigma}(G)$ to $W_2^{-\sigma}(G, \mathrm{b})$. Its domain $W_2^2(G, \mathrm{b})$ is dense in $L_2(G)$. Moreover, in the negative space $W_2^{2-\sigma}(G)$, its range coincides with $L_2(G)$ and is dense in $W_2^{-\sigma}(G, \mathrm{b})$. By virtue of (1.25), this operator can be closed to the whole space $W_2^{2-\sigma}(G)$ and, thus, it will realize the required homeomorphism between $W_2^{2-\sigma}(G)$ and $W_2^{-\sigma}(G, \mathrm{b})$.

It follows from the smoothness conditions formulated in Theorem 1.1 for $s = -2, -3, \dots$ that, for $\mathcal{L}^+$, estimates (1.12) with $s = \sigma - 2$ are true and, therefore, the argument presented above is correct.

III. It remains to prove the theorem in the case of $s = -1$. As before, it is necessary to establish the estimate

$$c_8\|u\|_{W_2^1(G)} \le \|\mathcal{L}u\|_{W_2^{-1}(G,\mathrm{b})} \le c_9\|u\|_{W_2^1(G)} \quad (u \in W_2^2(G, \mathrm{b})). \tag{1.26}$$

First, let us prove the first inequality in (1.26). For this purpose, we choose $q \ge 0$ as indicated in Lemma 1.2. By virtue of (1.13) and the Cauchy-Buniakowski inequality, we obtain

$$c_3\|u\|^2_{W_2^1(G)} \le \mathrm{Re}\big((\mathcal{L} + q\mathbb{1})u, u\big)_{L_2(G)} \le \|(\mathcal{L} + q\mathbb{1})u\|_{W_2^{-1}(G)}\|u\|_{W_2^1(G)}. \tag{1.27}$$

Therefore,

$$\begin{aligned} c_3\|u\|_{W_2^1(G)} &\le \|(\mathcal{L} + q\mathbb{1})u\|_{W_2^{-1}(G)} \le \|\mathcal{L}u\|_{W_2^{-1}(G)} + q\|u\|_{W_2^{-1}(G)} \\ &\le \|\mathcal{L}u\|_{W_2^{-1}(G)} + q\|u\|_{L_2(G)} \quad (u \in W_2^2(G, \mathrm{b})). \end{aligned}$$

Assume that the first inequality in (1.26) is not satisfied. Then there exists a sequence $(u_n)_{n=1}^{\infty} \subset W_2^2(G, \mathrm{b})$, $\|u_n\|_{W_2^{-1}(G)} = 1$, such that $\|\mathcal{L}u_n\|_{W_2^{-1}(G,\mathrm{b})} < n^{-1}$ ($n \in \mathbb{N}$). Since the embedding operator $W_2^1(G) \subseteq L_2(G)$ is compact (see assertion 1 in Subsection 1.1), one can extract a subsequence $(u_{n_k})_{k=1}^{\infty}$ such that, in $L_2(G)$, u_{n_k} tends to some function $\varphi \in L_2(G)$ as $k \to \infty$. By applying (1.27) to $u_{n_k} - u_{n_l}$, we conclude that $(u_{n_k})_{k=1}^{\infty}$ is fundamental in $W_2^1(G)$ and, therefore, $\varphi \in W_2^1(G)$; $u_{n_k} \xrightarrow[k\to\infty]{} \varphi$ in $W_2^1(G)$ and, since $\|u_{n_k}\|_{W_2^1(G)} = 1$, we have $\|\varphi\|_{W_2^1(G)} = 1$.

By using (1.21), we establish that

$$(\varphi, \mathcal{L}^+ v)_{L_2(G)} = \lim_{k\to\infty} (u_{n_k}, \mathcal{L}^+ v)_{L_2(G)} = \lim_{k\to\infty} (\mathcal{L}u_{n_k}, v)_{L_2(G)} = 0 \tag{1.28}$$

for any $v \in W_2^2(G, \mathrm{b})$). Here, we have taken into account that

$$|(\mathcal{L}u_{nk}, v)_{L_2(G)}| \le \|\mathcal{L}u_{nk}\|_{W_2^{-1}(G,\mathrm{b})} \|v\|_{W_2^1(G,\mathrm{b})} \le n_k^{-1} \|v\|_{W_2^1(G,\mathrm{b})} \to 0$$

as $k \to \infty$. But, according to step I, the functions $\mathcal{L}^+ v (v \in W_2^2(G, \mathrm{b}))$ run through the whole space $L_2(G)$. This and (1.28) imply that $\varphi = 0$ and, hence, we arrive at a contradiction. Thus, the first inequality in (1.26) is proved.

Let us prove the second inequality in (1.26). Let $u \in W_2^2(G, \mathrm{b})$ and $v \in W_2^1(G, \mathrm{b})$. By transferring the derivatives D_j from u to v in the terms of the form $(a_{jk}(x) D_j D_k u, v)_{L_2(G)}$ in the expression $(\mathcal{L}u, v)_{L_2(G)}$ and performing obvious estimation, we obtain $|(\mathcal{L}u, v)_{L_2(G)}| \le c_{10} \|u\|_{W_2^1(G)} \|v\|_{W_2^1(G)}$. Here, v is an arbitrary function from $W_2^1(G, \mathrm{b})$ and, therefore, this inequality yields $\|\mathcal{L}u\|_{W_2^{-1}(G,\mathrm{b})} \le c_{10} \|u\|_{W_2^1(G)} (u \in W_2^2(G, \mathrm{b}))$. Estimate (1.26) is proved.

The proof can be completed as in step II. Namely, we treat $\Lambda_0(\mathrm{b})$ as an operator acting from the space $W_2^1(G, \mathrm{b})$ to $W_2^{-1}(G, \mathrm{b})$. Its domain $W_2^2(G, \mathrm{b})$ is dense in $W_2^1(G, \mathrm{b})$ (equipped with the metric of $W_2^1(G)$) and its range coincides with $L_2(G)$ and is dense in $W_2^{-1}(G, \mathrm{b})$. By virtue of (1.26), the closure of this operator gives the required homeomorphism between $W_2^1(G, \mathrm{b})$ and $W_2^{-1}(G, \mathrm{b})$.

The smoothness conditions formulated for $s = -1$ in Theorem 1.1 are sufficient for the correctness of the proof presented above. □

2 Local Smoothing of Generalized Solutions of Elliptic Equations

In this section, we use the isomorphism theorem presented in Section 1 in proving one of the principal facts in the theory of elliptic equations — the theorem on smoothing of generalized solutions.

2.1 Generalized Solutions Inside a Domain

Let us first introduce a general definition: Let $\mathcal{L}$ be a linear differential expression of the form (1.1) in a bounded domain G with coefficients $a_\alpha \in C^\infty(G)$. Consider a differential equation in G

$$\mathcal{L}u = f. \tag{2.1}$$

Let $f \in W_2^s(G)$, where $s \in \mathbb{Z}$. Thus, f may be either an ordinary function (for $s \ge 0$) or a generalized function ($s < 0$).

A function (ordinary or generalized) $u \in W_2^t(G)$, where $t \in \mathbb{Z}$, is called a generalized solution of equation (2.1) in the domain G if the equality

$$(u, \mathcal{L}^+ v)_{L_2(G)} = (f, v)_{L_2(G)}. \tag{2.2}$$

holds for any $v \in C_0^\infty(G)$.

Note that if $u \in C^r(G)$ (r is the order of the expression $\mathcal{L}$), then by integrating by parts, the expression $\mathcal{L}^+$ on the left-hand side of (2.2) can be transferred to u and, as a result, relation (2.2) takes the form $(\mathcal{L}u, v)_{L_2(G)} = (f, v)_{L_2(G)}$ ($v \in C_0^\infty(G)$). Due to the arbitrariness of v, this implies that $(\mathcal{L}u)(x) = f(x)$, i.e., u is a classical solution of equation (2.1) (it is obvious that, in this case, the right-hand side f must be an ordinary function). In the general case, relation (2.2) defines, in a certain sense, a generalized solution. It should be also emphasized that "test" functions v appearing in (2.2) are finite and, therefore, equality (2.2) does not affect the boundary (on ∂G) properties of the function u.

If $G' \subseteq G$ is a subdomain of G with a sufficiently smooth boundary and relation (2.2) holds for any $v \in C_0^\infty(G')$, then we say that u is a generalized solution of equation (2.1) in G'.

Also note that the condition of infinite differentiability of the coefficients a_α can be replaced by the condition of finite differentiability, i.e., a_α must belong to some $C^{l(\alpha)}(G)$. The choice of the numbers $l(\alpha) \in \mathbb{Z}_+$ is governed by the sole condition that both the left-hand and right-hand sides of (2.2) should be meaningful (recall that u and f are, generally speaking, generalized functions). The reader can easily calculate these numbers without assistance.

Let us introduce natural and useful notions frequently used in what follows. In a fixed bounded domain $G \subset \mathbb{R}^N$ with sufficiently smooth boundary, we consider the negative Sobolev space $W_2^{-l}(G)(l \in \mathbb{N})$. Let $\alpha \in W_2^{-l}(G)$ and $\chi \in C^\infty(\tilde{G})$. The product $\chi\alpha$ is naturally defined as an element of $W_2^{-l}(G)$ by the formula

$$(\chi\alpha, u)_{L_2(G)} = \left(\alpha, \overline{\chi(x)}u(x)\right)_{L_2(G)} \quad (u \in W_2^l(G)).$$

Let $G' \subseteq G$ be a subdomain of G with sufficiently smooth boundary, $\alpha \in W_2^{-l}(G)$ ($l \in \mathbb{N}$). It may happen that, for any $\chi \subset C^\infty(\tilde{G})$ vanishing in a neighbourhood of the set $G \setminus G'$, the product $\chi\alpha$ belongs not only to $W_2^{-l}(G)$ but also to $W_2^k(G)$, where k is an element of the sequence $-l+1, -l+2, \ldots$.

In this case, we say that α belongs to $W_2^k(G)$ inside G' and write $\alpha \in W_{2,\text{loc}}^k(G')$.

Note that χ vanishes in a certain neighbourhood of the boundary $\partial G'$; this is why we speak about the inclusion of α inside G'. It is clear that, for ordinary functions α, this definition leads to a greater smoothness of α in G'. Since χ vanishes in a neighbourhood of $S \setminus G'$, the inclusion $\chi\alpha \in W_2^k(G)$ can be changed with a natural inclusion $\chi\alpha \in W_2^k(G')$.

By using the decomposition of the unit (see Section 11.1), one can prove the following natural localization lemma:

Lemma 2.1. *Assume that, for every point $x \in G'$, one can indicate its spherical neighbourhood $U(x) \subseteq G'$ such that $\alpha \in W^k_{2,\mathrm{loc}}(U(x))$. Then $\alpha \in W^k_{2,\mathrm{loc}}(G')$.*

Proof. In view of the local compactness of $\mathbb{R}^N$, we now select a countable subcovering $U(x_1), U(x_2), \dots$ from a covering of G' by the neighbourhoods $U(x)$ ($x \in G'$). Let $(\chi_j(x))_{j=1}^{\infty}$ be the corresponding decomposition of the unit, i.e., $\chi_j \in C_0^\infty(G)$ are nonnegative, vanish outside $U(x_j)$, and $\sum_{j=1}^{\infty} \chi_j(x) = 1$ ($x \in G'$). For $\chi \in C^\infty(\tilde{G})$ vanishing in a neighbourhood of $G \setminus G'$, one can write $\chi(x) = \sum_{j=1}^{n(x)} \chi_j(x)\chi(x)$, where $n(\chi) \in \mathbb{N}$ depends on χ. Therefore, for any $u \in W_2^l(G)$, we can write

$$(\chi\alpha, u)_{L_2(G)} = \big(\alpha, \overline{\chi(x)}u(x)\big)_{L_2(G)} = \sum_{j=1}^{n(x)} \big(\alpha, \chi_j(x)\overline{\chi(x)}u(x)\big)_{L_2(G)}$$

$$= \sum_{j=1}^{n(x)} (\chi_j\chi\alpha, u)_{L_2(G)} = \Big(\big(\sum_{j=1}^{n(x)} \chi_j\chi\alpha\big), u\Big)_{L_2(G)}.$$

Hence, $\chi\alpha = \sum_{j=1}^{n(x)} \chi_j\chi\alpha$. At the same time, $\chi_j\chi\alpha$ belongs to $W_2^k(G)$ by the condition. Therefore, $\chi\alpha \in W_2^k(G)$ and, consequently, $\alpha \in W^k_{2,\mathrm{loc}}(G')$. □

2.2 Smoothing Inside a Domain

We now formulate a theorem on smoothing of solutions of an elliptic equation inside a domain. As in the case of the isomorphism theorem, we present its formulation (Theorem 1.1) for second-order elliptic expressions, i.e., for $r = 2$.

Theorem 2.1. *Suppose that $\mathcal{L}$ is a second-order elliptic expression (1.1) with sufficiently smooth coefficients (the derivatives have real-valued coefficients) and $\varphi \in W_2^t(G)$ ($t \in \mathbb{Z}$) is a generalized solution of equation (2.1) whose right-hand side $f \in W_2^s(G)$ ($s \in \mathbb{Z}$) inside a domain G. Then the solution φ indeed belongs to the space $W_2^{2+s}(G)$ inside G, i.e., $\varphi \in W^{2+s}_{2,\mathrm{loc}}(G)$.*

The requirements imposed on the smoothness of coefficients have not been written explicitly with an intention not to make the formulation of the theorem too cumbersome; we formulate these requirements below in Remark 2.1. Thus, roughly speaking, the generalized solution of equation (2.1) inside the domain G is "smoother" than its right-hand side f exactly by the order of the equation (in this case, $r = 2$). This effect is typical of elliptic equations and can be used as a criterion for distinguishing these equations from the other partial differential equations.

Before proving the theorem, we establish an important property of the boundary-value problem (1.5). This result is formulated as Lemma 2.3. To prove it, we need the following lemma:

Lemma 2.2. *Let $G \subset \mathbb{R}^N$ be a bounded domain with sufficiently smooth boundary and let $d = \sup\{|x-y| \mid x, y \in G\}$ be its diameter. Then the estimate*

$$\|u\|_{L_2(G)} \le d\,\|u\|_{W_2^1(G)} \qquad (u \in W_2^1(G)) \tag{2.3}$$

holds for any $u \in W_2^1(G)$ vanishing on the boundary ∂G.

Proof. We restrict ourselves to the case of a convex domain G (in what follows, we use only convex domains). Consider a point $x = (x_1, \dots, x_N)$. Assume that, for fixed $x_2, \dots, x_N$, it lies in $\tilde{G}$ whenever x_1 varies within the limits $a_1(x_2, \dots, x_N) = a_1(x')$ and $b_1(x_2, \dots, x_N) = b_1(x')$ (x' denotes the point $(x_2, \dots, x_N)$). It is evident that, for $u \in C^1(\tilde{G})$ vanishing on ∂G, we have

$$u(x) = u(x_1, x') = \int_{a_1(x')}^{x_1} (D_1 u)(\xi_1, x')d\xi_1 \qquad (x_1 \in [a_1(x'), b_1(x')]). \tag{2.4}$$

By virtue of (2.4) and the Cauchy-Buniakowski inequality, we get

$$\begin{aligned}
\int_{a_1(x')}^{b_1(x')} |u(x_1, x')|^2 dx_1 &= \int_{a_1(x')}^{b_1(x')} \Big| \int_{a_1(x')}^{x_1} (D_1 u)(\xi_1, x')d\xi_1 \Big|^2 dx_1 \\
&\le \int_{a_1(x')}^{b_1(x')} (x_1 - a_1(x')) \int_{a_1(x')}^{x_1} \big|(D_1 u)(\xi_1, x')\big|^2 d\xi_1 dx_1 \\
&\le \frac{1}{2}\big(b_1(x') - a_1(x')\big)^2 \int_{a_1(x')}^{b_1(x')} \big|(D_1 u)(\xi_1, x')\big|^2 d\xi_1 \\
&\le d^2 \int_{a_1(x')}^{b_1(x')} \big|(D_1 u)(\xi_1, x')\big|^2 d\xi_1.
\end{aligned}$$

Integrating this inequality with respect to x' over the corresponding projection of the domain G, we obtain

$$\int_G |u(x)|^2 dx \le d^2 \int_G |(D_1 u)(x)|^2 dx \le d^2 \|u\|^2_{W_2^1(G)}. \qquad \square$$

Lemma 2.3. *Let $\mathcal{L}$ be a second-order elliptic expression (1.1) with real-valued coefficients of the derivatives and let $a^\alpha \in C^{|\alpha|}(\tilde{G}), \partial G \in C^2$. Consider a subdomain $G' \subseteq G$ with boundary $\partial G' \in C^2$ and the boundary-value problem (1.5) in this subdomain. It is stated that if the diameter of G' is sufficiently small, then zero is not an eigenvalue of this problem.*

Proof. The smoothness requirements of Lemma 2.3 coincide with those of Theorem 1.1 with $s = 0$. Thus, Lemma 1.2 is applicable in the domain G' and one can write the following estimate valid for all $u \in W_2^1(G')$ vanishing on $\partial G'$:

$$\operatorname{Re}\big((\mathcal{L} + q\mathbb{I})u, u\big)_{L_2(G')} \ge c_3 \|u\|^2_{W_2^1(G')}. \tag{2.5}$$

As follows from Remark 1.1, one can choose the same numbers $q > 0$ and $c_3 > 0$ in (2.5) for all $G' \subseteq G$, since they are governed only by the behaviour of the coefficients of $\mathcal{L}$ in G'. We fix these q and c_3. For indicated u, it follows from (2.5) that

$$c_3 \|u\|^2_{W_2^1(G')} \leq \left|\left((\mathcal{L} + q\mathbb{I})u, u\right)_{L_2(G')}\right| \leq \left|(\mathcal{L}u, u)_{L_2(G')}\right| + q\|u\|^2_{L_2(G')}. \tag{2.6}$$

Assume that the diameter of G' does not exceed $\left(c_3/(2q)\right)^{1/2}$. Then, by virtue of (2.3), $\|u\|^2_{L_2(G')} \leq \frac{c_3}{2q}\|u\|^2_{W_2^1(G')}$ and (2.6) implies that

$$|(\mathcal{L}u, u)_{L_2(G')}| \geq \frac{c_3}{2}\|u\|^2_{W_2^1(G')}.$$

It follows from the last inequality that if $\mathcal{L}u = 0$, then $u = 0$. □

Proof of Theorem 2.1. According to Lemma 2.1, it suffices to prove the theorem in the following local formulation: Let $x_0 \in G$. Then there exists its spherical neighbourhood $U(x_0) \subseteq G$ such that $\varphi \in W^{2+s}_{2,\mathrm{loc}}(U(x_0))$.

We take the radius of an open ball V centered at x_0 to be so small that zero is not an eigenvalue of the boundary-value problem (1.5) for $\mathcal{L}^+$ in the ball V. In view of Lemma 2.3, this is possible. We fix V and apply Theorem 1.1 on isomorphisms to $\mathcal{L}^+$ in V under the assumption that $|s| \leq m$, where m is a certain sufficiently large number (its choice is clarified in what follows). Below, (b) stands for the trivial boundary conditions on ∂V and $W_2^l(V, \mathrm{b})$ is the corresponding subspace of $W_2^l(V)$.

The proof is split into several steps. Steps 1–6 are aimed at proving the fact that if $t < 2 + s$, then u indeed lies in $W^{t+1}_{2,\mathrm{loc}}(V)$. Thus, in steps 1–6, we assume that $t < 2+s$. In what follows, χ always denotes a function from $C^\infty(\tilde{G})$ vanishing in a neighbourhood of $G \setminus V$.

(1) Here, we consider the case where $t = -1, -2, \ldots$ and prove that if $\varphi \in W_2^t(G)$ satisfies inside G the equation $\mathcal{L}\varphi = f$, where $f \in W_2^s(G)$, then $\chi\varphi \in W_2^{t+1}(V)$.

Indeed, let $w \in W_2^{2-t}(V, \mathrm{b})$. Then $\chi w \in W_2^{2-t}(G)$ and is finite. Therefore, one can substitute this function in (2.2). Note that, by the relevant limit transition, (2.2) can be extended to finite functions from $W_2^l(G)$, where $l = 2, 3, \ldots$ is sufficiently large so that both sides of equation (2.2) are meaningful. Hence,

$$\left(\varphi, \mathcal{L}^+(\chi w)\right)_{L_2(G)} = (f, \chi w)_{L_2(G)} \qquad \left(w \in W_2^{2-t}(V, \mathrm{b})\right); \tag{2.7}$$

$$\left(\mathcal{L}^+(\chi w)\right)(x) = \chi(x)(\mathcal{L}^+ w)(x) + (\mathcal{L}_\chi w)(x) \qquad (x \in G), \tag{2.8}$$

where $\mathcal{L}_\chi$ is a differential expression of the first order. Since χ vanishes in a neighbourhood of $G \setminus V$, the space $L_2(G)$ in (2.7) can be replaced by $L_2(V)$. The operator $W_2^{1-t}(V) \ni v \mapsto Av = \mathcal{L}_\chi v \in W_2^{-t}$ is clearly a continuous operator acting between these positive Sobolev spaces. Let A^+ be an operator adjoint to A that acts between the corresponding negative spaces (see Section 14.1).

Thus,

$$(\forall \alpha \in W_2^t(V))(\forall v \in W_2^{1-t}(V))\colon (\alpha, \mathcal{L}_\chi v)_{L_2(V)} = (\alpha, Av)_{L_2(V)} = (A^+\alpha, v)_{L_2(V)};$$

here, $A^+\alpha \in W_2^{t-1}(V)$. In particular,

$$(\varphi, \mathcal{L}_\chi w)_{L_2(V)} = (\psi, w)_{L_2(V)} \tag{2.9}$$

$$\big(\psi = A^+\varphi \in W_2^{t-1}(V),\ w \in W_2^{2-t}(V, \mathrm{b}) \subset W_2^{1-t}(V)\big).$$

By substituting (2.8) in (2.7) with $L_2(G)$ replaced by $L_2(V)$ and using the relations thus obtained, we find

$$\begin{aligned}(\chi\varphi, \mathcal{L}^+ w)_{L_2(V)} &= (\varphi, \chi\mathcal{L}^+ w)_{L_2(V)} = (f, \chi w)_{L_2(V)} - (\varphi, \mathcal{L}_\chi w)_{L_2(V)} \\ &= (\chi f - \psi, w)_{L_2(V)} = (\theta, w)_{L_2(V)} \quad \big(w \in W_2^{2-t}(V, \mathrm{b}),\end{aligned} \tag{2.10}$$

where $\theta = \chi f - \psi \in W_2^{t-1}(V)$.

By virtue of the fact that the mapping $W_2^{1-t}(V, \mathrm{b}) \ni w \mapsto \mathcal{L}^+ w \in W_2^{-t-1}(V)$ is a homeomorphism (according to Theorem 1.1), we can write

$$|(\theta, w)_{L_2(V)}| \le \|\theta\|_{W_2^{t-1}(V)} \|w\|_{W_2^{1-t}(V)} \le c_1 \|\theta\|_{W_2^{t-1}(V)} \|\mathcal{L}^+ w\|_{W_2^{-t-1}(V)} \tag{2.11}$$

$$\big(w \in W_2^{1-t}(V, \mathrm{b})\big).$$

Inequality (2.11) yields the existence of $\mu \in W_2^{t+1}(V)$ such that

$$(\theta, w)_{L_2(V)} = (\mu, \mathcal{L}^+ w)_{L_2(V)} \qquad \big(w \in W_2^{1-t}(V, \mathrm{b})\big) \tag{2.12}$$

(this fact is proved below). Inserting (2.12) in (2.10), we obtain

$$(\chi\varphi, \mathcal{L}^+ w)_{L_2(V)} = (\mu, \mathcal{L}^+ w)_{L_2(V)} \qquad \big(w \in W_2^{2-t}(V, \mathrm{b})\big).$$

Here, $\mathcal{L}^+ w$, by virtue of Theorem 1.1, runs through the whole $W_2^{2-t}(V)$. Therefore, the last equality means that $\chi\varphi - \mu \in W_2^{t+1}(V)$ as required.

Let us now establish representation (2.12). It follows from estimate (2.11) that, in fact, the expression $(\theta, w)_{L_2(V)}$ is a linear function of $\mathcal{L}^+ w$ but not of w (θ is fixed): $(\theta, w)_{L_2(V)} = l(\mathcal{L}^+ w)$. For the functional l, by virtue of (2.10), we have the following estimate:

$$|l(\mathcal{L}^+ w)| \le c_2 \|\mathcal{L}^+ w\|_{W_2^{-t-1}(V)} \qquad \big(w \in W_2^{1-t}(V, \mathrm{b})\big),$$

where $\mathcal{L}^+w$ runs through the whole $W_2^{-t-1}(V)$. In other words, it is continuous in $W_2^{-t-1}(V)$ and, therefore, admits the following representation in terms of an element $\mu \in W_2^{t+1}(V)$: $l(\mathcal{L}^+w) = (\mu, \mathcal{L}^+w)_{L_2(V)}$ $(w \in W_2^{1-t}(V, \mathrm{b}))$. Equality (2.12) is thus proved.

It is easy to see that, in proving this assertion, we have, in fact, used the requirement that $a_\alpha \in C^{|\alpha|-t}(G)$.

(2) Let us make some remarks necessary to consider the case $t = 0, 1, \dots$. Let $\varphi \in W_2^t(V)$ $(t = 1, 2, \dots)$, $w \in W_2^2(V)$. Integrating by parts, we arrive at the following formula for the expression $\mathcal{L}_\chi$ introduced above:

$$\begin{aligned}(\varphi, \mathcal{L}_\chi w)_{L_2(V)} &= \sum_{|\alpha|\le 1} \left(\varphi, c_\alpha(x) D^\alpha w\right)_{L_2(V)} \\ &= \sum_{|\alpha|\le 1} \left(d_\alpha(x) D^\alpha \varphi, w\right)_{L_2(V)} \end{aligned} \tag{2.13}$$

(the integrals over ∂V do not appear in this expression, since the coefficients c_α of the expression $\mathcal{L}_\chi$ vanish on ∂V together with all their derivatives due to the factor χ; d_α are new coefficients). Thus, if $t = 1, 2, \dots$, then one can write the following equality similar to (2.9):

$$(\varphi, \mathcal{L}_\chi w)_{L_2(V)} = (\psi, w)_{L_2(V)} \tag{2.14}$$
$$\left(\psi \in W_2^{t-1}(V), \quad \varphi \in W_2^t(V), \quad w \in W_2^2(V)\right).$$

Let us show that (2.14) is also true for $t = 0$. We fix the index $\beta, |\beta| \le 1$ $(\alpha = 0$ in (2.13)), and consider a continuous operator

$$W_2^1(V) \ni v \mapsto A_{0\beta} v = c_{0\beta}(x) D^\beta v \in L_2(V);$$

$A_{0\beta}^+$ is a continuous operator from $L_2(V)$ to $W_2^{-1}(V)$. In particular, we have

$$(\varphi, c_{0\beta} D^\beta w)_{L_2(V)} = (\varphi, A_{0\beta} w)_{L_2(V)} = (A_{0\beta}^+ \varphi, w)_{L_2(V)} \ (\varphi \subset L_2(V), w \in W_2^2(V)).$$

Substituting these expressions in (2.13) (for $t = 0$), we arrive at (2.14), where $\psi = \sum_{|\beta|\le 1} A_{0\beta}^+ \varphi \in W_2^{-1}(V)$.

(3) Below, we prove that (1) also holds for $t = 0, 1, \dots$. To do this, we first establish an analogue of (2.10).

By repeating the proof of (1) and using (2.14), we obtain

$$(\chi\varphi, \mathcal{L}^+ w)_{L_2(V)} = (\theta, w)_{L_2(V)} \tag{2.15}$$
$$\left(w \in W_2^2(V, \mathrm{b}), \ \theta = \chi f - \psi \in W_2^{t-1}(V)\right).$$

(4) Let us establish (1) for $t = 0$. According to Theorem 1.1, the closure of the mapping $W_2^2(V, \mathrm{b}) \ni w \mapsto \mathcal{L}^+ w \in L_2(V)$ in the relevant norms is a homeomorphism between $W_2^1(V, \mathrm{b})$ and $W_2^{-1}(V, \mathrm{b})$. Here, $\theta \in W_2^{-1}(V)$. Therefore, for any $w \in W_2^2(V, \mathrm{b})$, we have

$$\begin{aligned} \left|(\theta, w)_{L_2(V)}\right| &\leq \|\theta\|_{W_2^{-1}(V)} \|w\|_{W_2^1(V)} \\ &= \|\theta\|_{W_2^{-1}(V)} \|w\|_{W_2^1(V,\mathrm{b})} \\ &\leq c_2 \|\theta\|_{W_2^{-1}(V)} \|\mathcal{L}^+ w\|_{W_2^{-1}(V,\mathrm{b})}. \end{aligned} \tag{2.16}$$

As when deriving (2.12) from (2.11), it follows from inequality (2.16) that there exists $\mu \in W_2^1(V, \mathrm{b})$ such that

$$(\theta, w)_{L_2(V)} = (\mu, \mathcal{L}^+ w)_{L_2(V)} \qquad \big(w \in W_2^2(V, \mathrm{b})\big). \tag{2.17}$$

Relations (2.15) and (2.17) and the fact that $\mathcal{L}^+ w$ runs through the whole $L_2(V)$ imply the inclusion $\chi\varphi = \mu \in W_2^1(V, \mathrm{b}) \subset W_2^1(V)$. This completes the proof of statement (1) for $t = 0$.

It is easy to show that, in the proof of this statement, it suffices to require that $a_\alpha \in C^{|\alpha|+1}(G)$.

(5) Let us prove (1) for $t = 1, 2, \dots$. As above, according to Theorem 1.1, the closure of the mapping $W_2^2(V, \mathrm{b}) \ni w \mapsto \mathcal{L}^+ w \in L_2(V)$ in the relevant norms is a homeomorphism between $W_2^{-t+1}(V)$ and $W_2^{-t-1}(V, \mathrm{b})$. In this case, $\theta \in W_2^{t-1}(V)$. Therefore,

$$\left|(\theta, w)_{L_2(V)}\right| \leq \|\theta\|_{W_2^{t-1}(V)} \|w\|_{W_2^{-t+1}(V)} \leq c_3 \|\theta\|_{W_2^{t-1}(V)} \|\mathcal{L}^+ w\|_{W_2^{-t-1}(V,\mathrm{b})}$$

for any $w \in W_2^2(V, \mathrm{b})$. This implies the existence of $\mu \in W_2^{t+1}(V, \mathrm{b})$ which satisfies equality (2.17). This, (2.15), and the fact that $\mathcal{L}^+ w$ runs through the whole $L_2(V)$ enables us to conclude that $\chi\varphi = \mu \in W_2^{t+1}(V, \mathrm{b}) \subset W_2^{t+1}(V)$. Statement (1) is thus proved in the case under consideration.

Here, we have the following requirements of smoothness: $a_\alpha \in C^{|\alpha|+|1-t|}(G)$.

(6) We have proved that if $\varphi \in W_2^t(G)$ is a generalized solution of equation (2.1) with $f \in W_2^s(G)$ $(t, s \in \mathbb{Z})$ inside G and $t < 2 + s$, then $\chi\varphi \in W_2^{t+1}(V)$ for any $\chi \in C^\infty(\tilde{G})$ vanishing in a neighbourhood of $G \setminus V$ (it is clear that this inclusion can also be written in the form $\chi\varphi \in W_2^{t+1}(G)$). It follows from the proof that, without any changes, it works in the case where φ is a solution of equation (2.1) only inside V.

(7) If $t+1 = 2+s$, then the proof is completed. For $t+1 < 2+s$, we proceed as follows: Let $V_1 = V_1(x_0)$ be a spherical neighbourhood of a point x_0 whose radius is smaller than the radius of $V = V(x_0)$ and let $\chi_1 \in C^\infty(\tilde{G})$ be vanishing in a neighbourhood $G \setminus V$ and equal to one in $\tilde{V}_1$. Then $\varphi = \chi_1\varphi \in W_2^{t+1}(G)$

is a generalized solution of the same equation (2.1) inside V_1. Indeed, for any $v \in C_0^\infty(V_1)$, we have, according to (2.2),

$$(\chi_1\varphi, \mathcal{L}^+ v)_{L_2(G)} = (\varphi, \chi_1\mathcal{L}^+ v)_{L_2(G)} = (\varphi, \mathcal{L}^+ v)_{L_2(G)} = (f, v)_{L_2(G)}$$

(here, we have used the fact that $(\forall x \in G)\colon \chi_1(x)(\mathcal{L}^+ v)(x) = (\mathcal{L}^+ v)(x)$, since $(\mathcal{L}^+ v)(x)$ differs from zero only for $x \in V_1$ and, in this case, $\chi(x) = 1$).

Let us apply the assertion established in step (6) to a generalized solution $\varphi_1 \in W_2^{t+1}(G)$ of equation (2.1) inside V. This gives the inclusion $\chi\varphi \in W_2^{t+2}(G)$ for any $\chi \in C^\infty(\tilde{G})$ vanishing in a neighbourhood of $G \setminus V$.

If $t+2 = 2+s$, then the proof of the theorem is completed. For $t+2 < 2+s$, we proceed by analogy with the previous step, i.e., choose a spherical neighbourhood $V_2 = V_2(x_0)$ of the point x_0 whose radius is smaller than the radius of $V_1 = V_1(x_0)$, construct the corresponding function $\chi_2 \in C^\infty(\tilde{G})$, form $\varphi_2 = \chi_2\varphi_1$, and so on. As a result, after finitely many steps, we arrive at the inclusion $\varphi \in W_{2,\text{loc}}^{2+s}(V_n)$, where V_n is the corresponding neighbourhood of the point x_0 which can be taken as $V(x_0)$. Thus, the proof of Theorem 2.1 in the local formulation is complete. According to Lemma 2.1, this implies that Theorem 2.1 is valid in the general case. □

REMARK 2.1. It is convenient to formulate the smoothness requirements for the the coefficients of $\mathcal{L}$ in Theorem 2.1 in the following way: Let $a_\alpha \in C^{|\alpha|+p}(G)(|\alpha| \leq 2)$, where $p \in \mathbb{Z}_+$ is fixed. Then one can take $t \in [-p, p+2)$ and the generalized solution φ would automatically be in $W_{2,\text{loc}}^{\max(2+s,p+2)}(G)$. The sufficiency of this smoothness requirement for proving the theorem can be established quite easily.

REMARK 2.2. It is easy to see that Theorem 2.1 can also be formulated in following "local" form:

If $G' \subseteq G$ is a subdomain of G with sufficiently smooth boundary and $\varphi \in W_2^t(G)$ $(t \in \mathbb{Z})$ is a generalized solution of equation (2.1) inside G' with the right-hand side $f \in W_2^s(G')$ $(s \in \mathbb{Z})$, then $\varphi \in W_{2,\text{loc}}^{2+s}(G')$.

REMARK 2.3. In the formulation presented above, Theorem 2.1 is actually true for elliptic expressions of an arbitrary order. In this case, $\varphi \in W_{2,\text{loc}}^{2+s}(G')$ automatically. This can be proved by using the proper generalization of Theorem 1.1.

2.3 Smoothing up to the Boundary

In this subsection, we study the smoothness of generalized solutions of elliptic equations up to the boundary of a domain. We stress that the fact that equality (2.2) holds for functions v finite with respect to G implies the inclusion in the corresponding Sobolev space not for the solution φ itself but for its product $\chi\varphi$ by a "cutoff" function χ (i.e., ensures smoothness inside the domain). However, in this case where equality (2.2) holds for a larger supply of functions v, we can also establish much stronger properties of the solution φ. As in Theorem 1.1 on isomorphisms, we consider here only trivial boundary conditions.

Let $G \subset \mathbb{R}^N$ be a bounded domain whose boundary contains a sufficiently smooth piece γ (i.e., $\gamma \subseteq \partial G$ is a domain in the topology of the surface ∂G with sufficiently smooth boundary $\partial\gamma$ on ∂G). Let $C_0^l(G,\gamma)$ $(l \in \mathbb{N} \cup \{\infty\})$ denote the class of functions finite "with respect to G outside γ", i.e., functions from $C^l(\tilde{G})$ vanishing in a neighbourhood of the set $\partial G \setminus \partial\gamma$ in $\mathbb{R}^N$; $C_0^l(G,\gamma,\mathrm{b})$ is a subclass of this class that consists of functions vanishing in γ.

Consider the same equation (2.1) but necessarily of the second order.

We say that $u \in W_2^t(G)$ is a generalized solution of (2.1) inside G up to the piece γ (where it satisfies trivial boundary conditions) if equality (2.2) is satisfied for all $v \in C_0^\infty(G,\gamma,\mathrm{b})$.

Let $G' \subseteq G$ be a subdomain of G with sufficiently smooth boundary $\partial G'$ and let $\gamma \subseteq \partial G'$. Then one can say that $u \in W_2^t(G)$ is a generalized solution of equation (2.1) up to γ if equality (2.2) is satisfied for all $v \in C_0^\infty(G',\gamma,\mathrm{b})$ (extended by zero to the whole of G).

The fact that a generalized function $\alpha \in W_2^{-l}(G)$ $(l \in \mathbb{N})$ belongs to $W_2^k(G)$ inside G' up to the piece γ can also be formulated in a natural way:

As above, $\chi\alpha \in W_2^k(G)$ but the cutoff function $\chi \in C^\infty(\tilde{G})$ annihilates only in $G \setminus G'$ and in a neighbourhood of the set $\partial G' \setminus \partial\gamma$ in $\mathbb{R}^N$. For this type of inclusion of α in $W_2^k(G)$, we use the following notation: $\alpha \in W_{2,\mathrm{loc}}^k(G',\gamma)$.

For an inclusion of this sort, one can formulate the following analogue of Lemma 2.1 on localization:

Lemma 2.4. *Consider G, G', and γ defined above. Let $\alpha \in W_2^{-l}(G)$ $(l \in \mathbb{N})$ and $k = -l+1, -l+2, \dots$. Assume that, for every point $x \in G' \cup \gamma$, there exists its spherical neighbourhood $U(x)$ in $\mathbb{R}^N$ such that $U(x) \subseteq G'$ for $x \in G'$ and $\alpha \in W_{2,\mathrm{loc}}^k\big(U(x) \cap (G' \cup \gamma), \gamma \cap U(x)\big)$. Then $\alpha \in W_{2,\mathrm{loc}}^k(G',\gamma)$.*

Proof. In its principal points, the proof coincides with the proof of Lemma 2.1. Thus, we extract from the covering of the locally compact space $G' \cup \gamma$ by the neighbourhoods $W(x) = U(x) \cap (G' \cap \gamma)$ $(x \in G' \cup \gamma)$ a countable subcovering $W(x_1), W(x_2), \dots$. Let $\big(\chi_j(x)\big)_{j=1}^\infty$ be the corresponding decomposition of the unit, i.e., $\chi_j \in C_0^\infty(G,\gamma)$ are nonnegative and vanish outside $W(x_j)$ and $\sum_{j=1}^\infty \chi_j(x) = 1$ $(x \in G' \cup \gamma)$. This decomposition of the unit is constructed by analogy with the standard construction presented in Section 11.1.

Further, let $\chi \in C^\infty(\tilde{G})$ be the cutoff function introduced above. Then $\chi(x) = \sum_{j=1}^{n(\chi)} \chi_j(x)\chi(x)$ $(n(\chi) \in \mathbb{N})$ and we can repeat simple reasoning used in the proof of Lemma 2.1. This completes the proof of the required result. □

The theorem on smoothing of generalized solutions up to the boundary of a domain is similar to Theorem 2.1 and can be formulated as follows:

Theorem 2.2. *Let $\mathcal{L}$ be a second-order elliptic expression (1.1) with sufficiently smooth coefficients (the coefficients of derivatives are real-valued) given in $G \cup \gamma$, where γ is a sufficiently smooth piece of ∂G. Consider a generalized solution $\varphi \in W_2^t(G)$ $(t \in \mathbb{Z})$ of equation (2.1) with the right-hand side $f \in W_2^s(G)$ $(s \in \mathbb{Z})$ inside*

G up to the piece γ. Then this solution, in fact, belongs to the space $W_2^{2+s}(G)$ inside G up to the piece γ, i.e., $\varphi \in W_{2,\mathrm{loc}}^{2+s}(G,\gamma)$. If $2+s \geq 1$, then $\varphi(x) = 0$ for $x \in \gamma$.

Smoothness requirements for the coefficients a_α and the piece γ are presented in Remark 2.4.

Proof. The theorem is proved by repeating the arguments used in the proof of Theorem 2.1 with certain modifications described below.

By virtue of Lemma 2.4, the proof is localized. If $x_0 \in G$ and a spherical neighbourhood $W(x_0) = U(x_0) \subseteq G$, then the inclusion $\varphi \in W_{2,\mathrm{loc}}^{2+s}(W(x_0))$ follows from Theorem 2.1 or, more precisely, from Remark 2.2. Consider the case where $x_0 \in \gamma$. As V, we take a domain with sufficiently small diameter and sufficiently smooth boundary ∂V which has a nonempty intersection with γ that contains the point x_0. If we establish that $\varphi \in W_{2,\mathrm{loc}}^{2+s}(V, \partial V \cap \gamma)$ and $\varphi(x) = 0$ for $x \in \partial V \cap \gamma$ when $2+s \geq 1$, then the theorem will be proved. Indeed, it suffices to apply the localization lemma (Lemma 2.4) choosing as $W(x_0) = U(x_0) \cap (G \cup \gamma)$ a sufficiently small neighbourhood of the indicated type that belongs to V.

Let $t < 2+s$ and let χ be a cutoff function from $C^\infty(\tilde{G})$ vanishing in $G \setminus V$ and in a neighbourhood of the set $\partial G \setminus \partial\gamma$ in $\mathbb{R}^N$. As above, we prove that $\chi\varphi \in W_2^{t+1}(V)$ (i.e., $\varphi \in W_{2,\mathrm{loc}}^{t+1}(V, \partial V \cap \gamma)$). As in the proof of Theorem 2.1, we assume that t runs from $-\infty$ to ∞.

In the case where $t = -1, -2, \ldots$, it suffices to repeat the reasoning of step (1) without changes. Smoothness requirements are as follows: $a_\alpha \in C^{|\alpha|-t}(G \cup \gamma)$ and $\gamma \in C^{2-t}$.

Step (2) should be somewhat modified. Indeed, since χ, generally speaking, does not vanish on $\partial V \cap \gamma$, the coefficients c_α of the differential expression $\mathcal{L}_x$ do not necessarily vanish on $\partial V \cap \gamma$ and, hence, the transfer of derivatives as in (2.13) is impossible in the case $t = 1, 2, \ldots$. However, one can realize this operation under the additional assumption that the generalized solution φ vanishes on $\partial V \cap \gamma$. Therefore, equalities (2.13) and, hence, (2.14) hold for $t = 1, 2, \ldots$ provided that $\varphi(\varphi \restriction V \in W_2^t(V))$ vanishes on $\partial V \cap \gamma$.

For $t = 0$, the reasoning used in step (2) remains unchanged and equality (2.14) is true.

Steps (3) and (4) also do not change. Recall that, in step (4) ($t = 0$), we prove the inclusion $\chi\varphi = \mu \in W_2^1(V, \mathrm{b})$. Therefore, if a cutoff function χ does not vanish for $x \in \partial V \cap \gamma$, then we have $\varphi(x) = 0$ ($x \in \partial V \cap \gamma$).

Step (5) ($t = 1, 2, \ldots$) also remains unchanged under the additional assumption that φ vanishes on $\partial V \cap \gamma$. As mentioned above, relations (2.13) and (2.14) remain true in this case.

The remarks made in step (6) remain true with proper modification. The final step (7) is also preserved. One should only take $V_1, V_2, \ldots$ in the form of V-type domains, i.e., "to slide" along the piece γ, and take into account the following consideration mentioned above: As t passes through zero in moving from $-\infty$ to ∞, one can write $\varphi \in W_{2,\mathrm{loc}}^1(V, \mathrm{b})$ and note that φ vanishes on $\partial V \cap \gamma$. Therefore, further increase in t is possible.

It is worth noting that if we increase the smoothness of φ only from the values $t = 1, 2, \ldots$, then it is impossible to pass directly to the smoothness $t+1$, since it is not assumed that φ vanishes on γ. But, in this case, $\varphi \in L_2(G)$ and, therefore, one can start moving from $t = 0$ and prove that φ vanishes on $\partial V \cap \gamma$ as far as in the first step. □

Remark 2.4. The restrictions imposed on the smoothness of the coefficients of $\mathcal{L}$ coincide with those in Remark 2.1. For the smoothness of the piece γ, we assume that $\gamma \in C^{2+p}$. Under these assumptions, it is possible to apply Theorem 1.1 on isomorphisms for all s under consideration.

Remark 2.5. In this case, an analogue of Remark 2.2 on the possibility of formulation of Theorem 2.2 in the local form holds for $G' \subseteq G$ such that $\gamma \subseteq \partial G'$.

Remark 2.6. The results of Theorems 2.1 and 2.2 formulated in the local form in Remarks 2.2 and 2.5 can be easily reformulated for the case where $G \subseteq \mathbb{R}^N$ is an unbounded domain. Thus, the corresponding generalization of Theorem 2.1 can be obtained as follows:

Consider an elliptic expression given in the domain $G \subseteq \mathbb{R}^N$ and satisfying the conditions of the indicated theorem. Let $G' \subseteq G$ be a bounded subdomain of G with sufficiently smooth boundary. Let $\varphi \in W_2^t(G, p(x)dx)$ $(t \in \mathbb{Z})$ be a generalized solution of equation (2.1) inside G' with the right-hand side $f \in W_2^s(G')$ $(s \in \mathbb{Z})$. Then $\varphi \in W_{2,\mathrm{loc}}^{2+s}(G')$.

In this case, the smoothness conditions coincide with the corresponding conditions in Remark 2.1 and $0 < p(x) \in C(\tilde{G})$ is a weight. The fact that the indicated φ is a generalized solution inside G' means that relation (2.2) holds for $v \in C_0^\infty(G')$.

Theorem 2.2 is modified similarly.

3 Elliptic Differential Operators in a Domain with Boundary

3.1 The Case of a Bounded Domain

Let $G = \mathbb{R}^N$ be a bounded domain with sufficiently smooth boundary ∂G. In G, we consider a second-order elliptic formally selfadjoint differential expression $\mathcal{L}$ of the form (1.1) with real-valued coefficients. Note that if $\mathcal{L}$ takes the form (1.3), then the condition $\mathcal{L}^+ = \mathcal{L}$ of formal selfadjointness is equivalent to the following relation:

$$a_j(x) = \sum_{k=1}^{N} (D_k a_{jk})(x) \qquad (x \in G). \tag{3.1}$$

For given $\mathcal{L}$, by applying the standard procedure, one can introduce in $L_2(G)$ the minimal operator L defined as the closure of the operator $L_2(G) \supset C_0^2(G) \ni$

$u \mapsto L'u = \mathcal{L}u \in L_2(G)$. Consider the extension A of the operator L that corresponds to trivial boundary conditions (b) on ∂G. It is constructed as the closure of an operator $L_2(G) \supset C^2(\tilde{G}, \mathrm{b}) \ni u \mapsto A'u = \mathcal{L}u \in L_2(G)$ acting on $L_2(G)$ (this operator is an extension of L'). Recall that $C^2(\tilde{G}, \mathrm{b}) = \{u \in C^2(\tilde{G}) \mid u \restriction \partial G = 0\}$. The operator A' introduced above and, hence, A are Hermitian operators.

To prove this fact, it is convenient to use the following general Green's formula valid for any second-order elliptic expression $\mathcal{L}$ of the form (1.1), (1.3):

For any $u, v \in C^2(\tilde{G})$, we have

$$(\mathcal{L}u, v)_{L_2(G)} - (u, \mathcal{L}^+ v)_{L_2(G)}$$
$$= \int_{\partial G} |\mathrm{A}(x)\nu(x)| \left(\left(\frac{\partial u}{\partial \mu}\right)(x)\overline{v(x)} - u(x)\overline{\left(\frac{\partial v}{\partial \mu}\right)(x)} \right) dx + \int_{\partial G} \alpha(x) u(x)\overline{v(x)} dx, \tag{3.2}$$

$$\alpha(x) = \sum_{j=1}^{N} \left(a_j(x) - \sum_{k=1}^{N} (D_k a_{jk})(x)\nu_j(x) \right) \quad (x \in \partial G),$$

where $\nu(x) = (\nu_1(x), \ \ldots \ \nu_N(x))$ denotes the unit vector of the outer normal to ∂G at the point $x \in \partial G$ and $\mu(x)$ denotes the unit vector of the conormal defined by the formula

$$\mu(x) = \frac{\mathrm{A}(x)\nu(x)}{|\mathrm{A}(x)\nu(x)|}, \quad \mathrm{A}(x) = (a_{jk}(x))_{j,k=1}^{N} \qquad (x \in \partial G). \tag{3.3}$$

Formula (3.2) can be easily established by integrating by parts. Note that definition (3.3) of the unit vector $\mu(x)$ is possible due to the fact that the matrix $\mathrm{A}(x)$ is nondegenerate, which follows from the ellipticity of $\mathcal{L}$ (see (1.2)). It is also worth noting that if $\mathcal{L}^+ = \mathcal{L}$, then, according to (3.1), $\alpha(x) = 0$. Relation (3.2), clearly, implies that the operator A' is Hermitian.

An equality similar to (3.2) also holds for a general second-order expression $\mathcal{L}$, not necessarily elliptic. In this case, the first integral on the right-hand side of (3.2) is carried out not over ∂G but over its part $\partial G \backslash X$, where $X = \{x \in \partial G \mid \mathrm{A}(x)\nu(x) = 0\}$.

It is easy to show that the domain $\mathcal{D}(A)$ of the operator A coincides with the subspace $W_2^2(G, \mathrm{b}) = \{u \in W_2^2(G) \mid u \restriction \partial G = 0\}$ of the space $W_2^2(G)$ (which may also be defined as the completion of the set $C^2(\tilde{G}, \mathrm{b})$ in the metric of the space $W_2^2(G)$. In this case, it is necessary to assume that $\partial G \in C^2$. The proof of this assertion follows from the inequality: $(\exists p \geq 0, \ c_1 \geq 0, \ c_2 \geq 0)$ such that

$$c_1 \|u\|_{W_2^2(G)} \leq \|\mathcal{L}u\|_{L_2(G)} + p\|u\|_{L_2(G)} \leq c_2 \|u\|_{W_2^2(G)} \tag{3.4}$$
$$(u \in C^2(\tilde{G}, \mathrm{b})),$$

where the left inequality in (3.4) coincides with the coercivity inequality (1.11) with $s = 0$, while the right inequality is an elementary estimate.

The operator A is not only Hermitian but also selfadjoint.

Theorem 3.1. *Let G be a bounded domain with boundary $\partial G \in C^4$ and let $\mathcal{L}$ be a formally selfadjoint second-order elliptic expression (1.1) whose coefficients $a_\alpha \in C^{|\alpha|+2}(\tilde{G})$.*

Under these smoothness conditions, the operator $A = L(\mathrm{b})$ corresponding to $\mathcal{L}$ with trivial boundary conditions is selfadjoint in $L_2(G)$ and semibounded (from below).

Proof. Let $g \in \mathcal{D}(A^*)$. Then, for any $f \in \mathcal{D}(A)$, we have $(Af, g)_{L_2(G)} = (f, A^*g)_{L_2(G)}$. In particular, for any $v \in C^2(\tilde{G}, \mathrm{b})$,

$$(\mathcal{L}v, g)_{L_2(G)} = (v, A^*g)_{L_2(G)}$$

or, in view of the fact that $\mathcal{L} = \mathcal{L}^+$,

$$(g, \mathcal{L}^+v)_{L_2(G)} = (A^*g, v)_{L_2(G)} \qquad (v \in C^2(\tilde{G}, \mathrm{b})). \tag{3.5}$$

Relation (3.5) shows that $g \in L_2(G)$ is a generalized solution of the equation $\mathcal{L}u = A^*g \in L_2(G)$ inside G up to $\gamma = \partial G$ (where the trivial boundary conditions are imposed). By applying Theorem 2.2 for $s = 0$ and $t = 0$, we conclude that $g \in W_2^2(G)$, $g \upharpoonright \partial G = 0$, and $\mathcal{L}g = A^*g$. In other words, $g \in \mathcal{D}(A)$ and $A^*g = Ag$, i.e., $A^* \subseteq A$. Hence, $A = (A^*)^* \supseteq A^*$, i.e., $A^* = A$.

As follows from Remark 2.2 and the smoothness assumptions made above, Theorem 2.2 is applicable in this case.

The semiboundedness of A follows from inequality (1.13). □

REMARK 3.1. In formulating Theorem 3.1, we did not try to make the smoothness of the coefficients and boundary as low as possible. Here, we only note that, for piecewise smooth boundaries, this theorem may be not true. For example, it does not hold in the case where $\mathcal{L} = -\Delta$ in a region $G \subset \mathbb{R}^2$ if ∂G contains a corner point whose interior angle is greater than π.

REMARK 3.2. Theorem 3.1 also holds for operators $A = L(\mathrm{b})$ that correspond to some other boundary conditions (b). For example, one can consider the Neumann problem $\frac{\partial u}{\partial \mu} \upharpoonright \partial G = 0$ or the third boundary-value problem

$$\left(\frac{\partial u}{\partial \mu} + \sigma(x)u\right) \upharpoonright \partial G = 0$$

(σ is real-valued) and define the operator A as above with the sole difference that $u \in \mathcal{D}(A')$ does not vanish on ∂G but satisfies the indicated conditions. As follows from (3.2), the operator A' and, hence, A are Hermitian. Under certain additional smoothness requirements, which are not formulated here, the operator A is selfadjoint. This is proved as in Theorem 3.1 by using the technique developed in Sections 1 and 2, which can be reformulated for these boundary conditions.

□

Consider the following problem: Let $\mathcal{L} = \mathcal{L}^+$ be a general second-order differential expression. As mentioned above, it satisfies relation (3.2) with ∂G replaced by $\partial G \backslash X$ (in the first integral on the right-hand side) and without the second integral. Thus,

$$(\mathcal{L}u, v)_{L_2(G)} - (u, \mathcal{L}v)_{L_2(G)}$$
$$= \int_{\partial G \backslash X} |\,\mathrm{A}(x)\nu(x)| \left(\left(\frac{\partial u}{\partial \mu}\right)(x)\overline{v(x)} - u(x)\overline{\left(\frac{\partial v}{\partial \mu}\right)(x)} \right) dx \quad (3.6)$$
$$(u, v \in C^2(\tilde{G})).$$

The question is how to find "formally selfadjoint" boundary conditions (b) for given $\mathcal{L}$ and G, i.e., conditions imposed on u and v under which the right-hand side of (3.6) vanishes and, moreover, if $u \in C^2(\tilde{G})$ satisfies these conditions and is arbitrary and $v \in C^2(\tilde{G})$ is such that the right-hand side vanishes, then v also satisfies these conditions. (The three types of boundary conditions discussed above meet these requirements in the elliptic case.)

It is easy to see that the selfadjoint operator generated by $\mathcal{L}$ may satisfy just these boundary conditions (b). In selecting these (b), an essential role is played by the fact that, as u runs through $C^2(\tilde{G})$, the boundary values $u \restriction \partial G$ and $\frac{\partial u}{\partial v} \restriction \partial G$ take arbitrary (with some restrictions) values dense in the space L_2 constructed on ∂G. Here, we do not study this problem in detail and only note that if we construct, by analogy with the elliptic case, for given $\mathcal{L}, G$, and (b) of the indicated type,an operator $A = L(\mathrm{b}) \supseteq L$ acting on $L_2(G)$, then this operator would not necessarily be selfadjoint.

The spectrum of the operator thus constructed is described by the following simple theorem:

Theorem 3.2. *Assume that the conditions of Theorem 3.1 are satisfied and $A = L(\mathrm{b})$ is the selfadjoint operator in $L_2(G)$ constructed according to the procedure described above. It is stated that the spectrum of A coincides with a sequence $(\lambda_n)_{n=1}^{\infty}$ of real eigenvalues approaching $+\infty$; each λ_n is associated with a finite-dimensional eigensubspace.*

Proof. The location of the spectrum $S(A)$ on a semiaxis of the form $[\alpha, +\infty)$ $(\alpha \in \mathbb{R})$ evidently follows from the selfadjointness of A and its semiboundedness. By using Lemmas 1.2 and 1.1, we conclude that there exist sufficiently large $k \geq 0$ and $c_1 > 0$ such that the inequality

$$\|(\mathcal{L} + k\mathbb{I})u\|_{L_2(G)} \geq c_1 \|u\|_{W_2^2(G)} \qquad (u \in W_2^2(\tilde{G}, \mathrm{b}))$$

holds. This implies that

$$\|(A + k\mathbb{I})f\|_{L_2(G)} \geq c_1 \|f\|_{W_2^2(G)} \qquad (f \in W_2^2(G))$$

and, therefore, $-k \notin S(A)$, there exists $(A + k\mathbb{I})^{-1} = R_{-k}$, and

$$\|R_{-k}g\|_{W_2^2(G)} \le c_1^{-1}\|g\|_{L_2(G)} \qquad (g \in L_2(G)).$$

The last inequality implies that R_{-k} maps the unit ball in $L_2(G)$ into a ball in $W_2^2(G)$. By the embedding theorems (see 14.3), this ball is compact in $L_2(G)$. Therefore, the operator R_{-k} is compact. It follows from the properties of compact operators (see Section 9.4) that the spectrum $S(R_{-k})$ coincides with a sequence of eigenvalues (with finite multiplicity) that approaches zero. But if $\lambda \in S(A)$, then $(\lambda + k)^{-1} \in S(R_{-k})$, and vice versa. Moreover, the corresponding eigenvectors coincide. Therefore, the indicated character of the spectrum of the operator R_{-k} yields the required properties of the spectrum of the operator A. □

3.2 The Case of an Unbounded Domain

In this subsection, we consider operators generated by an elliptic formally selfadjoint second-order differential expression $\mathcal{L}$ with real-valued coefficients given in an unbounded domain $\tilde{G} \subset \mathbb{R}^N$ which does not coincide with $\mathbb{R}^N$ and has a sufficiently smooth boundary ∂G.

In this case, the operator $A = L(\mathrm{b})$ in $L_2(G)$ is defined as above. Indeed, we construct the operator $L_2(G) \supset C_0^2(\tilde{G}, \mathrm{b}) \ni u \mapsto A'u = \mathcal{L}u \in L_2(G)$, where $C_0^2(\tilde{G}, \mathrm{b})$ denotes the class of finite functions $C_0^2(\mathbb{R}^N)$ restricted to $\tilde{G}$ and taking zero values on ∂G. Relation (3.2) implies that the operator A' is Hermitian. The operator A is defined as the closure of A'. It is clear that A is an Hermitian extension of the minimal operator L corresponding to $\mathcal{L}$ and G.

It is clear that inequality (3.4) remains true for the functions $u \in C_0^2(\tilde{G}, \mathrm{b})$ but the constants p, c_1, and c_2 depend on the domain outside which the finite function u is identically equal to zero. Therefore, one can only say that $u \in \mathcal{D}(A)$ locally belongs to W_2^2 and vanishes on ∂G. More precisely, this means that if we denote by G_R $(R > 0)$ the intersection of G with an open ball with radius R centered at the origin, then $u \restriction \tilde{G}_R \in W_2^2(G_R)$ for $u \in \mathcal{D}(A)$ and vanishes on ∂G. Certainly, in this case, we also assume that $\partial G \in C^2$.

The selfadjointness of A depends on the behaviour of its coefficients as $|x| \to \infty$. Here, we do not cite the corresponding results because they are similar to the theorems proved in Section 4 for $\mathcal{L}$ defined in the whole of the space $\mathbb{R}^N$ and restrict ourselves to the proof of the fact that the "variation of $\mathcal{L}$ within a bounded domain" does not affect the selfadjointness of the corresponding operator A. More precisely, we prove the following theorem:

Theorem 3.3. *Let G be an unbounded domain with boundary $\partial G \in C^4$ and let $\mathcal{L}$ be a formally selfadjoint second-order elliptic expression (1.1) whose coefficients $a_\alpha \in C^{|\alpha|+2}(\tilde{G})$ and are real. Suppose that the corresponding operator in $L_2(G)$ $A = L(\mathrm{b})$ constructed for given $\mathcal{L}$ with trivial boundary conditions on ∂G is selfadjoint.*

Let $\mathcal{M}$ be another expression of the same form as $\mathcal{L}$ which coincides with $\mathcal{L}$ for $|x| \ge R$, where $R > 0$, and let $B = M(\mathrm{b})$ be the operator similar to A but constructed in terms of $\mathcal{M}$. It is stated that B is selfadjoint.

Proof. It suffices to show that the adjoint operator B^* is Hermitian (since, in this case, $B^* \subseteq (B^*)^* = B$ which, together with $B \subseteq B^*$, gives: $B^* = B$). By analogy with Theorem 3.1, one can prove that $\mathcal{D}(B^*)$ consists of the functions u which locally belong to W_2^2 and vanish on ∂G; $B^*u = \mathcal{M}u \quad (u \in \mathcal{D}(B^*))$. It should be noted that $g \in \mathcal{D}(B^*)$ is a generalized solution of the equation $\mathcal{M}u = B^*g$ inside G' up to $\partial G'$ (with the trivial boundary conditions), where G' is an arbitrary bounded subdomain of G with boundary $\partial G' \in C^4$. This follows directly from the definition of generalized solutions. Then we apply Theorem 2.2 with $s = 0$ in G'.

As above, let G_r be the intersection of G with the ball $\{x \in \mathbb{R}^N \mid |x| < r\}$. Then, by virtue of what has already been proved and (3.2), for any $u, v \in \mathcal{D}(B^*)$, we obtain

$$
\begin{aligned}
(B^*u, v)_{L_2(G)} - (u, B^*v)_{L_2(G)} &= (\mathcal{M}u, v)_{L_2(G)} - (u, \mathcal{M}v)_{L_2(G)} \\
&= \lim_{r\to\infty} \big((\mathcal{M}u, v)_{L_2(G_r)} - (u, \mathcal{M}v)_{L_2(G_r)}\big) \\
&= \lim_{r\to\infty} \int_{\gamma_r} |\,\mathrm{A}(x)\nu(x)| \left(\left(\frac{\partial u}{\partial \mu}\right)(x)\overline{v(x)} - u(x)\overline{\left(\frac{\partial v}{\partial \mu}\right)(x)} \right) dx,
\end{aligned}
\tag{3.7}
$$

$$\gamma_r = \{x \in \mathbb{R}^N \mid |x| = r\} \cap G$$

(here, we have used the fact that u, v vanish on ∂G). Note that, for $r > R$, the unit conormal vector $\mu(x)$ which appears in the square brackets in (3.7) has the same form as in the case of the expression $\mathcal{L}$. To prove the theorem, it is necessary to show that the limit on the right-hand side of (3.7) is equal to zero.

Acting absolutely similarly, one can also write equality (3.7) with $\mathcal{M}$ replaced by $\mathcal{L}$ and B^* replaced by $A^* = A$. In this case, in view of the fact that the operator A is Hermitian, the left-hand side of the corresponding equality is equal to zero. Finally, for any $u_1, v_1 \in \mathcal{D}(A)$, we get

$$\lim_{r\to\infty} \int_{\gamma_r} |\,\mathrm{A}(x)\nu(x)| \left(\left(\frac{\partial u_1}{\partial \mu}\right)(x)\overline{v_1(x)} - u_1(x)\overline{\left(\frac{\partial v_1}{\partial \mu}\right)(x)} \right) dx = 0.$$

This implies that the fact that the limit on the right-hand side of (3.7) is equal to zero will be established if we prove that, for any $v \in \mathcal{D}(B^*)$ and a nonnegative cutoff function $\chi \in C^\infty(\tilde{G})$ equal to zero for $|x| \le R_1$ and to one for $|x| \ge R_2$ ($R < R_1 < R_2$), the product $\chi(x)v(x)$ belongs to $\mathcal{D}(A)$. Let us prove this assertion.

Let $u \in C_0^2(\tilde{G}, \mathrm{b})$. Then

$$\mathcal{M}(\chi u) = \chi \mathcal{M}u + \mathcal{M}_\chi u = \chi \mathcal{L}u + \mathcal{M}_\chi u.$$

Here, $\mathcal{M}_\chi$ is a differential expression of the first order all coefficients of which contain (as factors) the first and second derivatives of χ and, therefore, vanish for

$|x| \geq R_2$. Taking this decomposition into account, for $v \in \mathcal{D}(B^*)$, according to the definition of B^*, we can write

$$\begin{aligned}(u, \chi B^* v)_{L_2(G)} &= (\chi u, B^* v)_{L_2(G)} = (B(\chi u), v)_{L_2(G)} \\ &= (\mathcal{M}(\chi u), v)_{L_2(G)} \\ &= (\chi \mathcal{L} u, v)_{L_2(G)} + (\mathcal{M}_\chi u, v)_{L_2(G)} \\ &= (\mathcal{L} u, \chi v)_{L_2(G)} + (u, (\mathcal{M}_\chi)^+ v)_{L_2(G)} \end{aligned} \tag{3.8}$$

for any $u \in C_0^2(\tilde{G}, \mathrm{b})$. In (3.8), $(\mathcal{M}_\chi)^+$ denotes the expression formally adjoint to $\mathcal{M}_\chi$; the required transfer is possible because u, v vanish on ∂G.

Since v locally belongs to W_2^2, we have $(\mathcal{M}_\chi)^+ v \in L_2(G)$ and, therefore, $\chi B^* v - (\mathcal{M}_\chi)^+ v = h \in L_2(G)$.

It follows from (3.8) that, for any $u \in C_0^2(\tilde{G}, \mathrm{b})$, we have $(\mathcal{L} u, \chi v)_{L_2(G)} = (u, h)_{L_2(G)}$, whence $\chi v \in \mathcal{D}(A^*) = \mathcal{D}(A)$. The required inclusion and, hence, the theorem are proved. □

All comments made in Remark 3.1 concerning smoothness are obviously true in the case of unbounded G. Theorem 3.3 can also be proved for (b) from Remark 3.2. At the same time, it is worth noting that if G is unbounded, then the operator A may be not semibounded and, certainly, its spectrum will be discrete only in special cases.

4 Differential Operators in $\mathbb{R}^N$

4.1 The Operator of Multiplication

In this section, an important role is played by the operators of multiplication (this becomes clear in Subsection 3). Therefore, it seems reasonable first to investigate the properties of these operators.

Let R be an abstract space of points x, let $\mathfrak{R}$ be a σ-algebra of its sets, and let $\mathfrak{R} \ni \alpha \mapsto \mu(\alpha) \in [0, \infty]$ be a σ-finite measure. In the space $H = L_2(R, \mathfrak{R}, d\mu)$, for a given complex-valued function a measurable with respect to $\mathfrak{R}$ and finite almost everywhere, we define the operator of multiplication by this function as follows:

$$\begin{gathered} H \supseteq \mathcal{D}(A) \ni f(x) \mapsto (Af)(x) = a(x) f(x) \in H, \\ \mathcal{D}(A) = \{ f \in H \mid a(x) f(x) \in H \}. \end{gathered} \tag{4.1}$$

Its domain $\mathcal{D}(A)$ is dense in H. Indeed, any function from H vanishing on $\alpha_n = \{x \in R \mid |a(x)| > n\}$ for some $n \in \mathbb{N}$ belongs to $\mathcal{D}(A)$. Furthermore, for any $f \in H$, $\mathcal{D}(A) \ni f(x) \chi_{R \setminus \alpha_n}(x) \to f(x)$ as $n \to \infty$ in H, since $\mu(\alpha_n) \to 0$ as $n \to \infty$ (χ_α is the indicator of the set α). It is easy to show that the operator A is normal and that A^* can be constructed similarly, according to the function $\overline{a(x)}$; it is bounded (selfadjoint) if and only if a is essentially bounded (real-valued).

The resolvent $R_z(A)$ is the operator of multiplication by the function $(a(x)-z)^{-1}$, where $z \in \mathbb{C}$ is such that this function is essentially bounded. This means that the resolution of the identity corresponding to A can be represented in the form

$$H \ni f(x) \mapsto (E(\alpha)f)(x) = \chi_\alpha(a(x))f(x) = \chi_{a^{-1}(\alpha)}(x)f(x),$$

where $a^{-1}(\alpha)$ denotes the complete preimage of the set α under the mapping a.

Let B be a normal operator similar to A constructed in terms of the function b. The operators A and B commute in the sense of Section 13.5, i.e., their resolutions of the identity are commuting. We suggest the reader to prove these simple assertions (see Exercises 13.4.1, 13.5.2, and 13.6.1).

4.2 Perturbation of an Operator

In this subsection, we proceed to the investigation of differential operators generated by an r-order expression $\mathcal{L}$ (1.1) defined on the whole $\mathbb{R}^N$ ($N \in \mathbb{N}$). Recall that the minimal operator L is defined in $L_2(\mathbb{R}^N)$ as the closure of the operator $L_2(\mathbb{R}^N) \supset C_0^r(\mathbb{R}^N) \ni u \mapsto L'u = \mathcal{L}u \in L_2$. Here, we consider only the case where $\mathcal{L}$ is formally selfadjoint. The operator L is Hermitian. Below, we present some conditions that should be imposed on the coefficients of $\mathcal{L}$ to ensure its selfadjointness.

For elliptic $\mathcal{L}$, a theorem similar to Theorem 3.3 can also be established in the case under consideration. For the second-order expressions, it can be formulated as follows:

Theorem 4.1. *Let $\mathcal{L}$ be a formally selfadjoint second-order elliptic expression (1.1) whose coefficients $a_\alpha \in C^{|\alpha|+2}(\mathbb{R}^N)$ are real. Suppose that the corresponding minimal operator L in $L_2(\mathbb{R}^N)$ is selfadjoint.*

Let $\mathcal{M}$ be another expression of the same form as $\mathcal{L}$ which coincides with $\mathcal{L}$ for $|x| \geq R$, where $R > 0$, and let M be the corresponding minimal operator. It is stated that M is selfadjoint.

Proof. It repeats the proof of Theorem 3.3 and appears to be even simpler because, instead of Theorem 2.2, one should clearly use Theorem 2.1 on smoothing of solutions inside a domain. The role of γ_r is now played by the sphere with radius r centered at the origin. □

4.3 Expressions with Constant Coefficients

Consider an r-order expression $\mathcal{L} = \mathcal{L}^+$ with constant coefficients defined on the space $\mathbb{R}^N$ ($N \in \mathbb{N}$). In this case, the minimal operator L is always selfadjoint: Indeed, speaking somewhat inaccurately one may say that after Fourier transformation, this operator turns into the operator of multiplication and the latter is selfadjoint as indicated in Subsection 1.

For given $\mathcal{L}$, we construct the polynomial

$$\mathcal{L}(\xi) = \sum_{|\alpha|\le r} a_\alpha (i\xi)^\alpha \tag{4.2}$$

$$((i\xi)^\alpha = (i\xi_1)^{\alpha_1} \dots (i\xi_N)^{\alpha_N},\ \xi = (\xi_1, \dots, \xi_N) \in \mathbb{R}^N).$$

Theorem 4.2. *The minimal operator L generated by a formally selfadjoint expression (1.1) with constant coefficients is selfadjoint. Its spectrum coincides with the closure of the set of values of polynomial (4.2)*

$$S(L) = \{\mathcal{L}(\xi) \mid \xi \in \mathbb{R}^N\}^\sim.$$

Proof. Consider the Schwartz space $\mathcal{S}(\mathbb{R}^N)$ which consists of the functions $\varphi \in C^\infty(\mathbb{R}^N)$ decreasing as $|x| \to \infty$ together with all their derivative faster than any power $|x|^{-n}$ $(n \in \mathbb{N})$ (see Section 11.3 and Section 14.4). Obviously, for any $\varphi \in \mathcal{S}(\mathbb{R}^N) \subset L_2(\mathbb{R}^N)$, we have $\mathcal{L}\varphi \in L_2(\mathbb{R}^N)$ and, therefore, parallel with L', one can consider an operator $L_2(\mathbb{R}^N) \supset \mathcal{S}(\mathbb{R}^N) \ni \varphi \mapsto L''\varphi = \mathcal{L}\varphi \in L_2(\mathbb{R}^N)$ in $L_2(\mathbb{R}^N)$ which is, in fact, an extension of the operator L'.

As is known, C_0^∞ is dense in $\mathcal{S}(\mathbb{R}^N)$ in the topology of this space (see Section 11.3). Recall that, for any $\varphi \in \mathcal{S}(\mathbb{R}^N)$, one can set $\varphi_n(x) = \chi_n(x)\varphi(x) \in C_0^\infty(\mathbb{R}^N)$, where $\chi_1(x)$ is a function from $C_0^\infty(\mathbb{R}^N)$ equal to one in a neighbourhood of zero and vanishing for $|x| \ge 1$ and $\chi_n(x) = \chi_1(n^{-1}x)$ $(n \in \mathbb{N})$. It is easy to show that $\varphi_n \to \varphi$ as $n \to \infty$ in the topology of $\mathcal{S}(\mathbb{R}^N)$, i.e., φ_n is the required approximating sequence. Note that convergence in $\mathcal{S}(\mathbb{R}^N)$ implies convergence in $L_2(\mathbb{R}^N)$. Therefore, in view of this denseness, we can write $\tilde{L}'' = \tilde{L}' = L$.

We now recall the well-known facts from Section 11.3. Consider the direct Fourier transformation $\mathcal{S}(\mathbb{R}^N) \ni \varphi(x) \mapsto \hat{\varphi}(\xi) \in \mathcal{S}(\mathbb{R}^N)$. It continuously maps the space $\mathcal{S}(\mathbb{R}^N)$ of functions of x onto the whole space $\mathcal{S}(\mathbb{R}^N)$ of functions of ξ. The inverse Fourier transformation denoted by $^\vee$ acts similarly. The direct and inverse Fourier transformations unitary map the corresponding spaces $L_2(\mathbb{R}^N)$ into each other. Further, for any $\varphi \in \mathcal{S}(\mathbb{R}^N)$, we have $(\mathcal{L}\varphi)^\wedge(\xi) = \mathcal{L}(\xi)\hat{\varphi}(\xi)$ $(\xi \in \mathbb{R}^N)$, where $\mathcal{L}(\xi)$ has the form (4.2).

Therefore, the operator of multiplication by the polynomial $\mathcal{L}(\xi)$ is the unitary image $\hat{L}''$ of the operator L'' in the space $L_2(\mathbb{R}^N)$ (in ξ) with the domain $\mathcal{D}(\hat{L}'') = \mathcal{S}(\mathbb{R}^N)$. This operator differs from the operator of multiplication defined by (4.1) (being its restriction), but, nevertheless, one can easily prove the selfadjointness of its closure. Thus, let $z \in \mathbb{C}\setminus\mathbb{R}$. Then $\mathcal{R}(\hat{L}'' - z\mathbb{1}) = \mathcal{S}(\mathbb{R}^N)$, by virtue of the fact that $(\mathcal{L}(\xi) - z)^{-1}\varphi(\xi) \in \mathcal{S}(\mathbb{R}^N)$ for any $\varphi \in \mathcal{S}(\mathbb{R}^N)$. Hence, $\mathcal{R}(\hat{L}'' - z\mathbb{1})$ is dense in $L_2(\mathbb{R}^N)$ and, therefore, $(\hat{L}'')^\sim$ is selfadjoint. But $(\hat{L}'')^\sim = (\tilde{L}'')^\wedge$, whence it follows that $\tilde{L}'' = L$ is selfadjoint.

The statement of the theorem concerning the spectrum of L follows from the fact that, for $z \in \mathbb{C}$, the function $\mathbb{R}^N \ni \xi \mapsto (\mathcal{L}(\xi) - z)^{-1} \in \mathbb{C}$ is bounded if and only if $z \notin \{\mathcal{L}(\xi) \mid \xi \subset \mathbb{R}^N\}^\sim$. □

By combining this theorem with Theorem 4.1, we conclude that if $\mathcal{L}$ is a second-order elliptic expression of the form indicated in Theorem 4.1 whose coefficients become constant outside a certain ball in $\mathbb{R}^N$, then the corresponding minimal operator L' is selfadjoint. In the next subsection, a much more general theorem is proved by using more powerful methods.

4.4 Semibounded Expressions

Theorem 4.3. *Let $\mathcal{L}$ be a formally selfadjoint second-order elliptic expression (1.1) whose coefficients $a_\alpha \in C^{2+[N/2]}(\mathbb{R}^N)$ are real. Suppose that $\mathcal{L}$ is semibounded below (on finite functions) and the leading coefficients $a_\alpha(x)$, $|\alpha| = 2$, are bounded for $x \in \mathbb{R}^N$. Then the minimal operator L corresponding to $\mathcal{L}$ is selfadjoint.*

Let us explain that the semiboundedness of the expression $\mathcal{L}$ from below is understood in the sense of the inequality

$$(\mathcal{L}u, u)_{L_2(G)} \geq \alpha \|u\|^2_{L_2(G)} \qquad (u \in C_0^\infty(\mathbb{R}^N)) \tag{4.3}$$

for some $\alpha \in \mathbb{R}$.

One can easily formulate sufficient conditions that should be imposed on the coefficients of $\mathcal{L}$ to guarantee the validity of (4.3). Thus, in the important case where $\mathcal{L}$ is the Schrödinger expression with real-valued potential q, i.e.,

$$(\mathcal{L}u)(x) = -(\Delta u)(x) + q(x)u(x) \qquad (x \in \mathbb{R}^N), \tag{4.4}$$

it suffices to require the semiboundedness from below of the potential q itself, i.e., the validity of the following inequality: There exists $c \in \mathbb{R}$ such that $q(x) \geq c$ ($x \in \mathbb{R}^N$) (we stress that this condition is sufficient but not necessary).

According to the general Theorem 4.3 as applied to the Schrödinger expression, the condition that should be imposed on the smoothness of the potential are rather restrictive, i.e., $q \in C^{2+[N/2]}(\mathbb{R}^N)$. At the same time, the proof of this theorem for the Schrödinger expression presented below works, e.g., for $q \in C(\mathbb{R}^N)$. Smoothness requirements can also be weakened for general $\mathcal{L}$.

Proof. It is based on the hyperbolic criterion of selfadjointness formulated above as Theorem 13.8.3 and on classical results on the solvability of the Cauchy problems for hyperbolic equations (see [Pet1]).

According to this criterion, to establish the selfadjointness of L, it is necessary to investigate the Cauchy problem for a vector function $\varphi(t)$ with values in $L_2(\mathbb{R}^N)$

$$\begin{gathered}\left(\frac{d^2\varphi}{dt^2}\right)(t) + L\varphi(t) = 0 \qquad (t \in [0, T]), \\ \varphi(T) = \varphi_0, \qquad \varphi'(T) = \varphi_1,\end{gathered} \tag{4.5}$$

and to prove that there exists a linear set Φ dense in $L_2(\mathbb{R}^N)$ such that, for some $b > 0$, any $T \in (0, b)$, and $\varphi_0, \varphi_1 \in \Phi$, problem (4.5) is strongly solvable. To do this, one must require the semiboundedness of the Hermitian operator L but, in the case under consideration, this restriction is already imposed as condition (4.3).

Consider the Cauchy problem for the hyperbolic equation

$$\begin{gathered}\left(\frac{d^2u}{dt^2}\right)(x,t) + (\mathcal{L}u)(x,t) = 0, \\ u(x,T) = \varphi_0, \quad \left(\frac{du}{dt}\right)(x,T) = \varphi_1 \qquad (\varphi_0, \varphi_1 \in C_0^\infty(\mathbb{R}^N))\end{gathered} \tag{4.6}$$

on a segment $[0,T]$ $(T > 0)$.

By virtue of the indicated classical results in the theory of partial differential equations, problem (4.6) possesses a solution $u \in C^2(\mathbb{R}^N \times [0,T])$ and this solution $u(x,t)$ is finite in x for any $t \in [0,T]$ (due to the finite rate of propagation of perturbations for (4.6)). Therefore, $u(\cdot,t)$ can be interpreted as a vector function $\varphi(t)$ with values in $L_2(\mathbb{R}^N)$ giving a strong solution of the Cauchy problem for (4.5) with $\Phi = C_0^\infty(\mathbb{R}^N)$. Thus, we have proved the required solvability of (4.5). The operator L is selfadjoint. □

The smoothness conditions imposed on the coefficients of $\mathcal{L}$ are sufficient for the solvability of (4.6) (they can be weakened for the Schrödinger expression as described above). The boundedness of the leading coefficients of $\mathcal{L}$ is ensured by (4.6) because one can easily show that, for finite initial data, the solution $u(x,t)$ is finite in x for any fixed t. The boundedness condition can also be weakened by assuming that the leading coefficients of $\mathcal{L}$ may increase as $|x| \to \infty$ and indicating an admissible growth rate.

4.5 Nonsmooth Potentials

In the theorems on the selfadjointness of L presented above, it was assumed that the coefficients of $\mathcal{L}$ are as smooth as required. At the same time, it is often necessary to establish selfadjointness in the case where these coefficients are not smooth. Below, we present several simple results for the operator L generated by the Schrödinger expression (4.4), i.e., for the Schrödinger operator.

The first result is absolutely elementary and applicable in many other cases (this will be evident). It is based on the following trivial remark:

Let A and B be, respectively, an operator with dense domain and a bounded operator acting on a Hilbert space H. Then $(A + B)^* = A^* + B^*$ (Theorem 12.3.2). Therefore, if we additionally require that A and B be selfadjoint, then $A + B$ $(\mathcal{D}(A + B) = \mathcal{D}(A))$ will also be selfadjoint.

Consider the Schrödinger differential expression $\mathcal{L}$ with real-valued potential q which locally belongs to L_2 $(q \in L_{2,\mathrm{loc}}(\mathbb{R}^N))$, i.e., the restriction of q to any ball in $\mathbb{R}^N$ belongs to L_2 on this ball. In this case, the ordinary definition of the Schrödinger operator as the minimal operator L given in Subsection 2 remains true, since $q(x)u(x) \in L_2(\mathbb{R}^N)$ for $u \in C_0^2(\mathbb{R}^N)$ (L' and L are clearly Hermitian).

Note that, obviously, the condition $q \in L_{2,\text{loc}}(\mathbb{R}^N)$ is also a necessary condition for this definition of the minimal operator.

For the Laplace expression $\mathcal{L}_0 = -\Delta$, the corresponding minimal operator L_0 is selfadjoint in $L_2(\mathbb{R}^N)$ as follows from Theorem 4.2. Let q be an essentially bounded function (i.e., $(\exists c > 0): |q(x)| \leq C$ for almost all $x \in \mathbb{R}^N$). Then $q \in L_{2,\text{loc}}(\mathbb{R}^N)$ and the definition of the Schrödinger operator L according to (4.4) is correct. However, in this case, the operator $L_2(\mathbb{R}^N) \ni f(x) \mapsto q(x)f(x) \in L_2(\mathbb{R}^N)$ is bounded and selfadjoint. Therefore, by using the reasonings presented above, we conclude that L is also selfadjoint.

Let us now consider a more complicated situation based on the application of the Rellich-Kato theorem (Theorem 13.10.1). The following theorem is true:

Theorem 4.4. *Let q be a real-valued locally square summable potential such that $q \in L_p(\mathbb{R}^N)$ with $p > N/2$. Then the Schrödinger operator is selfadjoint.*

In proving this theorem, we use the following generalization of the Hölder inequality: Let $p, p_1, p_2 \in [1, \infty]$ be such that

$$\frac{1}{p} = \frac{1}{p_1} + \frac{1}{p_2}. \tag{4.7}$$

If $f_1 \in L_{p_1},\ f_2 \in L_{p_2}$, then $f_1(x)f_2(x) \in L_p$ and we can write the inequality

$$\|f_1 f_2\|_{L_p} \leq \|f_1\|_{L_{p_1}} \|f_2\|_{L_{p_2}}, \tag{4.8}$$

where $L_r = L_r(R, \mathfrak{R}, d\mu)$ and μ is a nonnegative σ-finite measure given on a σ-algebra $\mathfrak{R}$ of subsets of the space R.

Indeed, by using the ordinary Hölder inequality applied to $|f_1(x)|^p$ and $|f_2(x)|^p$, we obtain

$$\begin{aligned}\|f_1 f_2\|_{L_p} &= \Big(\int_R |f_1(x)f_2(x)|^p d\mu(x)\Big)^{1/p} \\ &\leq \Big[\Big(\int_R |f_1(x)|^{pp_1p^{-1}} d\mu(x)\Big)^{p/p_1} \Big(\int_R |f_2(x)|^{p\cdot(p_1p^{-1})'} d\mu(x)\Big)^{((p_1p^{-1})')^{-1}}\Big]^{p^{-1}} \\ &= \|f_1\|_{L_{p_1}} \|f_2\|_{L_{p_2}}.\end{aligned} \tag{4.9}$$

Here, $1 = (p_1p^{-1})^{-1} + ((p_1p^{-1})')^{-1}$ is the standard relation between the original index $p_1p^{-1} \in [1, \infty]$ and its dual $(p_1p^{-1})'$. In (4.9), we have used the easily verified equality $p(p_1p^{-1})' = p_2$. □

Note that, as a result of a series of subsequent applications of (4.8), we arrive at the following generalization of this equality:

Let $p_1, \ldots, p_n,\ p \in [1, \infty]$ be such that $p^{-1} = p_1^{-1} + \cdots + p_n^{-1}$ $(n = 2, 3, \ldots)$. If $f_1 \in L_{p_1}, \ldots, f_n \in L_{p_n}$, then $f_1 \ldots f_n \in L_p$ and the following inequality holds:

$$\|f_1 \ldots f_n\|_{L_p} \leq \|f_1\|_{L_{p_1}} \cdots \|f_n\|_{L_{p_n}}. \tag{4.10}$$

In what follows, we also use the following Hausdorff-Young inequality (see, e.g., [ReS2]) introduced for the Fourier transformation $\mathcal{S}(\mathbb{R}^N) \ni f(x) \mapsto \hat{f}(\xi) \in \mathcal{S}(\mathbb{R}^N)$ and valid for $p \in [2,\infty]$:

$$\|f\|_{L_p(\mathbb{R}^N)} \le (2\pi)^{N/p-N/p'}\|\hat{f}\|_{Lp'(\mathbb{R}^N)} \quad (p_{-1}+(p')^{-1}=1, \quad f\in\mathcal{S}(\mathbb{R}^N)). \tag{4.11}$$

(This inequality generalizes the well-known Parseval equality valid for $p = 2$.)

Proof of Theorem 4.4. As already mentioned, we use Theorem 13.10.1, taking the operator L_0 generated by the Laplace expression $\mathcal{L}_0 = -\Delta$ as A and the operator of multiplication by the potential q as B. Let us prove that B is arbitrarily small as compared to L_0 in the sense of Section 13.10, i.e., for any $a > 0$, there exists $b = b(a) > 0$ such that

$$\begin{gathered}\|qf\|_{L_2(\mathbb{R}^N)} \le a\|L_0 f\|_{L_2(\mathbb{R}^N)} + b\|f\|_{L_2(\mathbb{R}^N)}\\ (f\in\mathcal{D}(\mathcal{L}_0)\subseteq\mathcal{D}(B)).\end{gathered} \tag{4.12}$$

This implies that the operator L is selfadjoint.

We prove (4.12). Let $f \in \mathcal{S}(\mathbb{R}^N)$. We choose α such that $1/2 = 1/p + 1/\alpha$. By using (4.8) with $p = 2$, $p_1 = p$, and $p_2 = \alpha$ and then (4.11) (here, $\alpha \ge 2$), we get, for fixed $t > 0$, that

$$\begin{aligned}\|qf\|_{L_2(\mathbb{R}^N)} &\le \|q\|_{L_p(\mathbb{R}^N)}\|f\|_{L_\alpha(\mathbb{R}^N)}\\ &\le c_1\|q\|_{L_p(\mathbb{R}^N)}\|\hat{f}\|_{L_{\alpha'}(\mathbb{R}^N)}\\ &\le c_2\left\|\frac{1+t|\xi|^2}{1+t|\xi|^2}\hat{f}(\xi)\right\|_{L_{\alpha'}(\mathbb{R}^N)}\\ &\le c_2\|(1+t|\xi|^2)^{-1}\|_{L_p(\mathbb{R}^N)}\|(1+t|\xi|^2)\hat{f}\|_{L_2(\mathbb{R}^N)}\\ &\quad (c_1=(2\pi)^{N/\alpha-N/\alpha'},\quad c_2=c_1\|q\|_{L_p}).\end{aligned} \tag{4.13}$$

Note that we have used here inequality (4.8) for the second time, setting $p = \alpha'$, $p_1 = p$, and $p_2 = 2$ (since $2^{-1} = p^{-1} + \alpha^{-1}$, $1 = \alpha^{-1} + (\alpha')^{-1}$, we have $(\alpha')^{-1} = p^{-1} + 2^{-1}$ and, therefore, equality (4.7) is satisfied).

Chahging the variables in the integral expression for the norm

$$\|(1+t|\xi|^2)^{-1}\|_{L_p(\mathbb{R}^N)}$$

according to the formula $\xi = \sqrt{t}\eta$, we obtain

$$t^{-N/2p}\|(1+|\eta|^2)^{-1}\|_{L_p(\mathbb{R}^N)} = t^{-N/2p}c_3$$

(due to the requirement $p > N/2$, we have $(1+|\eta|^2)^{-1} \in L_p(\mathbb{R}^N)$. By substituting this expression in (4.13) and using the Parseval equality, we can continue estimate

(4.13) as follows:

$$\begin{aligned}\|qf\|_{L_2(\mathbb{R}^N)} &\le c_2c_3t^{-N/2p}\|((1-t\Delta)f)^\frown\|_{L_2(\mathbb{R}^N)}\\ &= c_2c_3t^{-N/2p}\|(1-t\Delta)f\|_{L_2(\mathbb{R}^N)}\\ &\le c_2c_3t^{-N/2p}(\|f\|_{L_2(\mathbb{R}^N)}+t\|\Delta f\|_{L_2(\mathbb{R}^N)})\\ &= c_2c_3t^{(2p-N)/2p}\|L_0f\|_{L_2(\mathbb{R}^N)}+c_2c_3t^{-N/2p}\|f\|_{L_2(\mathbb{R}^N)}.\end{aligned} \tag{4.14}$$

Taking here sufficiently small $t > 0$, we arrive at inequality (4.12) for the functions $f \in \mathcal{S}(\mathbb{R}^N)$ (since $2p - N > 0$). Further, by constructing the closure of L_0 from its restriction to $\mathcal{S}(\mathbb{R}^N)$ and passing to the limit in (4.12), we conclude that $\mathcal{D}(B) \supseteq \mathcal{D}(L_0)$ and, hence, (4.12) is satisfied. □

4.6 The Schrödinger Operator as a Form Sum

As mentioned above, the inclusion $q \in L_{2,\text{loc}}(\mathbb{R}^N)$ is a necessary requirement for the possibility of the construction of the minimal operator L for given expression (4.4). At the same time, we often encounter Schrödinger expressions with more singular potentials q that do not belong to $L_{2,\text{loc}}(\mathbb{R}^N)$. In this case, it is necessary to solve the problem of realization of expression (4.4) (a "formal operator") as a selfadjoint operator in the space $L_2(\mathbb{R}^N)$. Sometimes, this can be done by using the theory of semibounded bilinear forms developed in Section 14.8. We now present a simple example to illustrate the possibility of a such realization on the basis of the KLMN theorem.

Consider the Schrödinger expression in $\mathbb{R}^3$ with the singular potential

$$q(x) = 1 - \frac{1}{|x|^\alpha}, \qquad \alpha > 0. \tag{4.15}$$

For the inclusion $q \in L_{2,\text{loc}}(\mathbb{R}^3)$, it is necessary to require that $\alpha < 3/2$. At the same time, the case where $\alpha \in [3/2, 2)$ and $q \notin L_{2,\text{loc}}(\mathbb{R}^3)$ is also of interest. We proceed as follows:

Denote by $C^\infty_{0,0}(\mathbb{R}^3)$ the collection of functions from $C^\infty_0(\mathbb{R}^3)$ which vanish in a certain neighbourhood of the origin. Consider the bilinear form

$$\begin{aligned}C^\infty_{0,0}(\mathbb{R}^3) \times C^\infty_{0,0}(\mathbb{R}^3) \ni \langle f, g\rangle \mapsto a(f,g)&\\ &= ((-\Delta + \mathbb{I})f, g)_{L_2(\mathbb{R}^3)}\\ &= \int_{\mathbb{R}^3}\Big(\sum_{j=1}^{3}(D_jf)(x)\overline{(D_jg)(x)} + f(x)\overline{g(x)}\Big)dx \in \mathbb{C}.\end{aligned}$$

This form is clearly positive and closable. Let $\tilde{a}$ be its closure.

Parallel with $\tilde{a}$, we consider an Hermitian nonpositive bilinear form defined by the equality

$$C_{0,0}^{\infty}(\mathbb{R}^3) \times C_{0,0}^{\infty}(\mathbb{R}^3) \ni \langle f, g\rangle \mapsto b'(f,g) = -\int_{\mathbb{R}^3} \frac{1}{|x|^{\alpha}} f(x)\overline{g(x)}dx \in \mathbb{C}.$$

The form b' is subordinate to a in the sense of (14.8.19), i.e., for any $p > 0$, there exists $q = q(p) > 0$ such that

$$|b'(f,f)| \le pa(f,f) + qb'(f,f) \qquad (f \in C_{0,0}^{\infty}(\mathbb{R}^3)). \tag{4.16}$$

We suggest the reader to prove the following inequality (by passing to spherical coordinates):

$$\int_{\mathbb{R}^3} \frac{1}{4|x|^2}|f(x)|^2 dx \le \int_{\mathbb{R}^3} (\sum_{j=1}^{3} |(D_j f)(x)|^2) dx \qquad (f \in C_{0,0}^{\infty}(\mathbb{R}^3)). \tag{4.17}$$

We fix $\alpha \in [3/2, 2)$. Then, for any $p > 0$, there exists $\varepsilon > 0$ such that $|x|^{-\alpha} \le p(4|x|^2)^{-1}$ for $|x| < \varepsilon$. Taking this estimate and (4.17) into account, we get

$$\begin{aligned} |b'[f]| = \int_{\mathbb{R}^3} \frac{1}{|x|^{\alpha}}|f(x)|^2 dx &= \int_{\{|x|<\varepsilon\}} \cdots + \int_{\{|x|\ge\varepsilon\}} \cdots \\ &\le p\int_{\{|x|<\varepsilon\}} \frac{1}{4|x|^2}|f(x)|^2 dx + \frac{1}{\varepsilon^{\alpha}}\int_{\{|x|\ge\varepsilon\}} |f(x)|^2 dx \\ &\le pa[f] + \frac{1}{\varepsilon^2}\|f\|^2_{L_2(\mathbb{R}^3)} \end{aligned}$$

for any $f \in C_{0,0}^{\infty}(\mathbb{R}^3)$.

Thus, (4.16) is established. By passing in (4.16) to the limit, we establish this inequality for $f \in \mathcal{D}(\tilde{a})$. Note that, in the inequality obtained, a is replaced by $\tilde{a}$ and b' is replaced by the form b obtained from b' by extending by continuity. As a result, we obtain an inequality of the form (14.8.19) sufficient for the application of Theorem 14.8.4. According to this theorem, the form $\tilde{a} + b$ is semibounded and closed and, therefore, it is possible to apply Theorem 14.8.2, which associates $\tilde{a}+b$ with a selfadjoint operator A. The latter is regarded as the operator associated with the Schrödinger expression (4.4) with potential (4.15). It is connected with Δ and q by a relation of the form (14.8.23).

In many cases, the theory of bilinear forms enables one to associate the Schrödinger expression with an operator even in the case where the role of potential q is played by a generalized function (such situations are typical, for example, of quantum field theory). We clarify this assertion by a simple example of a Schrödinger operator with δ-shaped potential in the one-dimensional case ($N = 1$). We follow the scheme used in of Section 14.8.4.

In the space $L_2(\mathbb{R})$, we consider the selfadjoint (by Theorem 4.2) minimal operator $A = L$ generated by the differential expression $(\mathcal{L}u)(x) = -u''(x)+u(x)$ $(x \in \mathbb{R})$. Given this operator, we construct the form $a(f,g) = (Af,g)_{L_2(\mathbb{R})}$ $(f,g \in \mathcal{D}(A))$ and denote its closure by $\tilde{a}$. On the basis of the arguments presented in Section 14.8.4, we conclude that $H_+ = \mathcal{D}(\tilde{a}) = \mathcal{D}(A^{1/2}) = W_2^1(\mathbb{R}) \subset C(\mathbb{R})$ (the last inclusion was established by using the embedding theorem (see Section 14.4)). Let us define an operator

$$H_+ = W_2^1(\mathbb{R}) \ni \varphi \mapsto \mathcal{B}\varphi = \varphi(0)\delta_0 \in W_2^{-1}(\mathbb{R}) = H_-,$$

where δ_0 is the δ-function concentrated at the origin, while $\mathcal{B}$ lies in $\mathcal{L}(H_+, H_-))$ and is nonnegative with respect to H_0.

It is not difficult to show that the form b associated with $\mathcal{B}$ satisfies estimate (14.8.22) with arbitrarily small $p \in (0,1)$ and some $q = q(p) \in [0,\infty)$. Indeed, due to the embedding theorem, $(\exists c \in (0,\infty))$ $(\forall u \in C_0^\infty(\mathbb{R}))$:

$$|u(0)| \leq c\|u\|_{W_2^1((-1,1))} \leq c\|u\|_{W_2^1(\mathbb{R})} = c\big(\|u'\|^2_{L_2(\mathbb{R})} + \|u\|^2_{L_2(\mathbb{R})}\big)^{1/2}.$$

In this inequality, we replace $u(x)$ by $u_\varepsilon(x) = u(\varepsilon x) \in C_0^\infty(\mathbb{R})$ $(\varepsilon \in (0,\infty))$. After differentiation and the change of the variables $\varepsilon x = y$, we obtain

$$|u(0)| = |u_\varepsilon(0)| \leq c\big(\varepsilon\|u'\|^2_{L_2(\mathbb{R})} + \varepsilon^{-1}\|u\|^2_{L_2(\mathbb{R})}\big)^{1/2}.$$

This inequality leads to (14.8.22) with arbitrarily small $p \in (0,1)$ and, therefore, the form sum $A\dot{+}B$ is defined. For this sum, one can write relation (14.8.23), which demonstrates that the point 0 is characterized by the compensation of the singularity of the operator $\mathcal{B}$ with the singularity of the operator $\mathcal{A}$.

The operator $A\dot{+}B$ thus constructed can be regarded as an operator realization of the differential expression $-u'' + u + \delta_0 u$. In this case, we have essentially used the fact that the Schrödinger operator is one-dimensional. Indeed, in the case where the dimension is greater than one, there are no required embedding theorems. We also stress that $B = 0$ in $\mathcal{D}(B) = \{\varphi \in W_2^1(\mathbb{R}) \mid \varphi(0) = 0\}$.

5 Expansion in Eigenfunctions and Green's Function of Elliptic Differential Operators

5.1 Generalized Eigenfunctions of Differential Operators

If a selfadjoint operator $A = L(\mathrm{b})$ is generated by an elliptic differential expression $\mathcal{L}$ with trivial boundary conditions in a bounded domain G, then, according to Theorem 3.2, its spectrum coincides with the sequence of real eigenvalues with finite-dimensional eigensubspaces. Thus, for the expansion in the eigenfunctions (vectors) of this operator, one can use relations (13.0.1) and (13.0.2) with the only difference that the sum in k is infinite.

However, in the case where a nondiscrete spectrum appears (an unbounded domain G, a nonelliptic expression $\mathcal{L}$, and more complicated boundary conditions), the situation becomes not so simple and expansions should be constructed according to the general scheme presented in Chapter 15.

Consider the general case of a formally selfadjoint r-order differential expression $\mathcal{L}$ (1.1) given on the whole space $\mathbb{R}^N$ whose coefficients are infinitely differentiable, i.e., $a_\alpha \in C^\infty(\mathbb{R}^N)$. Suppose that the corresponding minimal operator L is selfadjoint. To construct the expansion in its generalized eigenfunctions, one may consider the nuclear rigging of the space $L_2(\mathbb{R}^N)$ of the form

$$\Phi' = \mathcal{D}'(\mathbb{R}^N) \supseteq L_2(\mathbb{R}^N) = H_0 \supseteq \mathcal{D}(\mathbb{R}^N) = \Phi, \tag{5.1}$$

where $\mathcal{D}(\mathbb{R}^N) = C_0^\infty(\mathbb{R}^N)$ is the classical space of test functions. According to Theorem 14.4.6, this space is nuclear. Chain (5.1) is associated with the operator L in a standard way because $\mathcal{L}u \in C_0^\infty(\mathbb{R}^N)$ for any $u \in C_0^\infty(\mathbb{R}^N)$ and the mapping $\mathcal{D}(\mathbb{R}^N) \ni u \mapsto \mathcal{L}u \in \mathcal{D}(\mathbb{R}^N)$ is continuous. It is now possible to apply the general facts about expansions presented in Section 15.2 to the operator L and chain (5.1). Actually, in this case, one should not expect any significant simplifications as compared with the general situation of Chapter 15.

If the coefficients a_α of the expression $\mathcal{L}$ are not infinitely differentiable, then instead of the nuclear chain (5.1), one must use a properly constructed quasinuclear chain. Thus, if ($\exists l \in \mathbb{N} : \ l > N/2$) such that ($\forall \alpha$): $a_\alpha \in C^l(\mathbb{R}^N)$, then the required chain can be chosen in the form

$$\begin{aligned} H_- = W_2^{-l}(\mathbb{R}^N, p(x)dx) \supseteq L_2(\mathbb{R}^N) = H_0 \supseteq H_+ \\ = W_2^l(\mathbb{R}^N, p(x)dx) \supseteq \mathcal{D}(\mathbb{R}^N) = D, \end{aligned} \tag{5.2}$$

where the weight $p \in C^l(\mathbb{R}^N)$ approaches $+\infty$ as $|x| \to \infty$ with a rate sufficiently high to guarantee the quasinuclearity of the embedding $W_2^l(\mathbb{R}^N, p(x)dx) \subseteq L_2(\mathbb{R}^N)$ (the possibility of finding the required weight is ensured by Theorem 14.4.3). In (5.2), $\mathcal{D}(\mathbb{R}^N)$ is densely and continuously embedded in $W_2^l(\mathbb{R}^N, p(x)dx)$ and the mapping $\mathcal{D}(\mathbb{R}^N) \ni u \mapsto \mathcal{L}u \in W_2^l(\mathbb{R}^N, p(x)dx)$ is continuous. Thus, the quasinuclear chain (5.2) and the operator L are connected in a standard way. Therefore, it is possible to apply the general properties of expansions established in Section 15.2.

If the coefficients of $\mathcal{L}$ are less smooth than indicated, then the construction of an analogue of chain (5.2) becomes more difficult and we do not consider this case.

If $\mathcal{L}$ is defined in a domain $G \subset \mathbb{R}^N$ (bounded or not) with boundary ∂G, then, instead of the space $\mathcal{D}(\mathbb{R}^N)$ in chains (5.1) and (5.2), one can take the properly topologized set $C_0^\infty(G)$, while the positive space H_+ in (5.2) is clearly replaced by $W_2^l(G, p(x)dx)$ with a weight p selected to ensure the quasinuclearity of the embedding $H_+ \subseteq H_0$.

However, to take the influence of the boundary condition into account, it is necessary to select D in (5.2) more carefully. We clarify this remark for a second-order expression with trivial boundary condition. Let $a_\alpha \in C^l(\tilde{G})$ ($|\alpha| \le 2$), where $l > N/2$. We now take the space $W_2^l(G, p(x)dx)$ as H_+, and the class $C_0^{l+2}(\tilde{G}, \mathrm{b})$ composed of all functions from $C_0^{l+2}(\mathbb{R}^N)$ restricted to $\tilde{G}$ and taking zero values on ∂G as D. It is topologized in the following way: $C_0^{l+2}(\tilde{G}, \mathrm{b}) \ni u_n \to u \in C_0^{l+2}(\tilde{G}, \mathrm{b})$ as $n \to \infty$ if all u_n vanish outside a certain sufficiently large ball and their convergence to u in this ball is uniform together with all their derivatives up to $(l+2)$th order, inclusively (one can easily construct a topology that corresponds to this convergence). It is thus clear that $D \subseteq H_+$; this embedding and the mapping $C_0^{l+2}(G, \mathrm{b}) \ni u \mapsto \mathcal{L}u \in W_2^l(G, \mathrm{b}, p(x), dx)$ are continuous but D is not dense in H_+. Therefore, according to Remark 15.2.1, the quasinuclear chain associated with $L(\mathrm{b})$ in a standard way can be chosen in the form

$$\begin{aligned} H_- = W_2^{-l}(G, p(x)dx) \supseteq H_0 = L_2(G) \supseteq H_+ \\ = W_2^l(G, \mathrm{b}, p(x)dx) \supseteq C_0^{l+2}(\tilde{G}, \mathrm{b}) = D, \end{aligned} \tag{5.3}$$

where $W_2^{-l}(G, p(x)dx)$ denotes the corresponding negative space.

The choice of chain (5.1)–(5.3) suitable for the construction of expansions in generalized eigenfunctions of the operator L or $L(\mathrm{b})$ is obviously ambiguous. In this connection, we only note that chain (5.3) with $G = \mathbb{R}^N$ and $D = C_0^{l+2}(G)$ is suitable for constructing the expansion of the operator L (it differs from (5.2) by the space D).

General results of the theory of expansions are significantly simplified for elliptic expressions $\mathcal{L}$. This is connected with the theorems of local smoothing of generalized solutions of elliptic equations established for $r = 2$ in Section 2. Let us clarify this statement. Let $\mathcal{L}$ be a formally selfadjoint elliptic expression of second order in $\mathbb{R}^N$ and let L be the corresponding minimal operator in $L_2(\mathbb{R}^N)$ which is assumed to be selfadjoint. As already mentioned, to construct the expansion in generalized eigenfunctions of the operator L, one may use the following chain (a special case of (5.3)):

$$\begin{aligned} H_- = W_2^{-l}(\mathbb{R}^N, p(x)dx) \supseteq H_0 = L_2(\mathbb{R}^N) \supseteq H_+ \\ = W_2^l(\mathbb{R}^N, p(x)dx) \supseteq C_0^{l+2}(\mathbb{R}^N) = D. \end{aligned} \tag{5.4}$$

The generalized eigenfunction $\varphi \in W_2^{-l}(\mathbb{R}^N, p(x)dx)$ of the operator L corresponding to an eigenvalue λ, according to (15.2.2) and in agreement with the form (5.4) of the chain, satisfies the equality

$$(\varphi, \mathcal{L}u)_{L_2(\mathbb{R}^N)} = \lambda(\varphi, u)_{L_2(\mathbb{R}^N)} \qquad (u \in C_0^{l+2}(\mathbb{R}^N)). \tag{5.5}$$

In other words, φ is the generalized solution of the elliptic equation $(\mathcal{L} - \lambda\mathbb{I})\varphi = 0$ in $\mathbb{R}^N$ and, according to Theorem 2.1 (see also Remark 2.2), is a smooth function. The degree of its smoothness is determined by the smoothness of the coefficients a_α of the expression $\mathcal{L}$ (in this case, $f = 0 \in C^\infty(\mathbb{R}^N)$). Hence, if $a_\alpha \in C^{|\alpha|+l}(\mathbb{R}^N)$ ($|\alpha| \leq 2$), then $\varphi \in W^{l+2}_{2,\text{loc}}(\mathbb{R}^N)$, i.e., for any $\chi \in C_0^\infty(\mathbb{R}^N)$, the product $\chi\varphi \in W_2^{l+2}(\mathbb{R}^N)$. Consequently, by virtue of the embedding theorems, we have $\varphi \in C^{[l+2-N/2]}(\mathbb{R}^N)$.

A similar situation also takes place in the case where $\mathbb{R}^N$ is replaced by a domain G (bounded or not) with boundary ∂G, where we fix the trivial boundary conditions (b). We now consider the corresponding operator $A = L(\text{b})$ (which is assumed to be selfadjoint). As already explained, it is connected with the quasinuclear chain (5.3) in a standard way. The definition of the generalized eigenfunction $\varphi \in W_2^{-l}(G, p(x)dx)$ corresponding to λ has the form of relation (5.5) in which u runs through $C_0^{l+2}(\tilde{G}, \text{b})$. By applying Theorem 2.2 (on local smoothing up to the boundary of a domain) instead of Theorem 2.1, we conclude that φ is a smooth function vanishing on ∂G. If $a_\alpha \in C^{|\alpha|+l}(\mathbb{R}^N)$ ($|\alpha| \leq 2$) and $\partial G \in C^{l+2}$, then $\varphi \in W^{l+2}_{2,\text{loc}}(G, \text{b})$, i.e., for any $\chi \in C_0^\infty(\mathbb{R}^N)$, the product $(\chi \restriction \tilde{G})\varphi \in W_2^{l+2}(G)$ and, hence, $\varphi \in C^{[l+2-N/2]}(\tilde{G})$; $\varphi \restriction \partial G = 0$.

It seems necessary to emphasize that the imposed smoothness requirements are excessive. It is also clear that, for fixed chain (5.4) (or (5.3)), the smoother the coefficients a_α, the smoother the function φ.

As a general conclusion, we can state that, for the operator L or $L(\text{b})$ generated by an elliptic differential expression and "good" boundary conditions, the generalized eigenfunctions are smooth functions satisfying these boundary conditions.

A similar situation is typical of ordinary differential expressions (it is discussed in Section 6) as well as of some special nonelliptic expressions (in particular, of the class of so-called hypoelliptic expressions, for which it is possible to establish theorems similar to the theorem on smoothing).

Thus, for selfadjoint operators L and $L(\text{b})$ generated by elliptic expressions $\mathcal{L}$, the theory of expansions developed in Section 15.2 becomes much simpler because generalized eigenfunctions turn, in this case, into ordinary functions satisfying (b). It seems also useful to consider another approach to this situation, namely, to prove that $L, L(\text{b})$ are Càrleman operators and apply the general results of Section 15.4. These results give a more complete picture of expansions in the elliptic case. This approach is studied in Subsection 3.

5.2 Green's Function (Kernel of the Resolvent)

It seems reasonable first to clarify the character of the resolvent of an elliptic operator (i.e., of an operator generated by an elliptic expression $\mathcal{L}$ with "elliptic" (b), e.g., trivial for $r = 2$). First, we consider an expression $\mathcal{L}$ given in the whole space $\mathbb{R}^N$ (and, clearly, by virtue of Section 2, only in the case of second-order expressions).

Thus, let $\mathcal{L}$ be an elliptic formally selfadjoint second-order expression in $\mathbb{R}^N$ with sufficiently smooth coefficients and let L be the corresponding minimal operator which is supposed to be selfadjoint. Its resolvent $R_z = (L - z\mathbb{I})^{-1}$, where z is nonreal or, more generally, lies outside the spectrum $S(L)$ of the operator L, is a bounded operator in $L_2(\mathbb{R}^N)$. Therefore, R_z satisfies the conditions of the kernel theorem in the form of Theorem 14.6.3 with Remark 14.6.2.

Thus, consider the chain

$$H_- = W_2^{-l}(\mathbb{R}^N,\ p(x)dx) \supseteq H_0 = L_2(\mathbb{R}^N) \supseteq H_+ = W_2^l(\mathbb{R}^N,\ p(x)dx), \tag{5.6}$$

where $l > N/2$ and the weight p is such that the embedding $H_+ \subseteq H_0$ is quasinuclear; in the spaces of the chain, we define an involution $L_2(\mathbb{R}^N) \ni f(x) \mapsto \overline{f(x)} \in L_2(\mathbb{R}^N)$. The bilinear form of the operator R_z admits representation (14.6.23), i.e.,

$$\begin{gathered}(R_z u, v)_{L_2(\mathbb{R}^N)} = (\Gamma_z, v(x)\overline{u(y)})_{L_2(\mathbb{R}^{2N})} \\ (u, v \in W_2^l(\mathbb{R}^N, p(x)dx)),\end{gathered} \tag{5.7}$$

where $\Gamma_z \in (W_2^{-l}(\mathbb{R}^N, p(x)dx)) \otimes (W_2^{-l}(\mathbb{R}^N, p(x)dx))$. Note that here we have taken into account the equalities $L_2(\mathbb{R}^N) \otimes L_2(\mathbb{R}^N) = L_2(\mathbb{R}^{2N})$ and $(f \otimes g)(x,y) = f(x)g(y)$ $(f, g \in L_2(\mathbb{R}^N))$.

The generalized kernel Γ_z $(z \notin S(L))$ is called the kernel of the resolvent or Green's function of the operator L.

It is often convenient to write this generalized function with indicating the variables upon which it acts, i.e., $\Gamma_z = \Gamma_z(x,y)$. Somewhat formally, relation (5.7) can be rewritten in the form

$$(R_z u, v)_{L_2(\mathbb{R}^N)} = \int_{\mathbb{R}^N}\int_{\mathbb{R}^N} \Gamma_z(x,y) u(y)\overline{v(x)}\,dx dy. \tag{5.8}$$

At the same time, relation (5.8) can be made meaningful in a certain sense.

First, we show that Γ_z can be regarded as an element of the negative Sobolev space the with respect to variables $x, y \in \mathbb{R}^N$.

Lemma 5.1. *The following embeddings are true:*

$$\begin{aligned}W_2^{-2l}(\mathbb{R}^N, p(x)p(y)dxdy) &\supseteq (W_2^{-l}(\mathbb{R}^N, p(x)dx)) \otimes (W_2^{-l}(\mathbb{R}^N, p(y)dy)) \\ &\supseteq L_2(\mathbb{R}^{2N}) \supseteq (W_2^l(\mathbb{R}^N, p(x)dx)) \otimes (W_2^l(\mathbb{R}^N, p(y)dy)) \\ &\supseteq W_2^{2l}(\mathbb{R}^{2N}, p(x)p(y)dxdy),\end{aligned} \tag{5.9}$$

where $l \in \mathbb{Z}_+$ and $p \in C(\mathbb{R}^N)$ is a nonnegative weight. In (5.9), each space is dense in its left neighbour in the chain and the corrresponding embedding is continuous.

Proof. Denote $H_1 = W_2^1(\mathbb{R}^N, p(x)dx)$ and $H_2 = W_2^{2N}(\mathbb{R}^{2N}, p(x)p(y)dxdy)$. Clearly, it suffices to show that $H_2 \subseteq H_1 \otimes H_1$ densely and continuously. Let $u_j, v_j \in C_0^\infty(\mathbb{R}^N)$ $(j = 1, \dots, n)$. We set

$$u(x,y) = \sum_{j=1}^n u_j(x)v_j(y) \in C_0^\infty(\mathbb{R}^{2N}) \qquad (x, y \in \mathbb{R}^N). \tag{5.10}$$

We have

$$\begin{aligned}
(u,u)_{H_1\otimes H_1} &= \sum_{j,k=1}^n (u_j,u_k)_{H_1}(v_j,v_k)_{H_1} \\
&= \sum_{j,k=1}^n \Big(\sum_{|\alpha|\le l}\int_{\mathbb{R}^N} (D_x^\alpha u_j)(x)\overline{(D_x^\alpha u_k)(x)}p(x)dx\Big)\Big(\sum_{|\beta|\le l}\int_{\mathbb{R}^N} (D_x^\beta v_j)(y)\overline{(D_y^\beta v_k)(y)}p(y)dy\Big) \\
&= \sum_{j,k=1}^n \sum_{|\alpha|,|\beta|\le l}\int_{\mathbb{R}^{2N}} (D_x^\alpha u_j)(x)(D_y^\beta v_j)(y)\overline{(D_x^\alpha u_k)(x)(D_y^\beta v_k)(y)}p(x)dxdy \\
&= \sum_{j,k=1}^n \sum_{|\alpha|,|\beta|\le l}\int_{\mathbb{R}^{2N}} (D_x^\alpha D_y^\beta u_j(x)v_j(y))\overline{(D_x^\alpha D_y^\beta u_k(x)v_k(y))}p(x)p(y)dxdy \\
&= \sum_{|\alpha|,|\beta|\le l} (D_x^\alpha D_y^\beta u, D_x^\alpha D_y^\beta u)_{L_2(\mathbb{R}^{2N}, p(x)p(y)dxdy)} \\
&\le \sum_{|\gamma|\le 2l} \|D_{x,y}^\gamma u\|^2_{L_2(\mathbb{R}^{2N}, p(x)p(y)dxdy)}
\end{aligned} \tag{5.11}$$

(here, as usual, D_x, D_y, and $D_{x,y}$ denote derivatives with respect to x, y, and x and y). As u_j, v_j, and $n \in \mathbb{N}$ change, functions of the form (5.10) run through the set dense in $H_1 \otimes H_1$ and H_2. Therefore, inequality (5.11) implies the embedding $H_2 \subseteq H_1 \otimes H_1$ and its continuity. It is also clear that H_2 is dense in $H_1 \otimes H_1$. □

Lemma 5.1 yields the inclusion

$$\begin{aligned}
\Gamma_z \in (W_2^{-l}(\mathbb{R}^N, p(x)dx)) &\otimes (W_2^{-l}(\mathbb{R}^N, p(x)dx)) \\
&\subseteq W_2^{-2l}(\mathbb{R}^{2N}, p(x)p(y)dxdy) \qquad (z \notin S(L)).
\end{aligned} \tag{5.12}$$

For a given second-order elliptic expression $\mathcal{L}$ acting upon the functions $u(x)$ $(x \in \mathbb{R}^N)$, we now construct a "double" expression $\mathcal{M}$ acting upon the functions $u(x,y)$ according to the law

$$(\mathcal{M}u)(x,y) = (\mathcal{L}_x u)(x,y) + (\mathcal{L}_y u)(x,y) \qquad (x, y \in \mathbb{R}^N). \tag{5.13}$$

For the leading part $\mathcal{M}_0$ of the expression $\mathcal{M}$, we have

$$(\mathcal{M}_0 u)(x,y) = (\mathcal{L}_{0,x} u)(x,y) + (\mathcal{L}_{0,y} u)(x,y)$$

and, therefore, by changing D_x^α and D_y^β in $\mathcal{M}_0$ by ξ^α and η^β $(\xi, \eta \in \mathbb{R}^N)$, respectively, we find $\mathcal{M}_0(\langle\xi,\eta\rangle) = \mathcal{L}_{0,x}(\xi) + \mathcal{L}_{0,y}(\eta)$. Since $\mathcal{L}_0(\xi) \le 0$ and is equal to zero only if $\xi = 0$ (due to the ellipticity condition, see (1.2)), we have $\mathcal{M}_0(\langle\xi,\eta\rangle) \le 0$ and $\mathcal{M}_0(\langle\xi,\eta\rangle) = 0$ only in the case where $\langle\xi,\eta\rangle = 0$. Thus, $\mathcal{M}$ is a formally selfadjoint elliptic expression of the type investigated in Section 2 given in $\mathbb{R}^{2N}$.

Lemma 5.2. *Let $\mathcal{L}$ be a formally selfadjoint second-order elliptic expression in $\mathbb{R}^N$ whose coefficients $a_\alpha \in C^{2l}(\mathbb{R}^N)$ $(l > N/2)$. Assume that the corresponding minimal operator L is selfadjoint in $L_2(\mathbb{R}^N)$. Then the Green's function Γ_z of this operator satisfies the relation*

$$(\Gamma_z, (\mathcal{M} - 2\bar{z}\mathbb{I})u)_{L_2(\mathbb{R}^{2N})} = 2\int_{\mathbb{R}^N} \overline{u(x,x)}dx$$
$$(u(x,y) \in C_0^{2l+2}(\mathbb{R}^{2N}), \quad z \notin S(L)). \tag{5.14}$$

Proof. If suffices to establish equality (5.14) for functions u of the form $u(x,y) = u_1(x)v_1(y)$, where $u_1, v_1 \in C_0^{2l+2}(\mathbb{R}^N)$. Indeed, in this case, it is also true for linear combinations of these functions and, by using uniformly finite linear combinations of this sort, one can approximate any function $u \in C_0^{2l+2}(\mathbb{R}^{2N})$ in the sense of $C^{2l+2}(\mathbb{R}^{2N})$.

Taking (5.13), (5.7), and the fact that the coefficients of $\mathcal{L}$ are real into account, we obtain

$$\begin{aligned}
&(\Gamma_z, (\mathcal{M} - 2\bar{z}\mathbb{I})u_1(x)v_1(y))_{L_2(\mathbb{R}^{2N})} \\
&= (\Gamma_z, ((\mathcal{L}_x - \bar{z}\mathbb{I})u_1(x))v_1(y))_{L_2(\mathbb{R}^{2N})} + (\Gamma_z, u_1(x)((\mathcal{L}_y - \bar{z}\mathbb{I})v_1(y)))_{L_2(\mathbb{R}^{2N})} \\
&= (R_z\bar{v}_1, (\mathcal{L} - \bar{z}\mathbb{I})u_1)_{L_2(\mathbb{R}^N)} + (R_z\overline{(\mathcal{L} - \bar{z}\mathbb{I})v_1}, u_1)_{L_2(\mathbb{R}^N)} \\
&= (\bar{v}_1, R_z^*(L - \bar{z}\mathbb{I})u_1)_{L_2(\mathbb{R}^N)} + (R_z(L - z\mathbb{I})\bar{v}_1, u_1)_{L_2(\mathbb{R}^N)} \\
&= 2(\bar{v}_1, u_1)_{L_2(\mathbb{R}^N)} = 2\int_{\mathbb{R}^N} \overline{u_1(x)v_1(x)}dx. \qquad \square
\end{aligned}$$

It follows from equality (5.14) that Γ_z is a generalized solution of a certain elliptic equation. Indeed, we equip $C_0^{2l+2}(\mathbb{R}^{2N})$ with a natural topology as follows: $u_n \to u$ as $n \to \infty$ in $C_0^{2l+2}(\mathbb{R}^{2N})$ if u_n are uniformly finite and uniformly converge to u together with all their derivatives up to the $(2l+2)$th order, inclusively. In $C_0^{2l+2}(\mathbb{R}^{2N})$, we now define an antilinear continuous functional D ("diagonal") by the formula

$$(\mathrm{D}, u)_{L_2(\mathbb{R}^{2N})} = \int_{\mathbb{R}^N} \overline{u(x,x)}dx \qquad (u(x,y) \in C_0^{2l+2}(\mathbb{R}^{2N})). \tag{5.15}$$

Then, in accordance with Section 2, relation (5.14) means that $\Gamma_z \in W_2^{-2l}(\mathbb{R}^{2N}, p(x)p(y)dxdy)$ is a generalized solution of the elliptic equation

$$(\mathcal{M} - 2z\mathbb{I})u = 2\,\mathrm{D} \tag{5.16}$$

inside $\mathbb{R}^{2N}$.

This and the result established in Section 2 enable us to prove the following theorem:

Theorem 5.1. *Let $\mathcal{L}$ be a formally selfadjoint second-order elliptic expression in $\mathbb{R}^N$ whose coefficients $a_\alpha \in C^{|\alpha|+2l}(\mathbb{R}^N)$ ($|\alpha| \le 2$), where $l > N/2$ is fixed. Assume that the corresponding minimal operator L is selfadjoint in $L_2(\mathbb{R}^N)$. Then its Green's function Γ_z ($z \notin S(L)$) outside the diagonal $\gamma = \{\langle x, y\rangle \in \mathbb{R}^{2N} \mid x = y\}$ is an ordinary function $\Gamma_z(x,y)$ which locally belongs to the space W_2^{2l+2} with respect to the variable $\langle x, y\rangle$. If the value of one of the variables y or x is fixed, then $\Gamma_z(x,y)$ satisfies the equations*

$$\begin{aligned} (\mathcal{L}_x - z\mathbb{I})\Gamma_z(x,y) &= \delta_y \qquad (y \in \mathbb{R}^N), \\ (\mathcal{L}_y - z\mathbb{I})\Gamma_z(x,y) &= \delta_x \qquad (x \in \mathbb{R}^N). \end{aligned} \tag{5.17}$$

Let us clarify this formulation. In the language of the definitions introduced in Section 2, the assertion that Γ_z locally belongs to W_2^{2l+2} outside γ means that the product $\chi\Gamma_z$ belongs to $W_2^{2l+2}(\mathbb{R}^{2N})$ for any cutoff function $\chi \in C_0^\infty(\mathbb{R}^{2N})$ vanishing in a neighbourhood of the diagonal γ. Further, in the first equations in (5.17), δ_y denotes the δ-function concentrated at a fixed point y and the equation itself is equivalent to the validity of the following equality:

$$(\Gamma_z(x,y),\ (\mathcal{L}_x - \bar{z}\mathbb{I})v)_{L_2(\mathbb{R}^N)} = \overline{v(y)} \qquad (v \in C_0^{l+2}(\mathbb{R}^N),\ y \in \mathbb{R}^N). \tag{5.18}$$

If $\mathbb{R}^N \ni x \neq y$, then $(\mathcal{L}_x - z\mathbb{I})\Gamma_z(x,y) = 0$. The second equation in (5.17) has exactly the same sense.

Proof. As follows from relation (5.15), the generalized function D is concentrated on the diagonal γ. Therefore, outside γ, equation (5.16) turns into $(\mathcal{M} - 2x\mathbb{I})u = 0$. By applying Theorem 2.1 to the elliptic expression $\mathcal{M} - 2z\mathbb{I}$ in $\mathbb{R}^{2N}\backslash\gamma$, we conclude that Γ_z locally belongs to W_2^{2l+2} outside γ.

To deduce the first equation in (5.17), we substitute the function $((\mathcal{L}_x - z\mathbb{I})v)(x)$, where $v \in C_0^{l+2}(\mathbb{R}^N)$, for v in (5.17).This gives

$$(\Gamma_z, ((\mathcal{L}_x - z\mathbb{I})v)(x)\overline{u(y)})_{L_2(\mathbb{R}^{2N})} = (R_z u, (L - z\mathbb{I})v)_{L_2(\mathbb{R}^N)}$$

$$(u, R_z^*(L - z\mathbb{I})v)_{L_2(\mathbb{R}^N)} = (u, v)_{L_2(\mathbb{R}^N)} \qquad (u \in W_2^1(\mathbb{R}^N, p(x)dx)). \tag{5.19}$$

The left-hand side of (5.19) can be rewritten in the form $((\Gamma_z, (\mathcal{L}_x - z\mathbb{I})v(x))_{L_2(\mathbb{R}^N)}, \overline{u(y)})_{L_2(\mathbb{R}^N)}$, since it is easy to see that if $A \in H_- \otimes H_-$, then, for any $v \in H_+$, there exists a vector from H_- denoted by $(A, v)_{H_0}$ such that

$$(A, v \otimes u)_{H_0 \otimes H_0} = ((A, v)_{H_0}, u)_{H_0} \ (u \in H_+).$$

As a result, we get

$$((\Gamma_z, (\mathcal{L}_x - z\mathbb{I})v(x))_{L_2(\mathbb{R}^N)}, \overline{u(y)})_{L_2(\mathbb{R}^N)} = (u, v)_{L_2(\mathbb{R}^N)},$$

whence, in view of the arbitrariness of u, we arrive at the desired inequality (5.18).

The second equation in (5.17) is deduced similarly. □

REMARK 5.1. Let us clarify the situation with the smoothness of the function $\Gamma_z(x,y)$ outside γ. First, we consider the case where integer $l > N/2$. If, in this case, $a_\alpha \in C^{|\alpha|+m}(\mathbb{R}^N)$ $(|\alpha| \le 2)$ for some $\mathbb{N} \ni m \ge 2l$, then $\Gamma_z(x,y)$, regarded as a generalized function, belongs to the spaces (5.12) and locally belongs to W_2^{m+2} with respect to the variable $\langle x,y\rangle$ outside γ. This can be easily seen from the proof of Theorem 5.1.

To investigate the singularities of the kernel $\Gamma_z(x,y)$ for $x = y$, one must consider relation (5.12), where only one requirement is imposed on l, namely, $l > N/2$ (it is clear that the weight p determines the character of singularity only at ∞). These singularities can be described more exactly by using the notion of fundamental solution. Here, we omit the corresponding consideration but cite the final result. For sufficiently small $|x-y|$, we have

$$\Gamma_z(x,y) = e_z(x,y) + F_z(x,y), \tag{5.20}$$

where $e_z(x,y)$ is a fundamental solution for the expression $\mathcal{L} - z\mathbb{I}$ and $F_z(x,y)$ is a smooth function of the point $\langle x,y\rangle$. Note that the fundamental solution and, hence, $\Gamma_z(x,y)$ have singularities of the form $|x-y|^{2-N}$ for $N > 2$ and $\log|x-y|$ for $N = 2$. As far as fundamental solutions are concerned, see [Ber].

Equality (5.20) is proved quite easily. Indeed, the difference $\Gamma_z(x,y) - e_z(x,y)$ is a generalized solution of the homogeneous equation $(\mathcal{M} - 2z\mathbb{I})u = 0$, since the δ-functions that appear on the right-hand side disappear as a result of subtraction. Then we apply Theorem 2.1 on smoothing.

It follows from Theorem 5.1 that formula (5.18) is true for the functions $u, v \in L_2(\mathbb{R}^N)$ whose supports are situated at a certain positive distance. Due to the indicated character of the singularities of the kernel $\Gamma_z(x,y)$ for $x = y$ implied by (5.20), this formula remains true for a larger supply of functions u and v.

Let us now present a brief analysis of the situation that appears if $\mathcal{L}$ is considered in a certain (bounded or unbounded) domain $G \subset \mathbb{R}^N$ with boundary ∂G and trivial boundary conditions. Let $L(\mathrm{b})$ be the corresponding selfadjoint operator in $L_2(G)$ and let R_z $(z \notin S(L(\mathrm{b})))$ be its resolvent. For this resolvent, one can repeat all arguments presented above provided that the spaces $W_2^l(\mathbb{R}^N, p(x)dx)$ are replaced by properly chosen $W_2^l(G, p(x)dx)$ (in this connection, see Section 14.4).

Equation (5.16) is now considered in the domain $G \times G$ with boundary $\partial(G \times G) = (\partial G \times \tilde{G}) \cup (\tilde{G} \times \partial G)$ and it is easy to see that equality (5.14) holds for the functions $u(x,y) \in C^{2l+2}(\tilde{G} \times \tilde{G})$ finite at ∞ and vanishing in a certain neighbourhood U of an "angular set" in $\partial G \times \partial G$ and in the remaining part of the boundary $\partial(G \times G)$. Thus, equation (5.16) is now satisfied up to the piece $\partial(G \times G) \cap (\mathbb{R}^{2N} \setminus U)$ of the boundary of the domain $G \times G$, where the trivial boundary conditions are imposed. This enables us to use Theorem 2.2 (on smoothness up to the boundary) instead of Theorem 2.1 and make the following conclusion:

In this case, Green's function $\Gamma_z(x,y)$ $(x,y \in \tilde{G}, x \neq y)$ has the same type of smoothness as indicated above and, in addition, satisfies (b), *i.e., vanishes if at least one of the points x and y lies on ∂G.*

Here, it is necessary to assume that $\partial G \in C^{2l+2}$.

Note that results similar to those mentioned in this subsection are also true for elliptic expressions of any order and more general boundary conditions, i.e., for situations where it is possible to prove analogues of the theorems presented in Sections 1 and 2.

In conclusion, we present a simple example. Consider the Laplace expression $\mathcal{L} = -\Delta$ in $\mathbb{R}^3$. The corresponding minimal operator L (the Laplace operator) is selfadjoint in $L_2(\mathbb{R}^3)$ by virtue of Theorem 4.2. According to the same theorem, $S(L) = [0, \infty)$. The Green's function of this operator has the form

$$\Gamma_z(x,y) = \frac{1}{4\pi} \frac{e^{i\sqrt{z}|x-y|}}{|x-y|} \qquad (x, y \in \mathbb{R}^3, x \neq y; \quad z \notin [0, \infty)). \tag{5.21}$$

In proving this fact, we use the following simple procedure, which is also applicable to other differential expressions with constant coefficients in $\mathbb{R}^N$. As in the proof of Theorem 4.2, we consider the Fourier transformation $\hat{}$ that maps the operator L into the operator $\hat{L}$ of multiplication by the function $|\xi|^2$ $(\xi \in \mathbb{R}^3)$ in the space $L_2(\mathbb{R}^3)$. The resolvent of the operator $\hat{L}$ is given by the operator of multiplication by $(|\xi|^2 - z)^{-1}$ $(z \notin [0, +\infty))$. Performing the inverse Fourier transformation, we arrive at relation (5.21).

5.3 The Càrleman Property of Elliptic Operators

As mentioned in Subsection 1, it is convenient to construct expansion in generalized eigenfunctions of elliptic operators according to the following procedure: First, it is necessary to show that the operator under consideration possesses the Càrleman property and then use the general properties of expansions constructed for Càrleman operators (Section 15.4). As in the case of Green's function, we first present the corresponding results for expressions defined in $\mathbb{R}^N$.

Let us prove that sufficiently high powers R_z^m ($z \notin S(L)$ is fixed) of resolvents of the considered operators L are integral Càrleman operators, i.e., for certain dense set of functions $f \in L_2(\mathbb{R}^N)$, we have the following representation with measurable kernel $K(x,y)$:

$$(R_z^m f)(x) = \int_{\mathbb{R}^N} K(x,y) f(y) dy \qquad (x \in \mathbb{R}^N), \tag{5.22}$$

where the Càrleman condition is satisfied in the following form:

$$\int_{\mathbb{R}^N} |K(x,y)|^2 dx \leq d_n < \infty \quad (|y| \leq n, \quad n \in \mathbb{N}). \tag{5.23}$$

The role of the dense set is played by the finite functions f from $L_2(\mathbb{R}^N)$.

Theorem 5.2. *Let $\mathcal{L}$ be a formally selfadjoint second-order elliptic expression in $\mathbb{R}^N$ whose coefficients $a_\alpha \in C^{|\alpha|+2m-2}(\mathbb{R}^N)$ $(|\alpha| \le 2)$, where $\mathbb{N} \ni m > N/4$ is fixed. Suppose that the corresponding operator L is selfadjoint in $L_2(\mathbb{R}^N)$ and let R_z $(z \notin S(L))$ be its resolvent. Then its power R_z^m $(z \notin S(L))$ is an integral Càrleman operator* (5.22) *with estimates* (5.23) *(and, consequently, L is a Càrleman operator).*

Proof. The proof is split in several steps.

(1) Let us prove that $R_z^m f \in W_{2,\text{loc}}^{2m}(\mathbb{R}^N)$ for any $f \in L_2(\mathbb{R}^N)$. Denote

$$f_0 = f,\ f_{j+1} = R_z f_j = R_z^j f \in L_2(\mathbb{R}^N)\ (j = 0, \dots, m-1).$$

The function f_{j+1} is a generalized solution of the equation $((\mathcal{L} - z\mathbb{1})u)(x) = f_j(x)$ $(x \in \mathbb{R}^N)$ inside $\mathbb{R}^N$, i.e.,

$$\begin{aligned}(f_{j+1}, (\mathcal{L} - \bar{z}\mathbb{1})v)_{L_2(\mathbb{R}^N)} &= (R_z f_j, (L - \bar{z}\mathbb{1})v)_{L_2(\mathbb{R}^N)} \\ &= (f_j, v)_{L_2(\mathbb{R}^N)} \qquad (j = 0, \dots m-1)\end{aligned}$$

for any $v \in C_0^2(\mathbb{R}^N)$. Therefore, as a result of a series of successive applications of Theorem 2.1, we obtain

$$f_0 \in L_2(\mathbb{R}^N) \Rightarrow f_1 \in W_{2,\text{loc}}^2(\mathbb{R}^N) \Rightarrow f_2 \in W_{2,\text{loc}}^4(\mathbb{R}^N) \Rightarrow \dots \Rightarrow f_m \in W_{2,\text{loc}}^{2m}(\mathbb{R}^N)$$

as required.

(2) We now fix a bounded domain $G' \subset \mathbb{R}^N$ with sufficiently smooth boundary. It is stated that $\exists c_1 = c_1(G') > 0$:

$$\|(R_z^m f) \restriction G'\|_{W_2^{2m}(G')} \le c_1 \|f\|_{L_2(\mathbb{R}^N)} \qquad (f \in L_2(\mathbb{R}^N)). \tag{5.24}$$

Indeed, consider an operator $H_1 = L_2(\mathbb{R}^N) \ni f \mapsto Af = (R_z^m f) \restriction G' \in W_2^{2m}(G') = H_2$ that maps the whole Hilbert space H_1 into the Hilbert space H_2. These operators satisfy the conditions of the Banach closed graph theorem. In the case where $H_1 = H_2$, this is Theorem 12.3.4. At the same time, its proof (without any changes) can be generalized to the case of different H_1 and H_2 (the notion of closeness is generalized in a natural way).

Inequality (5.24) means that A is bounded. Therefore, by virtue of the indicated theorem, this inequality will be proved if we show that A is closed, i.e., that if $H_1 \ni f_n \to f$ and $Af_n \to g$ in H_2 as $n \to \infty$, then $Af = g$. In other words, one must prove that if $f_n \to f$ in $L_2(\mathbb{R}^N)$ and $(R_z^m f_n) \restriction G' \to g$ in $W_2^{2m}(G')$, then $(R_z^m f) \restriction G' = g$. But this is evident because, by virtue of the continuity of the operator R_z^m in $L_2(\mathbb{R}^N)$, we have $R_z^m f_n \to R_z^m f$ and, hence, $(R_z^m f_n) \restriction G' \to (R_z^m f) \restriction G'$ in $L_2(G')$. Thus, (5.24) is true.

(3) Since $2m > N/2$, according to the embedding theorems, we have

$$W_2^{2m}(G') \subset C(\tilde{G}') \quad \text{and} \quad \|u\|_{C(\tilde{G}')} \le c_2 \|u\|_{W_2^{2m}(G')}$$
$$(u \in W_2^{2m}(G'), \quad c_2 = c_2(G')).$$

Therefore, it follows from these relations and (5.24) that

$$|(R_z^m f)(x)| \le c_2 \|(R_z^m f) \restriction G'\|_{W_2^{2m}(G')} \le c_2 \|f\|_{L_2(\mathbb{R}^N)}$$
$$(x \in \tilde{G}', \quad c_3 = c_3(G')). \tag{5.25}$$

We fix $x \in \tilde{G}'$ and consider the linear functional

$$L_2(\mathbb{R}^N) \ni f \mapsto l_x(f) = (R_z^m f)(x) \in \mathbb{C}.$$

By virtue of (5.25), it is well-defined and continuous and, moreover, $\|l_x\| \le c_3$. Therefore, according to the Riesz theorem, it admits the representation $l_x(f) = (f, h_x)_{L_2(\mathbb{R}^N)}$ $(f \in L_2(\mathbb{R}^N))$, where $h_x \in L_2(\mathbb{R}^N)$ and $\|h_x\|_{L_2(\mathbb{R}^N)} = \|l_x\| \le c_3$. In other words, we have

$$(R_z^m f)(x) = \int_{\mathbb{R}^N} \overline{h_x(y)} f(y)\,dy,$$
$$\int_{\mathbb{R}^N} |h_x(y)|^2 dy \le c_3^2 = c_3^2(G') \qquad (f \in L_2(\mathbb{R}^N),\ x \in \tilde{G}'). \tag{5.26}$$

(4) If instead of G' we take a bounded domain $G'' \supset G'$, then it is also representable in the form (5.26) with a new function $h_x(y)$ which is constructed unambiguously and, therefore, coincides with the old function for $x \in \tilde{G}'$. By taking a sequence of balls with radii n centered at the origin and applying to each of these balls the procedure described above, we finally arrive at a function $h_x(y)$ $(x, y \in \mathbb{R}^N)$ such that

$$(R_z^m f)(x) = \int_{\mathbb{R}^N} \overline{h_x(y)} f(y)\,dy \qquad (f \in L_2(\mathbb{R}^N), \quad x \in \mathbb{R}^N),$$
$$\int_{\mathbb{R}^N} |h_x(y)|^2 dy \le d_n \qquad (|x| \le n; \quad n \in \mathbb{N}). \tag{5.27}$$

(5) It is now convenient to rewrite (5.27) with z, f, x, and y replaced by $\bar{z}, g, y$, and x, respectively. As a result, we obtain

$$(R_{\bar{z}}^m g)(y) = \int_{\mathbb{R}^N} \overline{h_y(x)} g(x)\,dx \qquad (g \in L_2(\mathbb{R}^N), \quad y \in \mathbb{R}^N),$$
$$\int_{\mathbb{R}^N} |h_y(x)|^2 dx \le d_n \qquad (|y| \le n; \quad n \in \mathbb{N}). \tag{5.28}$$

We introduce a kernel $K(x, y)$ by setting $K(x, y) = h_y(x)$ $(x, y \in \mathbb{R}^N)$. Then (5.28) yields

$$(R_{\bar{z}}^m g)(y) = \int_{\mathbb{R}^N} \overline{K(x, y)} g(x) dx \qquad (g \in L_2(\mathbb{R}^N), \quad y \in \mathbb{R}^N),$$
$$\int_{\mathbb{R}^N} \int_{G'} |K(x, y)|^2 dx dy < \infty \tag{5.29}$$

for any bounded domain $G' \subset \mathbb{R}^N$.

Assume that $f \in L_2(\mathbb{R}^N)$ and is finite and let $g \in L_2(\mathbb{R}^N)$. By using (5.29), we obtain

$$(R_z^m f, g)_{L_2(\mathbb{R}^N)} = (f, R_{\bar{z}}^m g)_{L_2(\mathbb{R}^N)} = \int_{\mathbb{R}^N} f(y) \Big(\int_{\mathbb{R}^N} K(x, y) \overline{g(x)} dx \Big) dy$$
$$= \int_{\mathbb{R}^N} \Big(\int_{\mathbb{R}^N} K(x, y) f(y) dy \Big) \overline{g(x)} dx \tag{5.30}$$

(the possibility of changing the order of integration in this relation is justified by the last relation in (5.29) and the fact that f is finite). In view of the arbitrariness of g in (5.30), we get representation (5.22) for finite $f \in L_2(\mathbb{R}^N)$. Since $K(x, y) = h_y(x)$, the second relation (inequality) in (5.28) yields (5.23). □

Thus, the elliptic operator L appearing in Theorem 5.2 is a Càrleman operator and, therefore, expansions in its generalized eigenfunctions can be constructed by applying the theory of expansions of Càrleman operators. It is not necessary to repeat here the relevant general facts from Section 15.4 as applied to the operator L. We only note that, by virtue of (5.23), the weight p may be chosen to be bounded in every bounded domain $\mathbb{R}^N$. This follows from the proof of Theorem 15.4.3.

It is worth noting that the generalized eigenfunction φ of the operator L corresponding to an eigenvalue λ is a smooth solution of the equation $(\mathcal{L}\varphi)(x) = \lambda\varphi(x)$ $(x \in \mathbb{R}^N)$, and its "generalized nature" manifests itself only in the fact that it belongs not to $L_2(\mathbb{R}^N)$ but to $L_2(\mathbb{R}^N, p^{-1}(x)dx)$, where $p(x) \geq 1$ is a weight that increases as $|x| \to \infty$ sufficiently rapidly. Hence, the function φ may increase as $|x| \to \infty$ and the character of its growth can be estimated. Here, we do not pay attention to this problem and only note that the proof of Theorem 15.4.3, in fact, establishes the required connection between the growth of the weight p and behaviour of the kernel K in representation (5.22) (the required type of its behaviour may be clarified). The smoothness of φ is governed by the smoothness of the coefficients of $\mathcal{L}$ and is described in Subsection 1.

The spectral kernel $P(x, y; \lambda)$ (the kernel of the operator $P(\lambda)$ of generalized projection (see (15.4.9)) plays a key role in the spectral problems for elliptic operators and, therefore, we investigate it in more detail.

Theorem 5.3. *Assume that the conditions of Theorem 5.2 are satisfied. Then the spectral kernel $P(x, y; \lambda)$ of the operator L for fixed λ regarded as a function of the point $\langle x, y\rangle \in \mathbb{R}^{2n}$ locally belongs to the space W_2^{2m} and satisfies the equations*

$$\begin{aligned} \mathcal{L}_x P(x, y; \lambda) &= \lambda P(x, y; \lambda) \qquad (x \in \mathbb{R}^N), \\ \mathcal{L}_y P(x, y; \lambda) &= \lambda P(x, y; \lambda) \qquad (y \in \mathbb{R}^N) \end{aligned} \tag{5.31}$$

provided that the value of one of the variables (y or x) is fixed.

Proof. By virtue of (15.4.10), we have $P(\cdot, \cdot; \lambda) \in L_{2,\text{loc}}(\mathbb{R}^{2N})$. It is not difficult to show that this function is a generalized solution inside $\mathbb{R}^{2N}$ of the equation

$$(\mathcal{M} - 2\lambda \mathbb{I})u = 0, \tag{5.32}$$

where $\mathcal{M}$ is the elliptic differential expression (5.13).

Indeed, it is necessary to show that the equality

$$(P, (\mathcal{M} - 2\lambda \mathbb{I})u)_{L_2(\mathbb{R}^{2N})} = 0 \tag{5.33}$$

holds for any $u \in C_0^2(\mathbb{R}^{2N})$; it is clear that the action of P upon a function from $C_0(\mathbb{R}^{2N})$ is described by an integral over $\mathbb{R}^{2N}$. It suffices to establish relation (5.33) for functions u of the form $u(x, y) = u_1(x)v_1(y)$, where $u_1, v_1 \in C_0^2(\mathbb{R}^{2N})$, since, by using linear combinations of these functions, one can approximate an arbitrary function $u \in C_0^2(\mathbb{R}^{2N})$ in the sense of $C^2(\mathbb{R}^{2N})$. For such u, we have

$$\begin{aligned} (P, (\mathcal{M}&-2\lambda\mathbb{I})u_1(x)v_1(y))_{L_2(\mathbb{R}^{2N})} \\ &= (P, (\mathcal{L}_x - \lambda\mathbb{I})u_1(x)v_1(y))_{L_2(\mathbb{R}^{2N})} \\ &\quad + (P, u_1(x)(\mathcal{L}_y - \lambda\mathbb{I})v_1(y))_{L_2(\mathbb{R}^{2N})} \\ &= \int_{\mathbb{R}^{2N}} P(x, y; \lambda)\big((\mathcal{L}_x - \lambda\mathbb{I})\overline{u_1(x)}\big)\overline{v_1(y)}\,dxdy \\ &\quad + \int_{\mathbb{R}^{2N}} \overline{P(x, y; \lambda)u_1(x)}\big((\mathcal{L}_y - \lambda\mathbb{I})v_1(y)\big)dxdy \\ &= (P(\lambda)\bar{v}_1, (L - \lambda\mathbb{I})u_1)_{L_2(\mathbb{R}^N)} \\ &\quad + ((L - \lambda\mathbb{I})\bar{v}_1, P(\lambda)u_1)_{L_2(\mathbb{R}^N)} = 0. \end{aligned} \tag{5.34}$$

Here, we have used equality (15.4.9) and the facts that the coefficients of $\mathcal{L}$ are real, that the kernel $P(x, y; \lambda)$ is Hermitian, and that $\mathcal{R}(P(\lambda))$ consists of generalized eigenvectors of the operator L corresponding to λ.

Thus, P is a generalized solution of equation (5.32) inside $\mathbb{R}^{2N}$ and, therefore, according to Theorem 2.1, belongs to $W_{2,\text{loc}}^{2m}(\mathbb{R}^{2N})$.

The first equality in (5.31) follows from the relation

$$\int_{\mathbb{R}^{2N}} P(x,y;\lambda)((\mathcal{L}_x - \lambda \mathbb{I})\overline{u_1(x)v_1(y)}dxdy$$
$$= (P(\lambda)\overline{v_1}, (L - \lambda\mathbb{I})u_1)_{L_2(\mathbb{R}^N)} = 0 \quad (5.35)$$

used in (5.34). According to what has been proved, $P(\cdot, \cdot; \lambda)$ in relation (5.35) is sufficiently smooth, while u_1 and v_1 are arbitrary functions from $C_0^2(\mathbb{R}^N)$. This gives $(\mathcal{L}_x - \lambda\mathbb{I})P(x,y;\lambda) = 0$. The second equality in (5.31) is established similarly.

□

Consider the situation where a differential expression $\mathcal{L}$ of the same form is defined in a (bounded or unbounded) domain $G \subset \mathbb{R}^N$ with boundary ∂G and, e.g., trivial boundary conditions (b) are given. Let $L(\mathrm{b})$ be the corresponding operator, which is supposed to be selfadjoint, and let R_z ($z \notin S(L(\mathrm{b}))$) be its resolvent. An analogue of Theorem 5.2 now takes the following form:

Theorem 5.4. *Assume that the expression $\mathcal{L}$ satisfies the conditions of Theorem 5.2 and is defined in a domain $G \subset \mathbb{R}^N$ with boundary $\partial G \in C^{2m}$. Then, for $m > N/4$, one can indicate a measurable kernel $K(x,y)$ such that*

$$(R_z^m f)(x) = \int_G K(x,y)f(y)dy \qquad (x \in G),$$
$$\int_G |K(x,y)|^2 dx \le d_n < \infty \qquad (y \in G, \quad |y| \le n, \quad n \in \mathbb{N}) \qquad (5.36)$$

for arbitrary $f \in L_2(G)$ vanishing inside a certain ball.

Proof. In step (1), it is similar to the proof of Theorem 5.2 with the following changes: Instead of Theorem 2.1, one should use Theorem 2.2, which states that $R_z^m f$ belongs to $W_2^{2m}(G)$ inside G up to an arbitrary bounded piece on the boundary ∂G. In the considerations of steps (2)–(4), G' and G'' should be chosen as intersection of G with balls centered at the origin. This would guarantee the convergence of the integral of $|K(x,y)|^2$ taken over $G \times G'$ which, in turn, enables us to perform estimation (5.30) for $f \in L_2(G)$ vanishing outside a certain ball.

□

Thus, in this case, $L(\mathrm{b})$ is also a Càrleman operator and, by virtue of the relation in (5.36), the weight p can be chosen to be bounded in every bounded part of G. As in the case of $\mathbb{R}^N$, the "generalized nature" of generalized eigenfunctions manifests itself only in the fact that it may be increasing as $|x| \to \infty$. On the boundary ∂G, it vanishes. This is implied by Theorem 2.2, which should be applied in this case instead of Theorem 2.1.

Theorem 5.3 remains true for the corresponding spectral kernel $P(x,y;\lambda)$ which, in this case, is, in addition, smooth in $\langle x, y\rangle$ up to the boundary of the domain $G \times G$ and vanish if at least one of the variables (x or y) lies on ∂G (this is an evident consequence of Theorem 2.2).

5.4 The Laplace Operator

Let us find explicit expressions for the spectral measure of the Laplace operator L in $\mathbb{R}^3$ and its spectral kernel. To do this, we use formula (5.21) for the Green's function of this operator (the kernel of its resolvent) and the general formula (13.6.17) that expresses the resolution of the identity in terms of the resolvent. This formula can now be rewritten in the following form:

$$E(\delta) = \lim_{\varepsilon\to 0+} \frac{1}{2\pi i} \int_{\delta+i\varepsilon} (R_z - R_{\bar{z}})dz \tag{5.37}$$

for any finite open interval $\delta \subset \mathbb{R}$. This representation of formula (13.6.17) is possible because here $E(\{a\}) = 0$ for any $a \in \mathbb{R}$. To establish this equality, it suffices, in accordance with the proof of Theorem 4.2, to pass from L to its Fourier image $\hat{L}$ equal to the operator of multiplication by $|\xi|^2$ in the space $\hat{L}_2(\mathbb{R}^3)$ of functions of ξ. Indeed, the resolution of the identity $\hat{E}$ of the operator L is such that $\hat{E}(\{a\}) = 0$ (in view of the formulas presented in Subsection 4.1).

By using (5.37), (5.21), and the general formulas (2.4) and (4.9), for any $u, v \in C_0(\mathbb{R}^3)$ and an arbitrary finite open interval δ, we get

$$\begin{aligned}
&\int_{\mathbb{R}^3}\int_{\mathbb{R}^3}\left(\int_\delta P(x,y;\lambda)d\rho(\lambda)\right)u(y)\overline{v(x)}dxdy \\
&= (E(\delta)u,v)_{L_2(\mathbb{R}^3)} = \lim_{\varepsilon\to 0+}\frac{1}{2\pi i}\left(\int_{\delta+i\varepsilon}(R_z - R_{\bar{z}})dzu, v\right)_{L_2(\mathbb{R}^3)} \\
&= \lim_{\varepsilon\to 0+}\frac{1}{2\pi i}\int_{\mathbb{R}^3}\int_{\mathbb{R}^3}\left(\int_{\delta+i\varepsilon}(\Gamma_z(x,y) - \Gamma_{\bar{z}}(x,y))dz\right)u(y)\overline{v(x)}dxdy \\
&= \lim_{\varepsilon\to 0+}\frac{1}{2\pi i}\int_{\mathbb{R}^3}\int_{\mathbb{R}^3}\left(\int_{\delta+i\varepsilon}\frac{1}{4\pi}\frac{e^{i\sqrt{z}|x-y|} - e^{i\sqrt{\bar{z}}|x-y|}}{|x-y|}dz\right)u(y)\overline{v(x)}dxdy \\
&= \int_{\mathbb{R}^3}\int_{\mathbb{R}^3}\left(\int_\delta \frac{1}{4\pi^2}\frac{\sin(\sqrt{\lambda}|x-y|)}{|x-y|}d\lambda\right)u(y)\overline{v(x)}dxdy.
\end{aligned} \tag{5.38}$$

In view of the arbitrariness of u, v, and δ, relation (5.38) yields the equality

$$P(x,y;\lambda)d\rho(\lambda) = \begin{cases} \dfrac{1}{4\pi^2}\dfrac{\sin(\sqrt{\lambda}|x-y|)}{|x-y|}d\lambda, & \lambda \geq 0, \\ 0, & \lambda < 0, \end{cases}.$$

valid for all $x, y \in \mathbb{R}^3$. This formula indicates that the (general, see Remark 15.1.3) spectral measure can now be chosen in the form $d\rho(\lambda) = d\lambda$ for $\lambda \geq 0$ and $d\rho(\lambda) = 0$ for $\lambda < 0$. Then the spectral kernel takes the form

$$P(x,y;\lambda) = \frac{1}{4\pi^2}\frac{\sin(\sqrt{\lambda}|x-y|)}{|x-y|} \qquad (x,y \in \mathbb{R}^3;\quad \lambda \geq 0).$$

It is easy to see that the Parseval equality (3.7) (or (3.1) if $\alpha = \mathbb{R}$) now takes the form of the Parseval equality for the classical three-dimensional Fourier transformation if the latter is written in spherical coordinates with respect to ξ.

6 Ordinary Differential Operators

6.1 Theorem on Smoothing of Solutions

This section contains a brief exposition of the theory of ordinary differential operators.

Consider an ordinary differential expression of the rth order ($r \in \mathbb{N}$) given in an open (finite or infinite) interval $G = (a, b) \subseteq \mathbb{R}$

$$(\mathcal{L}u)(x) = \sum_{\alpha=0}^{r} a_\alpha(x)(D^\alpha u)(x) \qquad \left(D = \frac{d}{dx}, \quad x \in G\right). \tag{6.1}$$

Assume that the coefficients a_α are complex-valued continuous functions defined in $\tilde{G}$. This means that we, in fact, study expression (1.1) with $N = 1$. The leading coefficient is assumed to be always nonzero, i.e., $a_r(x) \neq 0$ for all $x \in \tilde{G}$.

Let us first prove a theorem on smoothing of generalized solutions of the equation $\mathcal{L}u = f$, which plays the same role as Theorems 2.1 and 2.2 in the elliptic case. In this case, the proof differs from the proof of the indicated theorems and is based on the notion of a fundamental solution for (6.1). We recall this classical definition (see, e.g., [Kam, CoL]).

Below, we assume that the interval $G = (a, b)$ is bounded.

A fundamental solution is defined as a continuous function $\tilde{G} \times \tilde{G} \ni \langle x, \xi \rangle \mapsto e(x, \xi) \in \mathbb{C}$ with the following properties:

(i) For every fixed $\xi \in \tilde{G}$ and $x \neq \xi$, there exist partial derivatives $(D_x^\beta e)(x, \xi)$, $\beta = 0, \dots, r$, continuous in $\langle x, \xi \rangle$ in the triangles $\{\langle x, \xi \rangle \in \tilde{G} \times \tilde{G} \mid x \leq \xi\}$ and $\{\langle x, \xi \rangle \in \tilde{G} \times \tilde{G} \mid x \geq \xi\}$. If $\beta = 0, \dots, r-2$, then these derivatives exist and are continuous in $\langle x, \xi \rangle$ in the whole of the square $\tilde{G} \times \tilde{G}$.

(ii) For $\xi \in G$,

$$(D_x^{r-1} e)(\xi + 0, \xi) - (D_x^{r-1} e)(\xi - 0, \xi) = \frac{1}{a_r(\xi)}.$$

(iii) The equality

$$\mathcal{L}\left(\int_G e(x, \xi) f(\xi) d\xi\right) = f(x) \qquad (f \in C(\tilde{G}), \quad x \in \tilde{G}) \tag{6.2}$$

holds, where the integral under the sign of $\mathcal{L}$ belongs to $C^r(\tilde{G})$.

(iv) A fundamental solution always exists. It is defined ambiguously and can be constructed for an arbitrary system of r linearly independent solutions $u_1(x), \dots, u_r(x)$ of the equation $(\mathcal{L}u)(x) = 0$ $(x \in \tilde{G})$ by the formula

$$e(x,\xi) = \frac{\operatorname{sign}(x-\xi)}{2a_r(\xi)W(\xi)} \begin{vmatrix} u_1(\xi) & \dots & u_r(\xi) \\ (Du_1)(\xi) & \dots & (Du_r)(\xi) \\ \dots\dots\dots & \dots & \dots\dots\dots \\ (D^{r-2}u_1)(\xi) & \dots & (D^{r-2}u_r)(\xi) \\ u_1(x) & \dots & u_r(x) \end{vmatrix},$$

$$W(\xi) = \begin{vmatrix} u_1(\xi) & \dots & u_r(\xi) \\ (Du_1)(\xi) & \dots & (Du_r)(\xi) \\ \dots\dots\dots & \dots & \dots\dots\dots \\ \dots\dots\dots & \dots & \dots\dots\dots \\ (D^{r-1}u_1)(\xi) & \dots & (D^{r-1}u_r)(\xi) \end{vmatrix} \qquad (x, \xi \in G).$$

It follows from this formula that if the coefficients a_α belong to $C^q(\tilde{G})$ $(\alpha = 0, \dots, r)$ with some $q \in \mathbb{Z}_+$, then, for $x \neq \xi$, the derivatives $(D_x^\beta D_\xi^\gamma e)(x,\xi)$ exist for $\beta = 0, \dots, r+q$ and $\gamma = 0, \dots, 1+q$. All these derivatives are continuous in $\langle x, \xi \rangle$ in the indicated triangles.

In order not to make smoothness requirements imposed on the coefficients of $\mathcal{L}$ too excessive, we first prove an assertion on smoothing of the solutions of homogeneous equations.

A function $\varphi \in W_2^t(G)$, where $t \in \mathbb{Z}$, is called a generalized solution of the equation $\mathcal{L}^+ u = 0$ inside G if

$$(\varphi, \mathcal{L}v)_{L_2(G)} = 0. \tag{6.3}$$

for all $v \in C_0^\infty(G)$. For $t < 0$, we assume that $a_\alpha \in C^{|t|}(\tilde{G})$ $(\alpha = 0, \dots, r)$. It is clear that this definition corresponds to definition (2.2), although the expression $\mathcal{L}^+$ may not exist in view of the fact that the coefficients a_α are not smooth enough.

Lemma 6.1. *Assume that $a_\alpha \in C^{r+p-1}(\tilde{G})$ $(\alpha = 0, \dots, r)$ with some $p \in \mathbb{Z}_+$. Then every generalized solution $\varphi \in W_2^t(G)$ of the equation $\mathcal{L}^+ u = 0$, where $|t| \leq r+p-1$, in fact, belongs to $W_{2,\mathrm{loc}}^{r+p}(G)$.*

Proof. According to Lemma 2.1, it suffices to establish this lemma locally, i.e., to prove that, for every point $x_0 \in G$, there exists its spherical neighbourhood $U(x_0) = (x_0 - \varepsilon, x_0 + \varepsilon) \subseteq G$ such that $\varphi \in W_{2,\mathrm{loc}}^{r+p}(U(x_0))$. The proof is split into several steps.

(1) We fix $x_0 \in G$ and choose $\varepsilon > 0$ so small that $(x_0 - 3\varepsilon, x_0 + 3\varepsilon) \subseteq G$. Assume that $k(t) \in C^\infty(\mathbb{R})$ vanishes for $|t| \geq \varepsilon$ and is equal to one in a certain neighbourhood of zero. On $\tilde{G}$, for given $w \in C_0^\infty(U(x_0))$, we construct a function

of the form

$$v(x) = \int_G e(x,\xi)k(|x-\xi|)w(\xi)d\xi$$
$$= \int_{U(x_0)} e(x,\xi)\left[k(|x-\xi|)-1\right]w(\xi)d\xi + \int_{U(x_0)} e(x,\xi)w(\xi)d\xi \quad (x \in \tilde{G}). \tag{6.4}$$

This function vanishes for $|x - x_0| \geq 2\varepsilon$ and, therefore, is finite with respect to G. It is sufficiently smooth and belongs to $C_0^{2r+p-1}(G)$. This can be proved by differentiation under the integral sign due to the fact that e possesses, for $x \neq \xi$, the derivatives $(D_x^\beta e)(x,\xi)$ $(\beta = 0,\ldots,2r+p-1)$ continuous in both triangles defined in (i). The integrals on the right-hand side of (6.4) have the same degree of smoothness. But this means that the function v (6.4) can be substituted in (6.3) with t indicated in the formulation of the lemma.

Indeed, in (6.3), one can pass to the limit (in v) from the functions v lying in the space $C_0^\infty(G)$ to less smooth finite functions. In the worst case where $\varphi \in W_2^{-(r+p-1)}(G)$, it suffices to require that the limiting v must be in $C_0^{2r+p-1}(G)$ but this is guaranteed. In the case of less singular φ, this substitution is all the more possible.

(2) By virtue of (iii), we have

$$(\mathcal{L}v)(x) = \int_{U(x_0)} \mathcal{L}_x\big(e(x,\xi)[k(|x-\xi|)-1]\big)w(\xi)d\xi + w(x) \quad (x \in \tilde{G}). \tag{6.5}$$

Consider the kernel $K(x,\xi) = \mathcal{L}_x\big(e(x,\xi)[k(|x-\xi|)-1]\big)$ $(x,\xi \in \tilde{G})$. In view of the smoothness requirements imposed on the coefficients of $\mathcal{L}$, the fact that the factor $k(|x-\xi|)-1$ vanishes in a neighbourhood of the diagonal $x = \xi$, and the indicated properties of the fundamental solution e, we conclude that the derivatives $(D_x^\beta D_\xi^\gamma K)(x,\xi)$ exist for $\beta = 0,\ldots,r+p-1$, $\gamma = 0,\ldots,r+p$, and all $x,\xi \in \tilde{G}$ and, moreover, they are continuous with respect to $\langle x,\xi\rangle \in \tilde{G}\times\tilde{G}$.

We define the operator A in $L_2(G)$ by setting

$$(Au)(x) = \int_G K(x,\xi)u(\xi)d\xi \qquad (u \in L_2(G)). \tag{6.6}$$

This operator can be extended by continuity to the continuous operator (also denoted by A) acting from $W_2^{-(r+p)}(G)$ to $W_2^{r+p-1}(G)$. Indeed, let $L_2(G) \ni u_n \to 0$ in $W_2^{-(r+p)}(G)$ as $n \to \infty$. Then, for any $x \in \tilde{G}$ and $\beta = 0,\ldots,r+p-1$, we obtain

$$\begin{aligned}|(D_x^\beta Au_n)(x)| &= \left|\int_G (D_x^\beta K)(x,\xi)u_n(\xi)d\xi\right| = \left|\left(u_n, \overline{(D_x^\beta K)(x,\cdot)}\right)_{L_2(G)}\right| \\ &\leq \|(D_x^\beta K)(x,\cdot)\|_{W_2^{r+p}(G)}\|u_n\|_{W_2^{-(r+p)}(G)} \\ &\leq c_1\|u_n\|_{W_2^{-(r+p)}(G)} \qquad (n \in \mathbb{N}).\end{aligned}$$

This enables us to conclude that $D_x^\alpha A u_n \to 0$ uniformly in $x \in \tilde{G}$ for any $\alpha = 0, \ldots, r{+}p{-}1$ and, therefore, $\|D_x^\beta A u_n\|_{W_2^{r+p-1}(G)} \to 0$. Thus, A is a continuous operator acting from $W_2^{-(r+p)}(G)$ into $W_2^{r+p-1}(G)$. But then the operator A^+ adjoint to A with respect to $L_2(G)$ (see Subection 14.1.2) is a continuous operator acting from $W_2^{-(r+p-1)}(G)$ into $W_2^{r+p}(G)$. This property is used in the next step.

(3) Thus, by virtue of (6.5) and (6.6), we have $\mathcal{L}v = Aw + w$, where $w \in C_0^\infty(G)$. As mentioned in step (1), this expression can be inserted in (6.3). As a result, we obtain

$$0 = (\varphi, \mathcal{L}v)_{L_2(G)} = (\varphi, Aw)_{L_2(G)} + (\varphi, w)_{L_2(G)} = (A^+\varphi, w)_{L_2(G)} + (\varphi, w)_{L_2(G)}.$$

Therefore, for any $w \in C_0^\infty(U(x_0))$, we have $(\varphi, w)_{L_2(G)} = (-A^+\varphi, w)_{L_2(G)}$, where $-A^+\varphi \in W_2^{r+p}(G)$. This means that $\varphi \in W_{2,\mathrm{loc}}^{r+p}(G)$. □

Note that, according to the lemma just proved, the degree of smoothness of a generalized solution φ of the equation $\mathcal{L}^+ u = 0$ depends only on the degree of smoothness of the coefficients of $\mathcal{L}$. This is natural, since the right-hand side of this equation is equal to $0 \in C^\infty(\tilde{G})$.

Lemma 6.2. *Assume that $a_\alpha \in C(\tilde{G})$ $(\alpha = 0, \ldots, r)$. Then, for any $f \in L_2(G)$,*

$$g(x) = \int_G e(x,\xi) f(\xi) d\xi \in W_2^r(G), \quad \mathcal{L}\Big(\int_G e(x,\xi) f(\xi) d\xi\Big) = f(x) \qquad (x \in G). \tag{6.7}$$

Proof. Consider a sequence of functions $f_n \in C(\tilde{G})$ such that $f_n \to f$ as $n \to \infty$ in $L_2(G)$. According to (i)–(iii), for any $n \in \mathbb{N}$, we get

$$\begin{aligned}(D_x^\beta g_n)(x) &= D_x^\beta\Big(\int_G e(x,\xi) f_n(\xi) d\xi\Big)\\ &= \int_G (D_x^\beta e)(x,\xi) f_n(\xi) d\xi \qquad (\beta = 0, \ldots, r-1).\end{aligned}$$

Moreover, the kernels $(D_x^\beta e)(x,\xi)$ are bounded for $\langle x, \xi\rangle \in \tilde{G} \times \tilde{G}$. By using the Cauchy-Buniakowski inequality, for any $\beta = 0, \ldots, r-1$, we can write

$$\begin{aligned}\|D_x^\beta g_n - D_x^\beta g_m\|_{L_2(G)}^2 &= \int_G \Big|\int_G (D_x^\beta e)(x,\xi)(f_n(\xi) - f_m(\xi)) d\xi\Big|^2 dx\\ &\le c_1 \|f_n - f_m\|_{L_2(G)}^2 \underset{n,m\to\infty}{\longrightarrow} 0.\end{aligned} \tag{6.8}$$

Further, it follows from the equality $\mathcal{L}g_n = f_n$ and (6.1) that

$$(D_x^r g_n)(x) = \frac{1}{a_r(x)}\Big(f_n(x) - \sum_{\alpha=0}^{r-1} a_\alpha(x)(D^\alpha g_n)(x)\Big) \qquad (x \in \tilde{G}, \quad n \in \mathbb{N}). \tag{6.9}$$

Estimate (6.8) ensures the fundamentality of the sequence $(D_x^\beta g_n)_{n=1}^\infty$ in $L_2(G)$ for any $\beta = 0, \ldots, r-1$, but then, in view of equality (6.9), the sequence $(D_x^r g_n)_{n=1}^\infty$ is also fundamental. In other words, the sequence $(g_n)_{n=1}^\infty$ is fundamental in $W_2^r(G)$ but this yields relations (6.7). □

The following definition is a particular case of (2.2): Consider an equation

$$\mathcal{L}u = f \in L_2(G), \tag{6.10}$$

where $G = (a,b)$ and $\mathcal{L}$ has the form (6.1). The function $\varphi \in W_2^t(G)$ $(t = \dots, -1, 0, \dots, r-1)$ is called a generalized solution of equation (6.10) inside G if

$$(\varphi, \mathcal{L}^+ v)_{L_2(G)} = (f, v)_{L_2(G)} \tag{6.11}$$

for any $v \in C_0^\infty(G)$.

In order that the expression $\mathcal{L}^+$ exist and (6.11) be meaningful for $t < 0$, it is required that the coefficients a_α of the expression $\mathcal{L}$ be sufficiently smooth. It suffices to assume that $a_\alpha \in C^{\alpha+|t|}(\tilde{G})$ $(\alpha = 0, \dots, r)$.

By combining the results of Lemmas 6.1 and 6.2, we easily arrive at the following assertion:

Theorem 6.1. *Consider equation (6.10); it is assumed that* $a_\alpha \in C^{\alpha+r-1+p}(\tilde{G})$ $(\alpha = 0, \dots, r)$ *where* $p \in \mathbb{Z}_+$ *is fixed. Let* $\varphi \in W_2^t(G)$, *where* $t \in [-(r+p-1), r)$ *is a generalized solution of this equation inside the domain* G. *Then, in fact,* $\varphi \in W_2^r(G)$. *If the right-hand side of (6.10) is equal to zero, one can assume that* $t \in [-(r+p-1), r+p)$ *and, therefore,* $\varphi \in C^{r+p}(G)$.

Let us mention an important distinction between this theorem and Theorem 2.1. In the case of elliptic equations, a generalized solution inside a domain remains smooth only inside the domain; in the considered case, the solution is automatically smooth up to the boundary of G. This means that results similar to Theorem 2.2 become useless for ordinary differential equations.

Also note that, for the problems of spectral theory investigated below, it suffices to consider equations with right-hand sides $f \in L_2(G)$ or equal to zero, and Theorem 6.1 is formulated just for equations of this sort.

Proof. It is easy to see that the smoothness requirements imposed on the coefficients a_α in Theorem 6.2 lead to the same restrictions for $\mathcal{L}^+$. Indeed, the coefficient of D^α in this expression belongs to $C^{\alpha+r-1+p}(\tilde{G})$ $(\alpha = 0, \dots, r)$. This means that all coefficients of this sort belong to $C^{r-1+p}(\tilde{G})$ and it is possible to apply Lemma 6.1 with $\mathcal{L}^+$ playing a role of $\mathcal{L}$. By using this lemma, we can conclude that the smoothness of the generalized solution of equation (6.10) with $f = 0$ inside G is characterized by the inclusion

$$\varphi \in W_{2,\mathrm{loc}}^{r+p}(G). \tag{6.12}$$

Now let $f \in L_2(G)$ be arbitrary. We reduce the problem to the investigation of a homogeneous equation. According to Lemma 6.2, $g \in W_2^r(G)$ and $\mathcal{L}g = f$. We set $\psi = \varphi - g$. Since $W_2^r(G) \subseteq W_2^t(G)$ $(t \in [-(r+p-1), r))$, we have $\psi \in W_2^t(G)$. Clearly, the function $\varphi = g$ satisfies relation (6.11) and, therefore,

$(\psi, \mathcal{L}^+ v)_{L_2(G)} = 0$ for any $v \in C_0^\infty(G)$. By using Lemma 6.1 (with $\mathcal{L}$ replaced by $\mathcal{L}^+$), we conclude that $\psi \in W_{2,\text{loc}}^{r+p}(G)$. Then

$$\varphi = \psi + g \in W_{2,\text{loc}}^r(G). \tag{6.13}$$

Let us now show how to remove the index 'loc' in inclusions (6.12) and (6.13). Consider, e.g., (6.13). According to the embedding theorems, $W_{2,\text{loc}}^r(G) \subseteq C^{r-1}(G)$ $(N = 1)$. We fix $c \in (a, b)$ and introduce, for φ, a solution w of the Cauchy problem on $\tilde{G} = [a, b]$:

$$(\mathcal{L}w)(x) = f(x) \quad (x \in \tilde{G}); \qquad w(c) = \varphi(c), \ldots, (D^{r-1}w)(c) = (D^{r-1}\varphi)(c).$$

By virtue of the classical theorems, this solution exists and belongs to $W_2^r(G)$. Moreover, the function $\varphi(x)$ $(x \in G)$ is also a solution of this Cauchy problem in G. Indeed, in view of the already established smoothness of this function, one can transfer the expression $\mathcal{L}^+$ in (6.11) to φ and use the fact that v is arbitrary. Since the Cauchy problem is uniquely solvable, we have $\varphi(x) = w(x)$ $(x \in G)$ and, hence, $\varphi \in W_2^r(G)$.

For $f = 0$, the solution of the Cauchy problem $w \in C^{r+p}(\tilde{G})$ and, thus, the same inclusion can be written for φ. □

REMARK 6.1 As in Remarks 2.2, 2.5, and 2.6, we can now formulate Theorem 6.1 in the local form:

If $G' \subseteq G$ is an open interval, $f \in L_2(G')$ or $f = 0$, and (6.11) holds for any $v \in C_0^\infty(G')$, then φ belongs to $W_2^r(G')$ or $C^{r+p}(G')$. The original interval G can also be unbounded. In this case, we assume that $\varphi \in W_2^t(G, p(x)dx)$ with a certain weight $0 < p(x) \in C(\tilde{G})$. □

This concludes our investigation of the problem of smoothing of solutions of ordinary differential equations. Note only that an assertion similar to Theorem 6.1 but relating only to the case of smoothness inside a domain also holds for elliptic $\mathcal{L}$. It is based on the use of fundamental solutions for $\mathcal{L}$ and can, in a certain sense, replace Theorem 2.1.

6.2 Selfadjointness of Differential Operators

Theorems 4.1–4.4 related to the selfadjointness of the minimal operator in $\mathbb{R}$ are also applicable to studying the selfadjointness of operators generated by an ordinary differential expression. In this case, in Theorems 4.1, 4.3, and 4.4, one must take $\mathcal{L}$ in the form of the Sturm-Liouville differential expression with real-valued potential $q \in C(\mathbb{R})$, namely,

$$(\mathcal{L}u)(x) = -u''(x) + q(x)u(x) \qquad (x \in \mathbb{R}) \tag{6.14}$$

(i.e., the one-dimensional Schrödinger expression (4.4), $N = 1$). Clearly, in the case under consideration, one must use Theorem 6.1 instead of Theorem 2.1 in the proof of Theorem 4.1.

Here, we only clarify how Theorem 6.1 is used for proving selfadjointness in the case where G has a boundary, i.e., $G = (a, b), (a, \infty)$. For simplicity, we assume that $G = (a, b)$ and consider expression (6.14) defined on it.

We assume that expression (6.14) with $q \in C(\tilde{G})$ and the following boundary conditions (b) are given on $G = (a, b)$:

$$u(a)\cos\alpha + u'(a)\sin\alpha = 0, \quad u(b)\cos\beta + u'(b)\sin\beta = 0. \tag{6.15}$$

Here, $\alpha, \beta \in [0, \pi)$ are fixed. In the space $L_2(G)$, we construct the operator $L_2(G) \supset C^2(\tilde{G}, \mathrm{b}) \ni u \mapsto A'u = \mathcal{L}u \in L_2(G)$, where $C^2(\tilde{G}, \mathrm{b})$ denotes the subspace of all functions from $C^2(\tilde{G})$ satisfying (6.15).

The operator A' is Hermitian. Indeed, $\mathcal{L}^+ = \mathcal{L}$ and the Green formula (3.2) has the following form in this case: For any $u, v \in C^2(\tilde{G})$,

$$\begin{aligned}(\mathcal{L}u, v)_{L_2(G)} &- (u, \mathcal{L}v)_{L_2(G)} \\ &= -\big(u'(b)\overline{v(b)} - u(b)\overline{v'(b)}\big) + \big(u'(a)\overline{v(a)} - u(a)\overline{v'(a)}\big) = \partial(u, v).\end{aligned} \tag{6.16}$$

It is easy to see that if u and v belong to $C^2(\tilde{G}, \mathrm{b})$, i.e., satisfy (6.15), then the boundary form $\partial(u, v) = 0$. This implies that A is Hermitian.

As in Section 3, we denote the closure of A' by $A = L(\mathrm{b})$. This operator is also Hermitian. Clearly, it is an extension of the minimal operator L constructed for given $\mathcal{L}$ and G.

Theorem 6.2. *In $G = (a, b)$, we consider the Sturm-Liouville expression (6.14) with real-valued potential $q \in C(\tilde{G})$. The operator $A = L(\mathrm{b})$ corresponding to this expression with boundary conditions (6.15) is selfadjoint and semibounded (from below).*

Proof. Certainly, the proof is close to that of Theorem 3.1. We indicate some specific features of the case under consideration.

Let $g \in \mathcal{D}(A^*)$. Then $(\forall f \in \mathcal{D}(A))\colon (Af, g)_{L_2(G)} = (f, A^*g)_{L_2(G)}$. In particular, $\forall v \in C_0^2(G)\colon (\mathcal{L}v, g)_{L_2(G)} = (v, A^*g)_{L_2(G)}$ or

$$(g, \mathcal{L}^+v)_{L_2(G)} = (A^*g, v)_{L_2(G)} \qquad (v \in C_0^2(G)). \tag{6.17}$$

This equality shows that $g \in L_2(G)$ is a generalized solution of the equation $\mathcal{L}u = A^*g \in L_2(G)$ inside G (see (6.11)) and, hence, we can use Theorem 6.1. However, in order to apply it directly, we must assume that $q = a_0 \in C^1(\tilde{G})$ ($r = 2$, $p = 0$). To avoid this restriction, by using (6.14), we rewrite relation (6.17) in the form $(g, -v'')_{L_2(G)} = (A^*g - qg, v)_{L_2(G)}$ $(v \in C_0^2(G))$. Thus, g is a generalized solution of the equation $-u'' = A^*g - qg = h$ inside G; here, $h \in L_2(G)$ because $g \in L_2(G)$ and q is bounded as a function from $C(\tilde{G})$. In view of Theorem 6.1, we can now conclude that $g \in W_2^2(G)$ and $\mathcal{L}g = A^*g$.

To prove the selfadjointness of A, one must verify that g satisfies (b) of the form (6.15) (note that $W_2^2(G) \subset C^1(\tilde{G})$ and these (b) have sense for g).

We have $(\forall f \in \mathcal{D}(A))$: $(Af, g)_{L_2(G)} = (f, A^*g)_{L_2(G)} = (f, \mathcal{L}g)_{L_2(G)}$. By setting $f = u \in \mathcal{D}(A') = C^2(\tilde{G}, \mathrm{b})$ and using (6.16), we get

$$\begin{aligned} 0 &= (\mathcal{L}u, g)_{L_2(G)} - (u, \mathcal{L}g)_{L_2(G)} \\ &= -\big(u'(b)\overline{g(b)} - u(b)\overline{g'(b)}\big) + \big(u'(a)\overline{g(a)} - u(a)\overline{g'(a)}\big). \end{aligned} \tag{6.18}$$

Assume, in addition, that u vanishes in a neighbourhood of the point b. Then (6.18) yields

$$u'(a)\overline{g(a)} - u(a)\overline{g'(a)} = 0, \tag{6.19}$$

where $u(a)$ and $u'(a)$ are arbitrary values satisfying the first equality in (6.15). By using (6.15), we determine the ratio $u'(a)/u(a)$ (or $u(a)/u'(a)$) and substitute it in (6.19). As a result, we find that g satisfies (6.15) at the point a. Similarly, it can be shown that g also satisfies (6.15) at the point b. Let us prove that A is semibounded. Integration by parts yields

$$\begin{aligned} (\mathcal{L}u, u)_{L_2(G)} &= \int_a^b (-u''(x) + q(x)u(x))\overline{u(x)}dx \\ &= \int_a^b (|u'(x)|^2 + q(x)|u(x)|^2)dx - u'(b)\overline{u(b)} + u'(a)\overline{u(a)} \end{aligned} \tag{6.20}$$

for any $u \in C^2(\tilde{G}, \mathrm{b})$.

Consider the term $u'(a)\overline{u(a)}$ in (6.20). It is equal either to zero (if $\alpha = 0$ in (6.15)) or to $-\cot\alpha|u(a)|^2$.

In the latter case, we use the following estimate, which is a consequence of Lemma 6.3 proved below: For any $\varepsilon > 0$, there exists $k(\varepsilon) \geq 0$ such that

$$|u(a)|^2 \leq \varepsilon\|u'\|^2_{L_2(G)} + k(\varepsilon)\|u\|^2_{L_2(G)} \qquad (u \in C^2(\tilde{G})). \tag{6.21}$$

The term $u'(b)\overline{u(b)}$ is treated analogously. As a result, we establish that the last two terms in (6.20) are bounded in modulus from above by the expression on the right-hand side of (6.21). Setting $\varepsilon < 1$ and taking into account that q is bounded, we estimate the right-hand side in (6.20) from below by $\alpha\|u\|_{L_2(G)^2}$. The estimate obtained means that A is semibounded.

It is also clear that, in the case of boundary conditions of the form (6.15) with $\alpha, \beta = 0, \pi/2$, estimate (6.21) is not necessary. □

Estimate (6.21) obviously follows from the general lemma presented below.

Lemma 6.3. *Let $c \in [a, b] = \tilde{G}$ and $p \in (1, \infty]$ be fixed. Then, for any $\varepsilon > 0$, there exists $k(\varepsilon) \geq 0$ such that*

$$|u(c)| \leq \varepsilon\|u'\|_{L_p(G)} + k(\varepsilon)\|u\|_{L_p(G)} \qquad (u \in C^1(\tilde{G})). \tag{6.22}$$

Proof. By integration by parts, one can easily verify that, for any $u \in C^1(\bar{G})$ and $n \in (0, \infty)$, the following equality is true:

$$u(c) = (u', f_n)_{L_2(G)} + (u, g_n)_{L_2(G)}. \tag{6.23}$$

Here,

$$f_n(x) = \frac{(x-a)^{n+1}}{(b-a)(c-a)^n}, \quad g_n(x) = \frac{(n+1)(x-a)^n}{(b-a)(c-a)^n} \qquad (x \in [a,c]),$$
$$f_n(x) = \frac{-(b-x)^{n+1}}{(b-a)(b-c)^n}, \quad g_n(x) = \frac{(n+1)(b-x)^n}{(b-a)(b-c)^n} \qquad (x \in (c,b])$$

for $c \in (a,b)$; for $c = a, b$, the functions f_n and g_n are given by obviously changed expressions. By direct calculation, we establish that, for any $p' \in [1, \infty)$, the following inequalities hold:

$$\|f_n\|_{L_{p'}(G)} \le \left(\frac{b-a}{np'+p'+1}\right)^{1/p'},$$
$$\|g_n\|_{L_{p'}(G)} \le \frac{n+1}{(b-a)^{1-1/p'}(np'+1)^{1/p'}}. \tag{6.24}$$

Let $1/p + 1/p' = 1$. By applying the Hölder inequality to representation (6.23) and using estimates (6.24), we arrive at (6.22). Indeed, according to (6.24), the value $\|f_n\|_{L_{p'}(G)}$ can be made as small as desired for $n \to \infty$. □

Thus, Theorem 6.2 is proved. □

REMARK 6.1 It follows from the proof of Theorem 6.2 that $\mathcal{D}(L(\mathrm{b})) = W_2^2(G, \mathrm{b})$ (we have shown that $\mathcal{D}(A^*) = W_2^2(G, \mathrm{b})$ and $A^* = A = L(\mathrm{b})$). The coercivity inequality is also valid for $\mathcal{L}$ of the form (6.14), namely, $(\exists p \ge 0)\,(\exists c > 0)(\forall u \in W_2^2(G, \mathrm{b}))$:

$$\|\mathcal{L}u\|^2_{L_2(G)} + p\|u\|^2_{L_2(G)} \ge c\|u\|^2_{W_2^2(G)}. \tag{6.25}$$

Indeed, let $\lambda \le 0$ be so large that $(A - \lambda\mathbb{I})^{-1}$ exists. According to the argument presented above, $\mathcal{R}((A - \lambda\mathbb{I})^{-1}) = \mathcal{D}(A) = W_2^2(G, \mathrm{b})$ and, therefore, $(A - \lambda\mathbb{I})^{-1}$ can be regarded as an operator from $L_2(G)$ to $W_2^2(G)$. This operator is closed, which can be easily established by analogy with the proof of closedness of the operator A in step (2) of the proof of Theorem 5.2. Therefore, according to the Banach theorem on closed graph, it is bounded, i.e., $\|(\mathcal{L} - \lambda\mathbb{I})u\|_{L_2(G)} \ge c_1\|u\|_{W_2^2(G)}$ $(u \in W_2^2(G, \mathrm{b}))$. By extending this inequality to the left, we arrive at (6.25).

For the constructed operator $A = L(\mathrm{b})$, analogues of Theorems 3.2 and 3.3 are true, which are formulated and proved in exactly the same way as in Section 3. For example, the spectrum of A consists of the sequence $(\lambda_n)_{n=1}^\infty$ of real eigenvalues tending to $+\infty$. *Each λ_n is associated with a one-dimensional eigensubspace.* This statement can be proved as follows: Let $\alpha \ne 0$ in (6.15) and let $\lambda = \lambda_n$ be associated with eigenfunctions $\varphi_1(x;\lambda)$ and $\varphi_2(x;\lambda)$; $\varphi_j'(a;\lambda) = -\cot\alpha\varphi_j(a;\lambda)$ $(j=1,2)$. Then the function

$$\psi(x) = \varphi_1(x;\lambda) - \frac{\varphi_1(0;\lambda)}{\varphi_2(0;\lambda)}\varphi_2(x;\lambda)$$

is a solution of the equation $(\mathcal{L}\psi)(x) = \lambda\psi(x)$ $(x \in [a,b])$ with trivial initial conditions. Therefore, it is identically equal to zero, i.e., φ_1 and φ_2 are proportional. The case of $\alpha = 0$ is considered similarly. □

The results of this subsection can be easily generalized to the case of formally selfadjoint expressions of the form (6.1) of arbitrary order r on (a,b). Only certain technical difficulties arise in this case in connection with writing the Green formula (3.2) and choosing corresponding boundary conditions. Note that boundary conditions may establish a relationship between the values of functions and derivatives in both the points a and b (clearly, this is also true for the Sturm-Liouville expression (6.14)); this is so, e.g., for conditions of periodic type. Functions from the domain of the corresponding operator $A = L(\mathrm{b})$ belong to the subspace $W_2^r(G,\mathrm{b})$ of the space $W_2^r(G)$ comprising functions satisfying b.

6.3 Green's Function

We now consider expression (6.1) on the whole axis $\mathbb{R}$ or on the semiaxis $G = (a,\infty)$ (or $(-\infty,a)$). In the case of the whole axis, this expression is associated with the minimal operator L; in the case of a semiaxis, (6.1) is associated with a certain extension $L(\mathrm{b})$ of the minimal operator L on the semiaxis which corresponds to the boundary conditions at the point a. For example, in the case of the Sturm-Liouville expression (6.14) with the first relation in (6.15) as b, the operator $L(\mathrm{b})$ is constructed as the closure of the operator $L_2(G) \supset C_0^2(\tilde{G},\mathrm{b}) \ni u \mapsto \mathcal{L}u \in L_2(G)$ in $L_2(G)$; here, $C_0^2(\tilde{G},\mathrm{b})$ consists of all functions from $C^2(G)$ which are finite at ∞ and satisfy b with fixed α at the point a.

Assume that the operators L and $L(\mathrm{b})$ are selfadjoint. One can easily apply to these operators the results of Section 5 related to the Green's function and the expansion in eigenfunctions. Let us indicate the differences between these cases.

The results of Subsection 5.1 are applicable to the one-dimensional case ($N = 1$). We can assume that $l = 1$ in chains (5.2)–(5.4). Clearly, Theorem 6.1 should be used instead of Theorems 2.1 and 2.2. It is now helpful to clarify why, in the case of the semiaxis $G = (a,\infty)$, the *generalized eigenfunction φ, which, according to Theorem 6.1, belongs to $C^r(\tilde{G})$, satisfies* b *at the point a.*

Assume, for example, that $\mathcal{L}$ is the Sturm-Liouville expression (6.14) and b is the first relation in (6.15). As chain (15.2.1), it is expedient to choose the following chain:

$$H_- = W_2^{-1}(G,p(x)dx) \supseteq H_0 = L_2(G) \supseteq H_+ = W_2^1(G,p(x)dx) \supseteq D, \qquad (6.26)$$

where the weight $p \in C^1(G)$ is such that the embedding $H_+ \to H_0$ is quasinuclear and $D = C_0^{1+r}(\tilde{G},\mathrm{b})$ with natural topologization. In this case, relation (15.2.2) takes the form $(\varphi, \mathcal{L}u)_{L_2(G)} = \lambda(\varphi, u)_{L_2(G)}$ $(u \in D)$. According to Theorem 6.1, it guarantees the required smoothness of φ and the equality $\mathcal{L}\varphi = \lambda\varphi$. It follows from this relation that $(\mathcal{L}u, \varphi)_{L_2(G)} - (u, \mathcal{L}\varphi)_{L_2(G)} = 0$ $(u \in C_0^2(\tilde{G}, \mathrm{b}))$. This implies that φ satisfies b at the point a; one should only repeat the simple argument used in the proof of Theorem 6.2 to show that g satisfies b (see (6.18)). □

Let us proceed to studying the kernel of the resolvent (Green's function) of the minimal operator L generated by expression (6.1) of order r in $\mathbb{R}$. *Theorem 5.1 remains true in this case.* In the case of even r, its proof is completely the same because Lemma 5.2 remains true and the expression $\mathcal{M}$ in $\mathbb{R}^2$ is now elliptic (the corresponding form $\mathcal{M}_0(x, \xi) = a_r(x_1)\xi_1^r + a_r(x_2)\xi_2^r$ is not equal to zero, $(x = \langle x_1, x_2\rangle$, $\xi = \langle \xi_1, \xi_2\rangle \in \mathbb{R}^2)$. For odd r, one must modify the proof of smoothness of $\Gamma_z(x, y)$ in the pair of variables $\langle x, y\rangle$ outside the diagonal. One can use here an argument similar to the the proof of Theorem 6.1 but applied to both x and y simultaneously; the corresponding proof is given in [Ber]. □

The fundamental solution $e(x, \xi)$ for $\mathcal{L}$ described in Subsection 1 can be modified so that it will also be fundamental in ξ for the expression $\mathcal{L}^+$. Such a fundamental solution $e_z(x, y)$ constructed for $\mathcal{L} - z\mathbb{1}$ admits representation (5.20).

For the operators $L(\mathrm{b})$ generated by $\mathcal{L}$ in finite or semiinfinite intervals, the Green's function $\Gamma_z(x, y)$ possesses all properties described above. In addition, it satisfies the boundary condition used to construct $L(\mathrm{b})$ in each of the variables x and y. The corresponding comments were given in Subsection 1.

For the ordinary differential expression $\mathcal{L}$ (unlike the elliptic expression), the Green's function can be quite easily constructed in the form of a determinant formula similar to that presented in (iv), Subsection 1. Let us consider, e.g., the Sturm-Liouville expression (6.14) in $G = (a, b)$ with boundary conditions (6.15) and denote by $u_1(x; z), u_2(x; z)$ the solutions of the equation $(\mathcal{L}u)(x) = zu(x)$ $(x \in [a, b])$, where $z \notin S(L(\mathrm{b}))$ is fixed, which satisfy the following initial conditions corresponding to (6.15):

$$\begin{aligned} u_1(a; z) &= \sin\alpha, \quad u_1'(a; z) = -\cos\alpha; \\ u_2(b; z) &= \sin\beta, \quad u_2'(b; z) = -\cos\beta. \end{aligned} \tag{6.27}$$

Then
$$\Gamma_z(x, y) = \begin{cases} \dfrac{1}{W(z)} u_1(x; z) u_2(x; z), & x \le y, \\ \dfrac{1}{W(z)} u_2(x; z) u_1(y; z), & x \ge y, \end{cases}$$

$$W(z) = u_1(x; z)u_2'(x; z) - u_1'(x; z)u_2(x; z) \tag{6.28}$$

(recall that, for the Sturm-Liouville equation, the Wronskian W does not depend on x). One can easily verify by direct differentiation that (6.28) is really the kernel of the resolvent R_z of the operator $L(\mathrm{b})$.

In the case of the semiaxis $G = (a, \infty)$, formula (6.28) remains valid with the same u_1 and with u_2 determined by a certain behaviour at ∞ (this solution must belong to $L_2(G)$). In the case of $\mathbb{R}$, the solutions u_1 and u_2 are distinguished by their behaviour at $-\infty$ and $+\infty$.

6.4 Expansion in Generalized Eigenfunctions

The expansion in eigenvectors of the selfadjoint operator $L(\mathrm{b})$ for $\mathcal{L}$ on (a, b) was described in Subsection 2. In the case of the semiaxis (a, ∞) or the whole axis $\mathbb{R}$ where, as a rule, the continuous spectrum appears, one can follow the scheme of Subsection 5.1 and use chains of the form (6.26). However, as in the case of elliptic operators, it is more convenient to prove first that the operator corresponding to the formally selfadjoint $\mathcal{L}$ of the form (6.1) is a Càrleman operator and then use the general theory of expansion of Càrleman operators.

The proof of the fact that L or $L(\mathrm{b})$ is a Càrleman operator coincides with the proof of Theorem 5.2. Note only that, in this case, representations (5.22), (5.23) (for $N = 1$), and (5.36) (for $G = (a, \infty)$, (a, b)), are also valid in the case of $m = 1$, i.e., for the resolvent R_z itself. This is a consequence of the embedding theorem, i.e., $W_2^2(G') \subset C(\tilde{G}')$ if G' is an open interval of the axis $\mathbb{R}$ (see step (3) of the proof of Theorem 5.2). Therefore, to complete step (1) of this proof, it suffices to use Theorem 6.1; indeed, in this case, we need not consider the powers of the resolvent and increase the smoothness of the solution of equation (6.10) for f from a positive Sobolev space. We need only to increase the smoothness of the generalized solution φ from L_2 (as in the proof of selfadjointness, see Subsection 2). Therefore, Theorem 6.1 is used only in the case of $t = 0$ and, since $p \in \mathbb{Z}_+$, it suffices to assume that $p = 0$ in the smoothness conditions for the coefficients a_α (in the case of the Sturm-Liouville operator, one must require that $q \in C(\tilde{G})$; see the proof of Theorem 6.2). Thus, we have arrived, in fact, at the following result:

Theorem 6.3. *Let $\mathcal{L}$ be a formally selfadjoint expression (6.1) in $G = \mathbb{R}$, (a, ∞) whose coefficients a_α belong to $C^{\alpha+r-1}(\tilde{G})$. Given $\mathcal{L}$, we construct the minimal operator L in the case of $G = \mathbb{R}$ or the operator $L(\mathrm{b})$ corresponding to certain boundary conditions at the point a in the case of $G = (a, \infty)$. Assume that the operators L and $L(\mathrm{b})$ are selfadjoint in $L_2(G)$.*

It is stated that L and $L(\mathrm{b})$ are Càrleman operators and the results of Section 15.4 are applicable to the expansions of these operators in their generalized eigenfunctions. In particular, their generalized eigenfunctions belong to $C^{2r-1}(\tilde{G})$ and, in the case of $L(\mathrm{b})$, satisfy b *at the point a. There exists the spectral kernel $P(x, y; \lambda)$ $(x, y \in \tilde{G})$ which belongs to $C^{2r-1}(\tilde{G} \times \tilde{G})$ with respect to $\langle x, y \rangle$ and, for $L(\mathrm{b})$, satisfies* b *in each of the variables x and y.*

In the case of the Sturm-Liouville expression (6.14), it suffices to require that the potential q belong to $C(\tilde{G})$; in this case, $C^{2r-1}(\tilde{G})$ and $C^{2r-1}(\tilde{G} \times \tilde{G})$ are replaced by $C^r(\tilde{G})$ and $C^r(\tilde{G} \times \tilde{G})$.

Note that the smoothness of $P(x, y; \lambda)$ in each of the variables x and y while the other is fixed follows from Theorem 6.1. The smoothness in $\langle x, y \rangle$ in the case of even r is proved as in Subsection 3 by using the elliptic expression $\mathcal{M}$. In the general case, some additional reasoning should be involved.

6.5 The Spectral Matrix

According to (15.3.1) and (15.4.9), the Parseval equality for the expansion in generalized eigenfunctions in the case of Càrleman operators under consideration has the following form:

$$(u,v)_{L_2(G)} = \int_{\mathbb{R}} \left(\int_G \int_G P(x,y;\lambda) u(y) \overline{v(x)} dx dy \right) d\rho(\lambda). \tag{6.29}$$

Here, $G = \mathbb{R}$, (a, ∞) and $u, v \in C_0(\tilde{G})$, i.e., u and v belong to $C(\tilde{G})$ and are finite at ∞ (actually, u and v can by taken from the corresponding broader positive space).

Equality (6.29) for ordinary differential operators is usually written in a different way. Namely, denote by $\psi_0(x;\lambda), \ldots, \psi_{r-1}(x;\lambda)$ the fundamental system of solutions of the equation $(\mathcal{L}u)(x) = \lambda u(x)$ $(x \in \tilde{G})$ which satisfy the initial conditions

$$(D^k \psi_j)(c;\lambda) = \delta_{jk} \qquad (j,k = 0,\ldots,r-1; \quad \lambda \in \mathbb{R}), \tag{6.30}$$

where c is a fixed point of $\tilde{G}$. Every solution u of this equation can be expressed in terms of ψ_j as follows:

$$u(x) = \sum_{j=1}^{r-1} (D^j u)(c) \psi_j(x;\lambda) \qquad (x \in \tilde{G}). \tag{6.31}$$

The spectral kernel $P(x,y;\lambda)$ satisfies the equations

$$\mathcal{L}_x P(x,y;\lambda) = \lambda P(x,y;\lambda), \quad \bar{\mathcal{L}}_y P(x,y;\lambda) = \lambda P(x,y;\lambda) \qquad (x, y \in \tilde{G}). \tag{6.32}$$

These equations are, in fact, equations (5.31) for the ordinary differential expression. The conjugation bar over $\mathcal{L}_y$ appears due to the fact that the coefficients of $\mathcal{L}$ are complex-valued (in Sections 3–5, we considered only real-valued a_α). By applying formula (6.31) twice, we get

$$\begin{aligned} P(x,y;\lambda) &= \sum_{j=1}^{r-1} (D_x^j P)(c,y;\lambda) \psi_j(x;\lambda) \\ &= \sum_{j,k=1}^{r-1} (D_x^j D_y^k)(c,c;\lambda) \psi_j(x;\lambda) \overline{\psi_k(y,\lambda)} \qquad (x,y \in \tilde{G}). \end{aligned} \tag{6.33}$$

By substituting this representation in (6.29), we obtain the Parseval equality in the form

$$(u,v)_{L_2(G)} = \sum_{j,k=0}^{r-1} \int_{\mathbb{R}} \hat{u}_k(\lambda) \overline{\hat{v}_j(\lambda)} d\sigma_{jk}(\lambda), \tag{6.34}$$

$$\begin{aligned} \hat{u}_k(\lambda) &= \int_G u(x) \psi_k(x;\lambda) dx, \\ d\sigma_{jk}(\lambda) &= (D_x^j D_y^k P)(c,c;\lambda) d\rho(\lambda) \\ &(u, v \in C_0(\tilde{G}); \ \ j,k = 0,\ldots,r-1). \end{aligned} \tag{6.35}$$

Thus, *each function* $u \in C_0(\tilde{G})$ *can be associated with its Fourier transform*

$$\hat{u}(\lambda) = (\hat{u}_0(\lambda), \dots, \hat{u}_{r-1}(\lambda)) \tag{6.36}$$

whose components are given by (6.35); in this case, the Parseval equality has the form (6.34) (the Krein-Kodair expansion).

The matrix $(d\sigma_{jk}(\lambda))_{j,k=0}^{r-1}$ is called *spectral. This matrix, more exactly,* $(\forall \alpha \in \mathfrak{B}(\mathbb{R}))$: $(\sigma_{jk}(\alpha))_{j,k=0}^{r-1}$, *is positive definite.* Indeed, let us choose r points $x_\alpha \in G$ $(\alpha = 0, \dots, r-1)$ so that the matrix $(\psi_j(x_\alpha; \lambda))_{j,\alpha=0}^{r-1}$ is nondegenerate; this can be easily done due to (6.30). Then every vector $\xi = (\xi_0, \dots, \xi_{r-1}) \in \mathbb{C}^r$ can be represented in the form

$$\xi_j = \sum_{\alpha=0}^{r-1} c_\alpha \psi_j(x_\alpha; \lambda) \qquad (j = 0, \dots, r-1).$$

By using this representation and equality (6.33), we obtain

$$\begin{aligned}
\sum_{j,k=0}^{r-1} (D_x^j D_y^k P)(c, c; \lambda)\xi_k \bar{\xi}_j &= \sum_{j,k=0}^{r-1} (D_x^j D_y^k P)(c, c; \lambda) \sum_{\alpha,\beta=0}^{r-1} c_\beta \bar{c}_\alpha \psi_k(x_\beta; \lambda)\overline{\psi_j(x_\alpha; \lambda)} \\
&= \sum_{\alpha,\beta=0}^{r-1} c_\beta \bar{c}_\alpha \left\{ \sum_{j,k=0}^{r-1} (D_x^j D_y^k P)(c, c; \lambda)\psi_k(x_\beta; \lambda)\overline{\psi_j(x_\alpha; \lambda)} \right\} \\
&= \sum_{\alpha,\beta=0}^{r-1} P(x_\beta, x_\alpha; \lambda) c_\beta \bar{c}_\alpha \geq 0.
\end{aligned}$$

Here, the last inequality follows from the positive definiteness of the kernel $P(x, y; \lambda)$. □

Equality (6.34) can be extended by continuity to $u, v \in L_2(G)$ by introducing properly the concept of an integral over a spectral matrix. We only mention this without presenting the corresponding results.

The Fourier transform (6.36), (6.35) and the Parseval equality (6.34) differ from analogous transforms (15.3.4), (15.3.5); the latter can be regarded as a "diagonalization" of the former. We stress that the Fourier transform (6.36) depends analytically on $\lambda \in \mathbb{C}$ (as well as $\psi_j(x; \lambda)$), whereas the dependence of the Fourier transform (15.3.4) on λ is fairly irregular.

Note that the fundamental system of solutions ψ_j of the equation $\mathcal{L}u = \lambda u$ can also be chosen in a different way, namely, with initial conditions other than (6.30). For example, in the case of $G = (a, \infty)$, it is convenient to choose the initial conditions at the point $c = a$ in accordance with b; a function that satisfies these conditions also satisfies b. Thus, for the equation $(\mathcal{L}u)(x) = \lambda u(x)$ $(x \in \tilde{G})$ with boundary conditions b, one can choose a certain number $s < r$ of initial conditions which enable one to write an expansion of the form (6.31) of any solution of the problem considered in the solutions ψ_j satisfying these initial conditions. Since the spectral kernel $P(x, y; \lambda)$ satisfies b in x and y, it admits an expansion of type (6.33) in which ψ_j are exactly these solutions. It is important that r in (6.31) is replaced by $s < r$, i.e., the dimensionality $s \times s$ of the spectral matrix (6.35) is reduced.

In the case of the Sturm-Liouville expression (6.14), we have $r = 2$ and the spectral matrix in expansions (6.34)–(6.36) associated with the operator L on the axis $\mathbb{R}$ is two-dimensional.

However, if we consider the semiaxis $G = (a, \infty)$ and the first condition in (6.15) at the point a, then, for the corresponding operator $L(\mathrm{b})$, we can write the "scalar" expansions (6.34)–(6.36) with $s = 1$, i.e., the spectral matrix will be an ordinary scalar spectral (general) measure. In this case, as $\psi_1(x; \lambda)$, one should take the solution of the equation $(\mathcal{L}u)(x) = \lambda u(x)$ $(x \in \tilde{G})$ satisfying the first initial condition in (6.27).

6.6 Classical Fourier Transformation

In conclusion, we consider the following example: Assume that $(\mathcal{L}u)(x) = -iu'(x)$ $(x \in \mathbb{R})$. The corresponding minimal operator L is the simplest example of a self-adjoint differential operator. The expansion in its eigenfunctions gives the classical one-dimensional Fourier transformation. We construct such an expansion without using the Fourier transformation.

The operator L is selfadjoint. Indeed, let $\varphi \in L_2(\mathbb{R})$ be such that $L^*\varphi = \bar{z}\varphi$ $(\operatorname{Im} z \neq 0)$. For $v \in C_0^\infty(\mathbb{R})$, we have $(-iv' - zv, \varphi)_{L_2(\mathbb{R})} = 0$, i.e., φ is a generalized solution of the equation $-iu' - zu = 0$; by virtue of Theorem 6.1, φ is smooth. Then $\varphi(x) = ce^{i\bar{z}x} \in L_2(\mathbb{R})$. The last equality is true only for $c = 0$. Thus, $\varphi = 0$ and $L^* = L$.

Let us determine the resolvent of the operator L. Assume that $\operatorname{Im} z > 0$, $f \in C_0^\infty(\mathbb{R})$. The solution of the equation $(L - z\mathbb{I})u = f$ coincides with the solution $u \in L_2(\mathbb{R})$ of the differential equation $-iu'(x) - zu(x) = f(x)$ $(x \in \mathbb{R})$. Let us determine this solution. We have

$$
\begin{aligned}
(R_z f)(x) = u(x) = ie^{izx} \int_{-\infty}^{x} e^{-izy} f(y)dy \\
= \int_{-\infty}^{\infty} i\chi_{(-\infty,x)}(y) e^{iz(x-y)} f(y)dy \qquad (x \in \mathbb{R}),
\end{aligned}
$$

where $\chi_{(a,b)}$ is the indicator of the interval (a, b). Thus, the kernel of the resolvent has the following form:

$$
\Gamma_z(x, y) = \begin{cases} i\chi_{(-\infty,x)}(y) e^{iz(x-y)}, & \operatorname{Im} z > 0, \\ -i\chi_{(x,+\infty)}(y) e^{iz(x-y)}, & \operatorname{Im} z < 0 \end{cases} \qquad (x, y \in \mathbb{R}) \tag{6.37}
$$

(for $\operatorname{Im} z < 0$, this formula follows from the relation $R_z = R_{\bar{z}}^*$).

The spectral kernel can be calculated by analogy with the example (the case of the Laplace operator) in Subsection 5.4. For this purpose, one should use the general formula (5.37) and the Green's function (6.37). As a result of a simple limit procedure, we get

$$P(x, y; \lambda)d\rho(\lambda) = \frac{1}{2\pi}e^{i\lambda(x-y)}d\lambda \qquad (\lambda, x, y \in \mathbb{R}).$$

Thus, expansion (6.33) now takes the form

$$P(x, y; \lambda) = (2\pi)^{-1/2}e^{i\lambda x}(2\pi)^{-1/2}e^{-i\lambda y}, \quad d\sigma(\lambda) = d\lambda$$

and equality (6.34) is the usual Parseval equality for the decomposition in the classical Fourier integral.

The results presented above can be regarded as a construction of the usual one-dimensional Fourier integral. To construct an n-dimensional Fourier integral, one must use the theorem on expansion in joint generalized eigenfunctions of family of commuting selfadjoint operators. The role of these operators is played by the minimal operators generated in the space $L_2(\mathbb{R}^n)$ by the expressions $-iD_1, \dots, -iD_n$.

Bibliographical Notes

I. *Chapters 12–13.* Some branches of the theory of unbounded operators and their spectral representations are considered in detail in [AkG], [BeK], [BiS], [Kat], [Mau], [Ple], [ReS1], [ReS2], [Rud], and [Sam]. In particular, the generalized resolutions of the identity and M. Krein's description of all selfadjoint extensions are presented in [AkG], the products of commuting resolutions of the identity are investigated in [BeK], [BiS], and [Ple]; the proof of the spectral theorem based on the theory of commutative Banach algebras can be found in [Yos], [Mau], and [Rud]. Some other methods for establishing the selfadjointness of operators are presented in [BeK] and [ReS2]. Spectral representations of the families of commuting operators and their applications to the theorems of Stone's type, harmonic analysis, and noncommuting operators with constraints are described in [BeK] and [Sam].

Detailed information on differential equations in Banach spaces can be found in [Kre1]. Quasianalytic classes of functions are studied in [Man].

II. *Chapters 14–15.* For a detailed exposition of the theory of rigged spaces, see [Ber], [BeK], [GeV], [GoG], [LySt], and [Mau]; the spaces of test functions are thoroughly investigated in [GeS2] in the case of finitely many variables and in [BeK] in the case of the infinitely many variables.

Different versions of the kernel theorem can be found in [Ber], [BeK], [GeV], [Mau], [ReS1]. Many results about bilinear forms are presented in [BeK], [Kat], [ReS1], and [ReS2].

Expansions in generalized eigenvectors are considered in detail in [Ber], [BeK], [BeS], [GeV], and [Mau]. In particular, the book [BeK] deals with an expansion in joint generalized eigenvectors for a family of commuting normal operators of any cardinality. Direct integrals of Hilbert spaces are studied in [Mau] and [Nai1].

III. *Chapter 16.* Theorem 1.1 on isomorphisms and the results concerning the local smoothing of generalized solutions of elliptic equations inside a domain and up to its boundary in fairly general situations can be found in [Ber] and in journal papers. Similar results are presented in [LiM]. All necessary information on elliptic problems is contained in [Ber], [Lad], [Mikha], [Pet1], [Tri], and [Vl1]. The problems of smoothing inside a domain are discussed in [Mau].

The selfadjointness of differential operators with partial derivatives is investigated in [Ber] and [ReS2]; in the case of infinitely many independent variables, these problems are considered in [BeK]. These books and [BeS], [Gla], and [ReS3] contain additional information about spectral properties of differential operators with partial derivatives (including ordinary differential operators). The books [AkG], [DuS2], [Lev], [LeS], [Mar], [MyO], and [Nai2] are devoted to the study of the spectral properties of ordinary differential operators. The spectral properties of the indicated operators with operator-valued coefficients are investigated in [GoG] and [LySt] (such operators include certain classes of operators with partial derivatives).

The theory on expansions in generalized eigenfunctions of elliptic operators and the properties of Green's function are presented in [Ber], [BeK] (in the general case) and in [AkG], [DuS2], [CoL], [KoS], [Lev], [LeS], [Mar], and [Nai2] (in the case of ordinary differential operators). The facts from the theory of ordinary differential equations which are used in the book can be found in [Kam] and [CoL].

IV. Some branches of functional analysis are not reflected in our textbook. The reader who wants to study these subjects may use the following general remarks as an orientation key:

- **(i)** the theory of semigroups: [HiP], [Kat], [Mau], [ReS2], and [Yos];
- **(ii)** scattering theory: [AkG] and [ReS4];
- **(iii)** perturbations of operators: [AkG] and [Kat];
- **(iv)** nonselfadjoint operators and operator bundles: [GoK] and [Sad];
- **(v)** semiordered spaces: [KaA], [KLS], and [Vu1];
- **(vi)** spaces with indefinite metrics: [AzI];
- **(vii)** Banach algebras (normed rings): [GRS], [HiP], [Loo], [Mau], [Nai1], [Rud], and [Yos];
- **(viii)** topological groups and their representations: [BaR], [Kir], [Loo], [Nai3], and [Pon];
- **(ix)** differential calculus in infinite-dimensional spaces and nonlinear functional analysis: [BeK], [Die], [KaA], [KoF], [KrZ], and [LySo];
- **(x)** approximate methods in functional analysis: [Col], [KaA], and [KVZ].

References

[AkG] Akhiezer, N.I. and Glazman, I.M., *The Theory of Linear Operators in Hilbert Spaces* (Russian), Nauka, Moscow, 1966.

English translation of 1st ed.: New York, Ungar, 1961.

[AlM] Aleksandryan, R.A. and Mirzakhanyan, E.A., *General Topology* (Russian), Moscow, Vysshaya Shkola, 1979.

[AKR] Antonevich, A.B., Knyazev, P.N., and Radyno, Ya.V., *Problems and Exercises on Functional Analysis* (Russian), Minsk, Vysheishaya Shkola, 1978.

[AnR] Antonevich, A.B. and Radyno, Ya.V., *Functional Analysis and Integral Equations* (Russian), Minsk, Universitetskoe, 1984.

[ArP] Arkhangelsky, A.V. and Ponomaryov, V.I., *Foundations of General Topology in Problems and Exercises* (Russian), Moscow, Nauka, 1974.

[AzI] Azizov, T.Ya. and Iokhvidov, I.S., *Foundations of the Theory of Linear Operators in Spaces with Indefinite Metrics* (Russian), Moscow, Nauka, 1986.

[Ban] Banach, S., *Théorie des opérations linéaires*, Warsaw, 1932.

[BaR] Barut, A. and Rączka, R., *Theory of Group Representations and Applications*, Warszawa, PWN, 1977.

[Ber] Berezansky, Yu.M., *Expansions in Eigenfunctions of Selfadjoint Operators* (Russian), Naukova Dumka, Kiev, 1965.

English translation: Amer. Math. Soc. Transl., vol. 17, Providence, 1968.

[BeK] Berezansky, Yu.M. and Kondratyev, Yu.G., *Spectral Methods in Infinite-Dimensional Analysis* (Russian), Naukova Dumka, Kiev, 1988.

English translation: Dordrecht, Kluwer, 1995.

[BeS] Berezin, F.A. and Shubin, M.A., *Schrödinger Equations* (Russian), Moscow, Moscow University, 1983.

[BiS] Birman, M.Sh. and Solomyak, M.Z., *Spectral Theory of Selfadjoint Operators in Hilbert Spaces* (Russian), Leningrad, Leningrad University, 1980.

[CoL] Coddington, E.A. and Levinson, N., *Theory of Ordinary Differential Equations*, New York, McGraw-Hill, 1955.

[Col] Collatz, L., *Functional Analysis and Numerical Mathematics*, New York, Academic Press, 1966.

[Day] Day, M.M., *Normed Linear Spaces*, Berlin–Göttingen–Heidelberg, Springer, 1958.

[Die] Dieudonné, J., *Foundations of Modern Analysis*, New York–London, Academic Press, 1964.

[DuS1] Dunford, N. and Schwartz, J.T., *Linear Operators*, vol. I, *General Theory*, New York–London, Interscience, 1958.

[DuS2] Dunford, N. and Schwartz, J.T., *Linear Operators*, vol. II, *Spectral Theory. Selfadjoint Operators in Hilbert Spaces*, New York–London, Interscience, 1963.

[Edw] Edwards, R.E., *Functional Analysis. Theory and Applications*, New York–Chicago–San Francisco–Toronto–London, Holt, Rinehart, and Winston, 1965.

[GeO] Gelbaum, B.R. and Olmsted, J.M.H., *Counterexamples in Analysis*, San Francisco–London–Amsterdam, Holden-Day, 1964.

[GRS] Gelfand, I.M., Raikov, D.A., and Shilov, G.E., *Commutative Normed Rings* (Russian), Fizmatgiz, Moscow, 1960.

English translation: New York, Chelsea, 1964.

[GeS1] Gelfand, I.M. and Shilov, G.E., *Generalized Functions*, vol. 1: *Generalized Functions and Operations on Them* (Russian), Fizmatgiz, Moscow, 1958.

English translation: New York, Academic Press, 1964.

[GeS2] Gelfand, I.M. and Shilov, G.E., *Generalized Functions*, vol. 2: *Spaces of Test and Generalized Functions* (Russian), Fizmatgiz, Moscow, 1958.

English translation: New York, Academic Press, Gordon and Breach, 1968.

[GeV] Gelfand, I.M. and Vilenkin, N.Ya., *Generalized Functions*, vol. 4: *Some Applications of Harmonic Analysis. Rigged Hilbert Spaces* (Russian), Fizmatgiz, Moscow, 1961.

English translation: New York, Academic Press, 1964.

[GiS] Gikhman, I.I. and Skorokhod, A.V., *Introduction to the Theory of Random Processes* (Russian), Nauka, Moscow, 1965.

English translation: Philadelphia, Saunders, 1969.

[Gla] Glazman, I.M., *Direct Methods of Qualitative Spectral Analysis of Singular Differential Operators* (Russian), Fizmatgiz, Moscow, 1963.

English translations: Jerusalem, Israel Program Sci. Transls., 1965; New York, Davey, 1966.

[GlL] Glazman, I.M. and Lyubich, Yu.I., *Finite-Dimensional Linear Analysis in Problems* (Russian), Moscow, Nauka, 1969.

[GoK] Gokhberg, I.Ts. and Krein, M.G., *Introduction to the Theory of Linear Nonselfadjoint Operators* (Russian), Moscow, Nauka, 1965.

[GoG] Gorbachuk, V.I. and Gorbachuk, M.L., *Boundary Value Problems for Operator Differential Equations* (Russian), Kiev, Naukova Dumka, 1984.

English translation: Dordrecht, Kluwer AP, 1991.

[Hal1] Halmos, P.R., *Measure Theory*, Princeton, Van Nostrand, 1950.

[Hal2] Halmos, P.R., *A Hilbert Space Problem Book*, Princeton–Toronto–London, Van Nostrand, 1967.

[HiP] Hille, E. and Phillips, R.S., *Functional Analysis and Semi-Groups*, Providence, Amer. Math. Soc., 1957.

[HuP] Hutson, V.C.L. and Pym, J.S., *Applications of Functional Analysis and Operator Theory*, New York, Academic Press, 1980.

[Kam] Kamke, E., *Differentialgleichungen, Lösungsmethoden und Lösungen*, Leipzig, Gewöhnliche Differentialgleichungen, 1959.

[KaA] Kantorovich, L.V. and Akilov, G.P., *Functional Analysis* (Russian), Moscow, Nauka, 1984.

[Kat] Kato, T., *Perturbation Theory for Linear Operators*, Berlin, Springer, 1966.

[Kir] Kirillov, A.A., *Elements of Representation Theory* (Russian), Moscow, Nauka, 1978.

[KiG] Kirillov, A.A. and Gvishiani, A.D., *Theorems and Problems of Functional Analysis* (Russian), Moscow, Nauka, 1988.

[KoF] Kolmogorov, A.N. and Fomin, S.V. *Elements of Function Theory and Functional Analysis* (Russian), Moscow, Nauka, 1989.

[KoS] Kostyuchenko, A.G. and Sargsyan, I.S., *Distribution of Eigenvalues. Self-adjoint Ordinary Differential Operators* (Russian), Moscow, Nauka, 1979.

[KLS] Krasnoselsky, M.A., Lifshits, E.A., and Sobolev, A.V., *Positive Linear Systems. Methods of Positive Operators* (Russian), Moscow, Nauka, 1985.

[KVZ] Krasnoselsky, M.A., Vainikko, G.M., Zabreiko, P.P. et al., *Approximate Solution of Operator Equations* (Russian), Moscow, Nauka, 1969.

[KrZ] Krasnoselsky, M.A. and Zabreiko, P.P., *Geometric Methods of Nonlinear Analysis* (Russian), Moscow, Nauka, 1975.

[KZPS] Krasnoselsky, M.A., Zabreiko, P.P., Pustylnik, E.I. and Sobolevsky, P.E., *Integral Operators in Spaces of Summable Functions* (Russian), Moscow, Nauka, 1966.

[Kre1] Krein, S.G., *Linear Differential Equations in Banach Spaces* (Russian), Moscow, Nauka, 1967.

English translation: Providence, Amer. Math. Soc., 1971.

[Kre2] Krein, S.G., *Linear Equations in Banach Spaces* (Russian), Moscow, Nauka, 1971.

[Kre3] Krein, S.G. (editor), *Functional Analysis* (Russian), Moscow, Nauka, 1972.

English translation: Noordhoff, 1972.

[Kut] Kutateladze, S.S., *Foundations of Functional Analysis* (Russian), Novosibirsk, Nauka, 1983.

[Lad] Ladyzhenskaya, O.A., *Boundary Value Problems of Mathematical Physics* (Russian), Moscow, Nauka, 1973.

[Lev] Levitan, B.M., *Inverse Sturm-Liouville Problems* (Russian), Moscow, Nauka, 1984.

[LeS] Levitan, B.M. and Sargsyan, I.S., *Sturm-Liouville and Dirac Operators* (Russian), Moscow, Nauka, 1988.

[LiM] Lions, J.-L. and Magenes, E., *Problèmes aux Limites Non Homogènes et Applications*, vol. 1–3, Paris, Dunod, 1970.

[Loo] Loomis, L.H., *An Introduction to Abstract Harmonic Analysis*, Toronto–New York–London, Van Nostrand, 1953.

[LySt] Lyantse, V.E. and Storozh, O.G., *Methods of Unbounded Operator Theory* (Russian), Kiev, Naukova Dumka, 1983.

[LySo] Lyusternik, L.A. and Sobolev, V.I., *A Brief Course of Functional Analysis* (Russian), Moscow, Vysshaya Shkola, 1982.

[Man] Mandelbrojt, S., *Séries Adhérentes, Régularisation des Suites, Applications*, Paris, Gauthier-Villars, 1952.

[Mar] Marchenko, V.A., *Sturm-Liouville Operators and Their Applications* (Russian), Kiev, Naukova Dumka, 1977.

[Mau] Maurin, K., *Methods of Hilbert Spaces*, Warsaw, PWN, 1959.

English translation: Warsaw, PWN, 1967.

[Maz] Mazya, V.G., *Sobolev Spaces* (Russian), Leningrad, Leningrad University, 1985.

[Mikha] Mikhailov, V.P., *Partial Differential Equations* (Russian), Moscow, Nauka, 1976.

[Mikhl] Mikhlin, S.G., *Lectures on Linear Integral Equations* (Russian), Moscow, Fizmatgiz, 1959.

[MyO] Mynbaev, K.T. and Otelbaev, M.O., *Weight Functional Spaces and the Spectrum of Differential Operators* (Russian), Moscow, Nauka, 1988.

[Nai1] Naimark, M.A., *Normed Rings* (Russian), Moscow, Nauka, 1968.

English translation: Noordhoff, 1972.

[Nai2] Naimark, M.A., *Linear Differential Operators* (Russian), Moscow, Nauka, 1969.

[Nai3] Naimark, M.A., *Theory of Group Representations* (Russian), Moscow, Nauka, 1976.

[Nat] Natanson, I.P., *Theory of Functions of a Real Variable* (Russian), Moscow, Nauka, 1974.

[Nik] Nikolsky, S.M., *Approximation of Functions of Finitely Many Variables and Embedding Theorems* (Russian), Moscow, Nauka, 1977.

[Pet1] Petrovsky, I.G., *Lectures on Partial Differential Equations* (Russian), Moscow, Fizmatgiz, 1961.

[Pet2] Petrovsky, I.G., *Lectures on the Theory of Integral Equations* (Russian), Moscow, Moscow University, 1984.

[Ple] Plesner, A.I., *Spectral Theory of Linear Operators* (Russian), Moscow, Nauka, 1965.

English translation: New York, Ungar, 1969.

[Pon] Pontryagin, L.S., *Continuous Groups* (Russian), Moscow, Nauka, 1984.

[ReS1] Reed, M. and Simon, B., *Methods of Modern Mathematical Physics*, vol. 1, New York–San Francisco–London, Academic Press, 1972.

[ReS2] Reed, M. and Simon, B., *Methods of Modern Mathematical Physics*, vol. 2, New York–San Francisco–London, Academic Press, 1975.

[ReS3] Reed, M. and Simon, B., *Methods of Modern Mathematical Physics*, vol. 3, New York–San Francisco–London, Academic Press, 1978.

[ReS4] Reed, M. and Simon, B., *Methods of Modern Mathematical Physics*, vol. 4, New York–San Francisco–London, Academic Press, 1979.

[Ric1] Richtmyer, R.D., *Principles of Advanced Mathematical Physics*, vol. 1, New York–Heidelberg–Berlin, Springer, 1978.

[Ric2] Richtmyer, R.D., *Principles of Advanced Mathematical Physics*, vol. 2, New York–Heidelberg–Berlin, Springer, 1981.

[RiS] Riesz, F. and Szökefalvi-Nagy, B., *Leçons d'Analyse Fonctionelle*, Budapest, Akad. Kiadó, 1952.

English translation: New York, Ungar, 1955.

[RoR] Robertson, A.P. and Robertson, W., *Topological Vector Spaces*, Cambridge, University Press, 1964.

[Rud] Rudin, W., *Functional Analysis*, New York, Mc-Graw Hill, 1973.

[Sad] Sadovnichy, V.A., *Operator Theory* (Russian), Moscow, Moscow University, 1986.

[Sam] Samoilenko, Yu.S., *Spectral Theory of Families of Self-Adjoint Operators* (Russian), Kiev, Naukova Dumka, 1984.

English translation: Dordrecht, Kluwer, 1991.

[Sch] Schaefer, H., *Topological Vector Spaces*, Macmillan, 1966.

[Shi1] Shilov, G.E., *Mathematical Analysis. Special Course* (Russian), Moscow, Fizmatgiz,1960.

[Shi2] Shilov, G.E., *Mathematical Analysis. Second Special Course* (Russian), Moscow, Moscow University, 1984.

[ShG] Shilov, G.E. and Gurevich, B.L., *Integral, Measure, and Derivative* (Russian), Moscow, Nauka, 1967.

[Sob1] Sobolev, S.L., *Introduction to the Theory of Cubature Formulas* (Russian), Moscow, Nauka, 1974.

[Sob2] Sobolev, S.L., *Some Applications of Functional Analysis in Mathematical Physics* (Russian), Moscow, Nauka, 1988.

[Tel] Telyakovsky, S.A., *Problem Book on the Theory of Real Variable Functions* (Russian), Moscow, Nauka,1980.

[Tre] Trenogin, V.A., *Functional Analysis* (Russian), Moscow, Nauka, 1980.

[TPS] Trenogin, V.A., Pisarevsky, B.M., and Soboleva, T.S., *Problems and Exercises on Functional Analysis* (Russian), Moscow, Nauka, 1984.

[Tri] Triebel, H., *Interpolation Theory. Function Spaces. Differential Operators*, Berlin, VEB Deutscher Verlag der Wissenschaften, 1978.

[Vl1] Vladimirov, V.S., *Equations of Mathematical Physics* (Russian), Moscow, Nauka, 1971.

[Vl2] Vladimirov, V.S., *Generalized Functions in Mathematical Physics* (Russian), Moscow, Nauka, 1979.

English translation: Moscow, Mir, 1979.

[VMV] Vladimirov, V.S., Mikhailov, V. P., Vasharin, A.A. et al., *Problem Book on Equations of Mathematical Physics* (Russian), Moscow, Nauka, 1982.

English translation: Moscow, Mir, 1982.

[Vu1] Vulikh, B.Z., *Introduction to Functional Analysis* (Russian), Moscow, Nauka, 1967.

[Vu2] Vulikh, B.Z., *A Brief Course on the Theory of Real Variable Functions* (Russian), Moscow, Nauka, 1973.

[Yos] Yosida, K., *Functional Analysis*, Berlin–Göttingen–Heidelberg, Springer, 1965.

Index

Titles previously published in the series

OPERATOR THEORY: ADVANCES AND APPLICATIONS

BIRKHÄUSER VERLAG

Edited by **I. Gohberg,**

School of Mathematical Sciences, Tel-Aviv University, Ramat Aviv, Israel

73. **A. Feintuch, I. Gohberg** (Eds): Nonselfadjoint Operators and Related Topics. Workshop on Operator Theory and Its Applications, Beersheva, February 24–28, 1994, (ISBN 3-7643-5097-0)

74. **R. Hagen, S. Roch, B. Silbermann**: Spectral Theory of Approximation Methods for Convolution Equations, 1994, (3-7643-5112-8)

75. **C.B. Huijsmans, M.A. Kaashoek, B. de Pagter** (Eds): Operator Theory in Function Spaces and Banach Lattices. The A.C. Zaanen Anniversary Volume, 1994, (ISBN 3-7643-5146-2)

77. **J. Lindenstrauss, V.D. Milman** (Eds): Geometric Aspects of Functional Analysis Israel Seminar GAFA 1992-94, 1995, (ISBN 3-7643-5207-8)

78. **M. Demuth, B.-W. Schulze** (Eds): Partial Differential Operators and Mathematical Physics: International Conference in Holzhau (Germany), July 3-9, 1994, 1995, (ISBN 3-7643-5208-6)

79. **I. Gohberg, M.A. Kaashoek, F. van Schagen**: Partially Specified Matrices and Operators: Classification, Completion, Applications, 1995, (ISBN 3-7643-5259-0)

80. **I. Gohberg, H. Langer** (Eds): Operator Theory and Boundary Eigenvalue Problems International Workshop in Vienna, July 27-30, 1993, 1995, (ISBN 3-7643-5259-0)

81. **H. Upmeier**: Toeplitz Operators and Index Theory in Several Complex Variables, 1995, (ISBN 3-7643-5280-5)

83. **A.B. Antonevich**: Linear Functional Equations. Operator Approach, 1995, (ISBN 3-7643-2931-9)

84. **L.A. Sakhnovich**: Integral Equations with Difference Kernels on Finite Intervals, 1996, (ISBN 3-7643-5267-1)